Contents

Expanded contents

Chapter 9   **How to work with fragments**

Section 3 The News Reader app

Chapter 10   **How to work with threads, files, adapters, and intents**

Introduction

Android is arguably the world's most widely used operating system for mobile devices such as smartphones and tablets. As of May 2013, over 900 million Android devices had been activated and over 48 billion apps had been installed from the Google Play store.

In the coming years, Android's popularity is likely to continue for a couple reasons. First, Google releases Android under an open-source license that allows it to be modified and freely distributed by device manufacturers. Second, since Android is open-source with a large community of developers, it is able to evolve according to the needs of the developers who use it.

Who this book is for

This book is for anyone who wants to learn professional skills for developing Android apps. The only prerequisite is a basic understanding of the Java programming language roughly equivalent to chapters 1 through 14 of our core Java book, *Murach's Java Programming*. Once you have the necessary Java skills, this book should work for you even if you have absolutely no experience developing mobile apps.

What version of Android this book supports

The apps presented in this book have been designed to take advantage of the latest features of Android 4.2 and have been tested on this version of Android. However, most of these apps have also been designed to work with earlier versions of Android going back to version 2.2, and they have been tested on Android 2.2.

Since Android is backwards-compatible, the apps presented in this book should continue to work for future versions of Android, though they might not take advantage of the latest features. For example, the apps in this book have been tested on Android 4.3, which was released just as this book went to press, and they work well with that version of Android.

What operating systems this book supports

The software you need for developing Android apps is available for Windows, Mac OS X, and Linux. If you're using Windows, you can use Appendix A to download and install this software. If you're using Mac OS X, appendix B shows how to download and install this software. If you're using another platform such as Linux, you probably already know how to install this software! If not, you can use appendix B as a general guide and search the Internet for more information if you need it.

What IDE this book supports

This book shows you how to use the Eclipse IDE (Integrated Development Environment) to code, test, and debug applications. Although there are other excellent IDEs for working with Android, I decided to present Eclipse because I think it's the most widely used IDE for Android development. In addition, it's available for free, and it runs on all operating systems.

Even if you want to use another IDE, I recommend that you use Eclipse to work with the applications presented in this book. Then, once you're done with this book, it should be easy for you to switch from Eclipse to another IDE. However, if you'd prefer to get started with another IDE right away, you can do that too. But first, you'll need to convert the Eclipse projects that are available from our web site so you can open them in your preferred IDE.

Since the IntelliJ IDE has been gaining in popularity in recent years, appendix C shows how to install and configure IntelliJ to work with this book. In addition, it shows how to use IntelliJ to import the Eclipse projects for this book. After that, you should be able to use this book with IntelliJ. If you hit any problems, you can search the Internet for solutions.

In May of 2013, IntelliJ announced that it is collaborating with Google to create a new IDE called Android Studio. This IDE is based on the IntelliJ source code but streamlined for Android development. As this book goes to press, Android Studio is only available as an early access preview. However, since it's backed by Google and IntelliJ, it might become very popular after a production version is released. As a result, it's worth keeping an eye on this IDE for the next few years.

How to get the software you need

You can download all of the software that you need for this book for free from the Internet. This software includes the source code for this book, the Java SDK (Software Development Kit), and the ADT (Android Developer Tools) bundle, which includes the Android SDK, Eclipse, and the ADT plugin for Eclipse.

What you'll learn in this book

As I wrote this book, I did my best to focus on the practical features that you'll need for developing professional Android applications. Here's a quick tour:

- In section 1, you'll quickly master the basics of Android programming. In chapter 1, you'll learn some background terms and concepts. In chapter 2, you'll learn how to use Eclipse for Android development. In chapter 3, you'll learn how to develop your first Android app, a simple but complete Tip Calculator app. And in chapter 4, you'll learn how to use Eclipse to thoroughly test and debug your apps.

- In section 2, you'll learn some essential Android skills by enhancing the Tip Calculator app that you developed in chapter 3. To start, in chapter 5, you'll learn how to use different layouts and widgets to develop a sophisticated user interface. In chapter 6, you'll learn several techniques for handling high- and low-level events. In chapter 7, you'll learn how to use themes and styles to improve the appearance of your app. In chapter 8, you'll learn how to use menus and preferences. And in chapter 9, you'll learn how to use fragments to allow your app to take advantage of large screens that are available from some mobile devices such as tablets.

- In section 3, you'll learn how to develop a News Reader app. Along the way, you'll learn more essential Android skills. In chapter 10, for example, you'll learn how to read an RSS feed from the Internet, save that data in a file, and display it on the user interface. In chapter 11, you'll learn how to use a service to download data for an app even when the app isn't running, and to notify a user when new data is available. And in chapter 12, you'll learn how to respond to actions that are broadcast by Android and its apps.

- In section 4, you'll learn how develop a Task List app that stores one or more to-do lists. In chapter 13, you'll learn how to create a database to store these lists. In chapter 14, you'll learn how to use tabs and a custom adapter to display these tasks on the user interface. In chapter 15, you'll learn how to use a content provider to allow other apps to work with the same data as this app. And in chapter 16, you'll learn how to create an app widget that can display some of this app's data on a device's Home screen.

- In section 5, you'll learn how to deploy apps to the Google Play store. Then, you'll learn how to create a Run Tracker app that uses the Google Maps API.

Why you'll learn faster and better with this book

Like all our books, this one has features to help you learn as quickly and easily as possible. Here are just three of those features.

- To help you develop applications at a professional level, this book presents complete, non-trivial apps. For example, chapter 12 presents the final version of the News Reader app, chapter 16 presents the final Task List app, and chapter 18 presents a Run Tracker app. Studying these apps is a great way to master Android development.

- All of the information in this book is presented in our unique paired-pages format, with the essential syntax, guidelines, and examples on the right page and the perspective and extra explanation on the left page. This helps you learn more while reading less, and it helps you quickly find the information that you need when you use this book for reference.

- The exercises at the end of each chapter give you a chance to try out what you've just learned and to gain valuable, hands-on experience with Android programming. They guide you through the development of some of the book's apps, and they challenge you to apply what you've learned in new ways. And because most of these exercises start from code that you can download from our web site, you'll spend your time practicing new skills instead of doing busywork.

How our downloadable files make learning easier

To make learning easier, you can download the source code for all the apps presented in this book from our web site (www.murach.com). Then, you can view the complete code for these apps as you read each chapter, you can run these apps to see how they work, and you can copy portions of code for use in your own apps.

You can also download the starting points and solutions for the exercises in this book. That way, you don't have to start every exercise from scratch. This takes the busywork out of doing these exercises. As a result, you get more practice in less time. In addition, if you encounter a problem, you can easily check the solution. This helps you to keep moving forward with less chance that you'll get hung up on a minor issue. For more information about these downloads, please see appendix A (Windows) and appendix B (Mac).

Support materials for trainers and instructors

If you're a corporate trainer or a college instructor who would like to use this book for a course, we offer an Instructor's CD that includes: (1) a complete set of PowerPoint slides, (2) test banks, (3) additional chapter exercises that aren't in this book, (4) solutions to those exercises, (5) projects that the students start from scratch, and (6) solutions to those projects.

To learn more about this Instructor's CD and to find out how to get it, please go to our web site at www.murach.com and click the Trainers link or the Instructors link. Or, if you prefer, you can call Kelly at 1-800-221-5528 or send an email to kelly@murach.com.

A companion book

As you read this book, you may discover that your Java skills aren't as strong as they ought to be. In that case, I recommend that you get a copy of our core Java book (*Murach's Java Programming*). This book will get you up to speed with Java and show you how to use all of the core skills you need for developing Android apps. In short, I think our Java book makes a great companion book for this Android book.

Please let us know how this book works for you

When I started writing this book, my goal was (1) to teach you Android programming as quickly and easily as possible and (2) to teach you the practical Android concepts and skills that you need for developing professional apps. Now, I hope I have succeeded. If you have any comments about this book, I would appreciate hearing from you at murachbooks@murach.com.

Thanks for buying this book. I hope you enjoy reading it, and I wish you great success with your Android programming.

Joel Murach, Author
joel@murach.com

Section 1

Get started fast with Android

This section gets you started fast with Android programming. Chapter 1 introduces you to some important terms, concepts, and background information. Chapter 2 shows you how to use Eclipse to work with Android projects. Chapter 3 introduces you to the basic skills that you need for developing Android apps. And chapter 4 shows you how to test and debug Android apps.

To illustrate these skills, this section shows how to write, test, and debug a simple Tip Calculator app. This app calculates the tip you would give based on the amount of a bill. When you complete these chapters, you'll be able to write, test, and debug simple applications of your own.

1

An introduction to Android

This chapter introduces you to the concepts and terms that you need for developing Android apps. To give you a clear idea of how the code for an Android app works, this chapter presents some of the code for a simple app that calculates tips. When you finish this chapter, you'll be ready to get started with Android development.

Android overview

Android is a Linux-based operating system designed primarily for touch-screen mobile devices such as smartphones and tablet computers. Before you begin developing apps that run on Android, it's helpful to take a moment to consider the types of Android devices and apps that are available today. In addition, it's helpful to understand the history, versions, and architecture of Android.

Types of devices

Figure 1-1 starts by showing two of the most popular Android devices, a smartphone and a tablet. However, the code for Android is open-source. As a result, it can be customized to work with other types of electronic devices such as eBook readers, cameras, home automation systems, home appliances, vehicle systems, game consoles, and so on.

An Android phone and tablet

Other types of Android devices

- Readers
- Cameras
- Home automation systems
- Home appliances
- Vehicle systems
- Game consoles

Description

- *Android* is a Linux-based operating system designed primarily for touchscreen mobile devices such as smartphones and tablet computers.
- Since the code for Android is open-source, it can be customized to work with other types of electronic devices.

Figure 1-1   Types of Android devices

Types of apps

Android has a large community of developers writing *applications*, more commonly referred to as *apps*, that extend the functionality of Android devices. Figure 1-2 lists some categories of different types of apps and describes some of the functionality available from each category. As you review this list, keep in mind that these categories are only intended to give you a general idea of the various types of Android apps. More categories exist, and the functionality that's available from Android apps is constantly expanding.

If you have used a smartphone or tablet, you should be familiar with some of the apps listed in this figure. For example, you probably use apps to send and receive text messages and email. You probably use apps to take pictures, listen to music, and watch video. You probably use apps to get directions and navigate to a location. You probably use apps to check the weather, read news, and browse the web.

In addition, you may use apps from social media companies like Facebook and Twitter to work with social media. You may use apps to play games like Angry Birds. The main point here is that there are many different types of apps and that application developers are constantly creating new apps that use the capabilities of Android phones in new ways.

Some apps come preinstalled on the device. For example, most phones include apps for managing contacts, using the phone, sending and receiving text messages and email, working with photos and video, and web browsing. Other apps are available for download through Google Play or third-party sites. Some apps are free. Other apps are expensive. All apps were created by somebody like you who learned how to develop Android apps.

Types of Android apps

Category	Functionality
Communications	Send and receive text messages, send and receive email, make and receive phone calls, manage contacts, browse the web.
Photography	Take photos, edit photos, manage photos.
Audio	Play audio, record audio, edit audio.
Video	Play video, record video, edit video.
Weather	View weather reports.
News	Read news and blogs.
Personalization	Organize home screen, customize ringtones, customize wallpaper.
Productivity	Manage calendar, manage task list, take notes, make calculations.
Finance	Manage bank accounts, make payments, manage insurance policies, manage taxes, manage investment portfolio.
Business	Read documents, edit documents, track packages.
Books	Read and search eBooks.
Reference	Get info from a dictionary, thesaurus, or wiki.
Education	Prepare for exams, learn foreign languages, improve vocabulary.
Shopping	Buy items online, use electronic coupons, compare prices, keep grocery lists, read product reviews.
Social	Use social networking apps such as Facebook and Twitter.
Fitness	Monitor and document workouts.
Sports	Track sport scores, manage fantasy teams.
Travel	Get directions, use GPS to navigate to a location, get information about nearby places to visit.
Games	Play games such as arcade games, action games, puzzles, card games, casino games, sports games.

Description

- Android has a large community of developers writing *applications*, more commonly referred to as *apps*, that extend the functionality of Android devices.
- Android apps are available for download through Google Play or third-party sites.

Figure 1-2   Types of Android apps

A brief history

Figure 1-3 summarizes the history of Android. In 2003, a handful of entre-preneurs in Palo Alto, California, founded Android, Inc. In 2005, Google bought Android, Inc. Then, in 2007, the *Open Handset Alliance* was announced. This alliance consists of a consortium of hardware, software, and telecommunication companies devoted to advancing open standards for mobile devices. That same year, Google released Android code as open source under the Apache License. In addition, Google helped to create the *Android Open Source Project* (*AOSP*) and put it in charge of the maintenance and further development of Android.

In 2008, the first version of the Android *Software Development Kit* (*SDK*) was released. This SDK contains all of the tools that Android developers need to develop apps. Later in 2008, the first Android phones became available.

Since 2008, new versions of the Android SDK have continued to be released, and Android devices have continued to proliferate. During that time, hundreds of thousands of apps have been developed, and billions of apps have been downloaded.

A brief history of Android

Year	Event
2003	Android, Inc. is founded in Palo Alto, California.
2005	Google buys Android, Inc.
2007	The *Open Handset Alliance* is announced. This alliance consists of a consortium of hardware, software, and telecommunication companies devoted to advancing open standards for mobile devices.
	Google releases Android code as open source under the Apache License.
	The *Android Open Source Project* (*AOSP*), led by Google, is tasked with the maintenance and further development of Android.
2008	Android *Software Development Kit* (*SDK*) 1.0 is released. This kit contains all of the tools needed to develop Android apps.
	The first Android phones become available.
2009-present	New versions of the Android SDK continue to be released.
	Android devices continue to proliferate.
	Hundreds of thousands of apps are developed.
	Billions of apps are downloaded.

Description

- Android has grown steadily since its release in 2008 until the present.

Figure 1-3   A brief history of Android

Versions

Figure 1-4 describes all major releases of Android starting with version 1.6 and ending with version 4.3. Here, each version of Android corresponds with an API (Application Programming Interface) number. For example, version 4.2 corresponds with API 17, and version 2.2 corresponds with API 8. In addition, each version has a code name that's based on something sweet. For example, version 2.2 carries the code name Froyo, while version 4.1 is known as Jelly Bean.

As you develop an Android app, you must decide the minimum API level that your app supports. As of April 2013, many developers choose 7 or 8 as the minimum API level to support since that covers a high percentage of all Android devices.

The distribution percentages shown here are from April 2013. As time progresses, more users will upgrade from older devices to newer ones. As a result, you should check the URL shown in this figure to get current percentages before you decide the minimum API level for your app.

As you review this figure, you may notice that it doesn't include some versions of Android such as 1.0 and 1.1. That's because there are virtually no devices left that run on these versions of Android. As time marches on, it's inevitable that the same fate will befall other older versions of Android too.

Android versions

Version	Code name	API	Release year	Distribution
1.6	Donut	4		0.1%
2.1	Eclair	7		1.7%
2.2	Froyo	8	2010	4.0%
2.3 - 2.3.2	Gingerbread	9		0.1%
2.3.3 - 2.3.7		10		39.7%
3.2	Honeycomb	13	2011	0.2%
4.0.3 - 4.0.4	Ice Cream Sandwich	15		29.3%
4.1	Jelly Bean	16	2012	23.0%
4.2		17		2.0%
4.3		18	2013	0%

A URL for current distribution percentages

`http://developer.android.com/about/dashboards`

Description

- Many new versions of Android were released from 2009 to 2012.

- The distribution percentages in this figure are from April 2013. To get current percentages, please visit the URL shown above.

- As you develop an Android app, you must decide the minimum API level that your app supports. As of April 2013, many developers choose Android 2.2 as the minimum API level to support since that covers a high percentage of all Android devices.

Figure 1-4   Android versions

System architecture

Figure 1-5 shows the Android system architecture, which is also known as the *Android stack*. This stack has four layers.

The bottom layer of the stack is Linux, an open-source operating system that's portable and secure. This operating system provides low-level drivers for hardware, networking, file system access, and inter-process communication (IPC).

The second layer up in the stack contains the native libraries. These libraries are written in C or C++. They include the *Dalvik virtual machine* (*VM*), which works similarly to the *Java virtual machine* (*JVM*). However, the Dalvik VM was designed specifically for mobile devices and their inherent limitations, such as battery life and processing power.

The third layer up in the stack contains the application framework. This layer is written mostly in Java, and it provides libraries that can be used by the top layer of the stack. In this book, you'll learn how to use some of these libraries, such as the libraries for the notification manager, content providers, and the location manager.

The top layer of the stack contains Android apps. These apps include pre-installed apps such as the apps that you can use to manage the home screen, manage your contacts, make and receive calls, browse the web, and so on. In addition, these apps include other apps that you can download and install. These types of apps are written in Java, and they are the type of apps that you'll learn to develop in this book.

Android stack

Description

- Linux is an open-source operating system that's portable and secure.
- The native libraries are written in C or C++. These libraries provide services to the Android application layer.
- The *Dalvik virtual machine* (*VM*) works similarly to the *Java virtual machine* (*JVM*). However, the Dalvik VM was designed specifically for mobile devices and their inherent limitations such as battery life and processing power.
- The application framework provides libraries written in Java that programmers can use to develop Android apps.

Figure 1-5   The Android system architecture

How apps are compiled and run

When you develop an Android app, you typically use an *IDE* (*Integrated Development Environment*) such as Eclipse to create a *project*. A project contains all of the files for the app including the files for the Java source code. In the next chapter, you'll learn how to use Eclipse to work with Android projects.

When you're ready to test a project, you can run it. Figure 1-6 shows how this works. When you run a project, the IDE typically compiles and packages the project automatically before running it. This is known as *building* the project.

When the IDE builds a project, it compiles the *Java source code* (.java files) into *Java bytecodes* (.class files). Then, it compiles the Java bytecodes into *Dalvik executable files* (.dex files) that can be run by the Dalvik virtual machine that's available from all Android devices.

When the IDE builds a project, it puts the .dex files and the rest of the files for the project into an *Android package* (.apk file). This file contains all of the files necessary to run your app including the .dex files and other compiled resources, uncompiled resources, and a binary version of the Android manifest. The *Android manifest* is a file that specifies some essential information about an app that the Android system must have before it can run the app. In its non-binary version, the Android manifest is stored in a file named AndroidManifest.xml.

For security reasons, all Android apps must be digitally *signed* with a certificate. During development, the IDE typically signs the app for you automatically using a special debug key. Then, it runs the app on the specified physical device such as a smartphone or tablet. Or, it runs the app on the specified *emulator*, which is a piece of software that runs on your computer and mimics an Android device. An Android emulator can also be called an *Android Virtual Device* (*AVD*).

The *Android debug bridge* (*ADB*) lets your IDE communicate with an emulator or a physical Android device. This is necessary to provide the debugging capabilities described in chapter 4.

When you are ready to release the app, you must sign the app in release mode, using your own private key. For more information about this, please see chapter 17.

Android system architecture

Description

- When you develop an Android app, you typically use an *IDE* (*Integrated Development Environment*) such as Eclipse to create a *project*, and you typically use Java as the programming language.

- When you develop an Android app, you can run it on a physical Android device, such as a smartphone or tablet. Or, you can run it on an *emulator*, which is a piece of software that runs on your computer and acts like an Android device. An Android emulator can also be called an *Android Virtual Device* (*AVD*).

- Before you can run a project, you must build the project. Typically, the IDE automatically builds a project before running it.

- When the IDE builds a project, it compiles the *Java source code* (.java files) into *Java bytecodes* (.class files) and then into *Dalvik executable files* (.dex files). Dalvik executable files can be run by the Dalvik virtual machine that's available from all Android devices.

- When the IDE builds a project, it puts the files for the project into an *Android package* (.apk file). This file contains all of the files necessary to run your app on a device or emulator including the .dex files, compiled resources (the resources.arsc file), uncompiled resources, and a binary version of the AndroidManifest.xml file.

- To run an app on an emulator or device, the app must be *signed* with a digital certificate that has a private key. During development, the IDE automatically signs the app for you in debug mode using a special debug key. Before you release an app, you must sign the app in release mode, using your own private key. For more information about this, please see chapter 17.

- The *Android debug bridge* (*ADB*) lets your IDE communicate with an emulator or a physical Android device.

Figure 1-6 How an Android app is compiled and run

A simple Android app

To give you a better idea of how an Android app works, this topic presents a basic Tip Calculator app. For a simple app like this one, the programmer only needs to modify three files: the XML file for the user interface, the XML file for the display text, and the Java source code for the activity. In addition, the programmer may sometimes need to view or modify the Android manifest file for the app.

The user interface

Figure 1-7 shows the user interface for the Tip Calculator app after it has been displayed in an emulator for Android 4.2 (API 17). Of course, this app can also be run on emulators for other Android platforms. Similarly, this app can also be run on a physical device that's configured for debugging.

The emulator shown in the figure includes a message that indicates that the hardware keyboard is not enabled in AVD and that you should use the on-screen keyboard. As a result, you can use the on-screen keyboard, known as a *soft keyboard*, to enter text as shown in this figure. To do that, you click on the Bill Amount text box. When you do, the emulator should display a soft keyboard that's optimized for entering a number with a decimal point. When you're done using that keyboard to enter an amount for the bill, you can click the Done key. When you do, the app calculates the tip and total for the bill.

If the Hardware Keyboard option is enabled for an emulator, the emulator should display a message that indicates that you can use your computer's keyboard to enter text. In that case, the emulator won't display a soft keyboard when you click on the Bill Amount text box. Instead, you use your computer's keyboard to enter text. Although this provides an easier way to enter text, it doesn't accurately emulate touchscreen devices.

By default, this Tip Calculator app uses a tip percent of 15%. However, if you want to increase or decrease the tip amount, you can click the Increase (+) or Decrease (-) buttons. When you do, the app recalculates the tip and total amounts for the bill.

The Tip Calculator app with the soft keyboard displayed

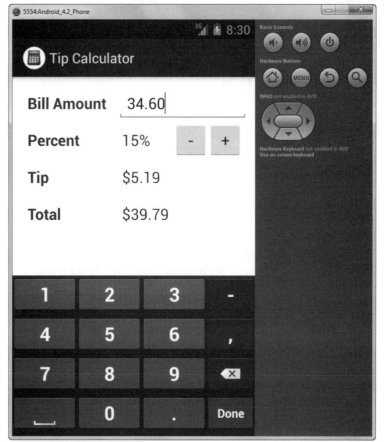

Description

- The Tip Calculator app shown in this figure is displayed in an emulator for Android 4.2 (API level 17).
- If the Hardware Keyboard option is not enabled for an emulator, you can use the on-screen keyboard, known as a *soft keyboard*, to enter text as shown in this figure.
- If the Hardware Keyboard option is enabled for an emulator, you can use your computer's keyboard to enter text.
- To calculate a tip, click the Bill Amount text box and enter the bill amount. When you're done, press the Done key.
- To increase or decrease the tip amount, click the Increase (+) or Decrease (-) buttons.
- The app automatically recalculates the tip and total amounts whenever the user changes the bill amount or tip percent.

Figure 1-7   The user interface for the Tip Calculator app

The XML for the user interface

Figure 1-8 shows the XML for the user interface. This code defines an *activity*, which is a screen that users can interact with, and it's stored in a file named activity_tip_calculator.xml.

If you have experience with XML or HTML, you may be able to understand much of this code already. If not, don't worry! You'll learn the details about how this code works in chapter 3. For now, here's a brief explanation.

An XML file consists of one or more *elements*. These elements can be nested within each other. In this file, the RelativeLayout element is the *parent element*, and it has ten *child elements*: seven TextView elements, one EditText element, and two Button elements.

Each of these elements has multiple *attributes*. For example, the RelativeLayout element has a padding attribute that determines how much space there is on the screen between the RelativeLayout element and its child elements. Many of these attributes are set automatically for you when you use a graphical tool like Eclipse to create the layout. As a result, you often don't need to set these attributes. However, you sometimes need to modify them.

The RelativeLayout element defines a *layout*, which is a container that determines where its child elements are displayed on the screen. In Android, there are several types of layouts that use different techniques to control where the child elements are displayed. The *relative layout* organizes its child elements in relation to one another.

The RelativeLayout element has several attributes that determine how it works. Here, the first two attributes are set automatically, and you can usually ignore them. Then, the layout_width and layout_height attributes determine that the layout should use the whole screen. Finally, the padding attribute determines that there should be some *padding*, or space, between the edge of the layout and its child elements.

To specify the amount of space, the padding attribute specifies a measurement of "10dp". Here, *dp* stands for *density-independent pixel*. In general, it's considered a best practice to use density-independent pixels for padding and margins whenever possible as it allows your app to scale properly on various devices.

The child elements define the *controls* for the bill amount. In Android, controls are typically referred to as *widgets*.

The first TextView element displays "Bill Amount" on the user interface. In Android terms, this element is often referred to as a *text view*. Since this widget displays text that labels another widget, it can also be referred to as a *label*.

The first attribute of this widget defines an ID for the widget. Here, the value for the id attribute begins with "@+id/". This indicates that the following text is the ID for the widget. In this case, the widget has an ID of billAmountLabel. The next three attributes for this widget set its height, width, and padding.

The activity_tip_calculator.xml   **Page 1**

```xml
<RelativeLayout xmlns:android="http://schemas.android.com/apk/res/android"
    xmlns:tools="http://schemas.android.com/tools"
    android:layout_width="match_parent"
    android:layout_height="match_parent"
    android:padding="10dp" >

    <!-- The bill amount -->

    <TextView
        android:id="@+id/billAmountLabel"
        android:layout_width="wrap_content"
        android:layout_height="wrap_content"
        android:padding="10dp"
        android:text="@string/bill_amount_label"
        android:textSize="20sp"
        android:textStyle="bold" />

    <EditText
        android:id="@+id/billAmountEditText"
        android:layout_width="wrap_content"
        android:layout_height="wrap_content"
        android:layout_alignBaseline="@+id/billAmountLabel"
        android:layout_marginLeft="5dp"
        android:layout_toRightOf="@+id/billAmountLabel"
        android:ems="8"
        android:inputType="numberDecimal"
        android:text="@string/bill_amount"
        android:textSize="20sp" >

        <requestFocus />
    </EditText>

    <!-- The tip percent -->

    <TextView
        android:id="@+id/percentLabel"
        android:layout_width="wrap_content"
        android:layout_height="wrap_content"
        android:layout_alignLeft="@+id/billAmountLabel"
        android:layout_below="@+id/billAmountLabel"
        android:padding="10dp"
        android:text="@string/tip_percent_label"
        android:textSize="20sp"
        android:textStyle="bold" />

    <TextView
        android:id="@+id/percentTextView"
        android:layout_width="wrap_content"
        android:layout_height="wrap_content"
        android:layout_alignBaseline="@+id/percentLabel"
        android:layout_alignLeft="@+id/billAmountEditText"
        android:padding="5dp"
        android:text="@string/tip_percent"
        android:textSize="20sp" />
```

Figure 1-8   The XML for the user interface (part 1 of 3)

The fifth attribute sets the text that's displayed on the widget to the value that's stored in the bill_amount_label string of the strings.xml file, which you'll learn about in a moment. To get a string from the strings.xml file, the value for this attribute begins with "@string/". This indicates that the following text is the name of the string element that contains the text.

To specify the size of the text, the textSize attribute specifies a measurement of "20sp". Here, *sp* stands for *scale-independent pixel*. In general, it's considered a best practice to use scale-independent pixels for text sizes whenever possible as it allows your text sizes to scale properly. Finally, the textStyle attribute sets the style of this text to bold.

The EditText element that follows displays the widget that allows the user to enter a bill amount. The Android documentation refers to this type of widget as an *editable text view*, but you can also refer to it as a *text box*. Most of the attributes of this EditText element work similarly to those of the previous TextView element. However, there are some differences. For example, the alignment attributes align the text box relative to its label. Then, the ems attribute specifies that the text box width should be 8 ems (8 of the letter *m*). Next, the inputType element specifies that the soft keyboard should only allow the user to enter decimal numbers.

To make it easy to test this app, the text attribute for the EditText element specifies a default value of "34.60". During development, you can leave this attribute set to a default value like this one. However, once the application is working, you can set this attribute to an empty string.

The next two TextView elements define the widgets that display the tip percent. By now, you should have a general idea of what most of the attributes do. The main point to notice is that the layout attributes align these controls relative to one another. For example, the label for the tip percent is aligned with the left side of the label for the bill amount. Similarly, the percent label is displayed below the bill amount label.

The Button elements define the buttons that allow the user to increase or decrease the tip percent. These widgets are 45dp wide and high and they're aligned to the right of the TextView widgets for the tip percent.

The last four TextView elements define the widgets that display the tip and total amounts. These elements work like the previous TextView elements shown in this figure.

The activity_tip_calculator.xml file   **Page 2**

```xml
<Button
    android:id="@+id/percentDownButton"
    android:layout_width="45dp"
    android:layout_height="45dp"
    android:layout_alignBaseline="@+id/percentTextView"
    android:layout_marginLeft="25dp"
    android:layout_toRightOf="@+id/percentTextView"
    android:text="@string/decrease"
    android:textSize="20sp" />

<Button
    android:id="@+id/percentUpButton"
    android:layout_width="45dp"
    android:layout_height="45dp"
    android:layout_alignBaseline="@+id/percentDownButton"
    android:layout_toRightOf="@+id/percentDownButton"
    android:text="@string/increase"
    android:textSize="20sp" />

<!-- the tip amount -->

<TextView
    android:id="@+id/tipLabel"
    android:layout_width="wrap_content"
    android:layout_height="wrap_content"
    android:layout_alignLeft="@+id/percentLabel"
    android:layout_below="@+id/percentLabel"
    android:padding="10dp"
    android:text="@string/tip_amount_label"
    android:textSize="20sp"
    android:textStyle="bold" />

<TextView
    android:id="@+id/tipTextView"
    android:layout_width="wrap_content"
    android:layout_height="wrap_content"
    android:layout_alignBaseline="@+id/tipLabel"
    android:layout_alignLeft="@id/billAmountEditText"
    android:padding="5dp"
    android:text="@string/tip_amount"
    android:textSize="20sp" />
```

Figure 1-8   The XML for the user interface (part 2 of 3)

The XML for the display text

Part 3 of figure 1-8 also shows the XML for the display text. This XML is stored in a file named strings.xml, and it defines the names and values for the display text. For example, the string element with a name of app_name has a value of "Tip Calculator". The string element with a name of bill_amount_label has a value of "Bill Amount". The string element with a name of bill_amount has a value of "34.60". And so on. To differentiate between the widgets that display labels and the widgets that display results, I have appended "_label" to the names of the labels.

At this point, you may be wondering why Android provides a mechanism for separating the display text from the user interface. The reason is because Android apps are likely to run on devices used around the world. As a result, Android attempts to make it easy to develop apps that are appropriate for different locales. This is known as *localization*. For example, you can provide one strings.xml file for English and another for French. Then, if Android determines that the locale for the device is French, it will use the strings.xml file that contains the French text. You'll learn more about how that works in chapter 3.

The activity_tip_calculator.xml file   **Page 3**

```xml
    <!-- the total -->

    <TextView
        android:id="@+id/totalLabel"
        android:layout_width="wrap_content"
        android:layout_height="wrap_content"
        android:layout_alignLeft="@+id/tipLabel"
        android:layout_below="@+id/tipLabel"
        android:padding="10dp"
        android:text="@string/total_amount_label"
        android:textSize="20sp"
        android:textStyle="bold" />

    <TextView
        android:id="@+id/totalTextView"
        android:layout_width="wrap_content"
        android:layout_height="wrap_content"
        android:layout_alignBaseline="@+id/totalLabel"
        android:layout_alignLeft="@+id/tipTextView"
        android:padding="5dp"
        android:text="@string/total_amount"
        android:textSize="20sp" />

</RelativeLayout>
```

The strings.xml file

```xml
<resources>
    <string name="app_name">Tip Calculator</string>
    <string name="title_activity_tip_calculator">Tip Calculator</string>

    <string name="bill_amount_label">Bill Amount</string>
    <string name="bill_amount">34.60</string>

    <string name="tip_percent_label">Percent</string>
    <string name="tip_percent">15%</string>
    <string name="increase">+</string>
    <string name="decrease">-</string>

    <string name="tip_amount_label">Tip</string>
    <string name="tip_amount">$0.00</string>

    <string name="total_amount_label">Total</string>
    <string name="total_amount">$0.000</string>
</resources>
```

Description

- In Android development, an *activity* is a screen that users can interact with.
- The activity_tip_calculator.xml file contains the XML that defines the user interface.
- The strings.xml file contains the XML for the text strings that are displayed in the user interface.

Figure 1-8   The XML for the user interface (part 3 of 3)

The Java source code

Figure 1-9 shows the Java source code that handles events that occur on the user interface, gets input from the user interface, performs calculations, and updates the user interface. This source code is stored in a file named TipCalculatorActivity.java.

If you have experience with Java, you may be able to understand much of this code already. If not, don't worry! You'll learn the details about how this code works in chapter 3. For now, here's a brief explanation.

This source code declares a class named TipCalculatorActivity. To start, the package statement stores this class in the com.murach.tipcalculator package. Then, it imports all the Java and Android classes and interfaces that are needed for this class. As you can see, most of these classes and interfaces are stored in the android.view or android.widget packages.

The TipCalculatorActivity class inherits the Activity class that's provided by Android. This class defines the methods that are called during the lifecycle of an activity. This allows the TipCalculatorActivity class to override these lifecycle methods to perform tasks at different points in the activity's life.

Of these methods, the most important one is the onCreate method, which is called when Android creates an activity. This method is the only lifecycle method that's used by the activity shown in this chapter. However, in chapter 3, you'll learn how to use two other activity lifecycle methods to save and restore data if the activity is interrupted by another activity such as a phone call.

The TipCalculatorActivity class also implements two interfaces. These interfaces allow this class to handle the events that occur on the user interface. First, the OnEditorActionListener interface makes the class a listener for the EditorAction event. As a result, this class must implement the onEditorAction method. Then, the OnClickListener interface makes the class a listener for the Click event. As a result, this class must implement the onClick method.

Within the class, the first six statements define the instance variables for the widgets that the class needs to work with. Then, the seventh statement defines an instance variable for the tip percent and sets its default value to .15. Since this statement uses the float type, this code appends an *f* to the number to indicate that the literal is of the float type, not the double type.

The onCreate method begins by passing its Bundle parameter to the superclass (the Activity class). Then, it uses the setContentView method that's available from the Activity superclass to display the user interface that's defined in the activity_tip_calculator.xml file. To do that, this method uses the R class to access the compiled resources that are available to the Android app. This class is created and compiled automatically when you build a project. In other words, Android converts the XML in the activity_tip_calculator.xml file into Java code, compiles it, and makes it available through the R class.

The TipCalculatorActivity.java file   **Page 1**

```java
package com.murach.tipcalculator;

import java.text.NumberFormat;

import android.os.Bundle;
import android.view.KeyEvent;
import android.view.View;
import android.view.View.OnClickListener;
import android.widget.Button;
import android.widget.EditText;
import android.widget.TextView;
import android.widget.TextView.OnEditorActionListener;
import android.app.Activity;

public class TipCalculatorActivity extends Activity
implements OnEditorActionListener, OnClickListener {

    // define instance variables for the widgets
    private EditText billAmountEditText;
    private TextView percentTextView;
    private Button   percentUpButton;
    private Button   percentDownButton;
    private TextView tipTextView;
    private TextView totalTextView;

    // define an instance variable for tip percent
    private float tipPercent = .15f;

    @Override
    public void onCreate(Bundle savedInstanceState) {
        super.onCreate(savedInstanceState);
        setContentView(R.layout.activity_tip_calculator);

        // get references to the widgets
        billAmountEditText = (EditText) findViewById(R.id.billAmountEditText);
        percentTextView = (TextView) findViewById(R.id.percentTextView);
        percentUpButton = (Button) findViewById(R.id.percentUpButton);
        percentDownButton = (Button) findViewById(R.id.percentDownButton);
        tipTextView = (TextView) findViewById(R.id.tipTextView);
        totalTextView = (TextView) findViewById(R.id.totalTextView);

        // set the listeners
        billAmountEditText.setOnEditorActionListener(this);
        percentUpButton.setOnClickListener(this);
        percentDownButton.setOnClickListener(this);

        // calculate the tip and display the results
        calculateAndDisplay();
    }
```

Figure 1-9   The Java source code (part 1 of 2)

After displaying the user interface, the onCreate method gets references to the six widgets that it declared earlier. To do that, it calls the findViewById method for each widget and uses the R class to pass the ID of the widget to that method. Then, it casts the widget object that's returned to the appropriate type. These types correspond with the names of the elements in the XML file for the user interface.

After getting references to the widgets, this code sets the listeners. First, it sets the current class (TipCalculatorActivity) as the listener for the EditorAction event on the bill amount text box. As a result, the onEditorAction method in this class is executed when the user uses the bill amount text box to enter an amount for the bill. Then, this code sets the current class as the listener for the Click event on both of the buttons. As a result, the onClick method is executed when the user clicks on either button.

After setting the listeners, this code calls the calculateAndDisplay method. This method calculates the tip and total amounts and displays all current data on the user interface. To start, it gets the string that was entered into the bill amount text box. Then, it checks whether the string is empty. If so, it sets the bill amount to zero. Otherwise, it converts the string to a float value. Next, it calculates the tip and total amounts. Finally, it formats and displays the results of the calculations. To do that, it uses the NumberFormat class to convert the float values into string values that have currency and percent formatting.

The onEditorAction method is executed whenever the user presses an action key on a soft keyboard such as the Done key. This method calls the calculateAndDisplay method to calculate the new tip amount and total and display it on the user interface.

The onClick method is executed whenever the user clicks on either of the buttons. This method begins by using a switch statement to check which button is clicked. Then, if the Decrease (-) button is clicked, this code decreases the tip percent by 1 percent and calls the calculateAndDisplay method to refresh the data that's on the user interface. Conversely, if the Increase (+) button is clicked, this code increases the tip percent by 1 percent and calls the calculateAndDisplay method to refresh the data.

The TipCalculatorActivity.java file   **Page 2**

```java
    public void calculateAndDisplay() {

        // get the bill amount
        String billAmountString = billAmountEditText.getText().toString();
        float billAmount;
        if (billAmountString.equals("")) {
            billAmount = 0;
        }
        else {
            billAmount = Float.parseFloat(billAmountString);
        }

        // calculate tip and total
        float tipAmount = billAmount * tipPercent;
        float totalAmount = billAmount + tipAmount;

        // display the results with formatting
        NumberFormat currency = NumberFormat.getCurrencyInstance();
        tipTextView.setText(currency.format(tipAmount));
        totalTextView.setText(currency.format(totalAmount));

        NumberFormat percent = NumberFormat.getPercentInstance();
        percentTextView.setText(percent.format(tipPercent));
    }

    @Override
    public boolean onEditorAction(TextView v, int actionId, KeyEvent event) {
        calculateAndDisplay();
        return false;
    }

    @Override
    public void onClick(View v) {
        switch (v.getId()) {
        case R.id.percentDownButton:
            tipPercent = tipPercent - .01f;
            calculateAndDisplay();
            break;
        case R.id.percentUpButton:
            tipPercent = tipPercent + .01f;
            calculateAndDisplay();
            break;
        }
    }
}
```

Figure 1-9   The Java source code (part 2 of 2)

The Android manifest

Figure 1-10 shows the AndroidManifest.xml file, which is known as the *Android manifest*. This file specifies some essential information about an app that the Android system must have before it can run the app. Within the manifest, there is always one application element. Within the application element, there may be one or more activity elements. However, for this app, there is only a single activity. This activity is identified as the main activity, and it is launched when the app starts.

Like the XML file for the user interface, the manifest file can access strings in the strings.xml file. For example, the label attribute of the application element specifies the app_name string in the strings.xml file. Similarly, the label attribute of the activity element specifies the title_activity_tip_calculator string in the strings.xml file.

This manifest file sets the screenOrientation attribute of the activity to "portrait". As a result, this application can only be displayed in portrait orientation, not landscape orientation. That's important because this activity doesn't include the code that's necessary to gracefully handle the switch between the portrait and landscape orientations. In chapter 3, you'll learn how to handle changes in orientation.

The AndroidManifest.xml file

```xml
<manifest xmlns:android="http://schemas.android.com/apk/res/android"
    package="com.murach.tipcalculator"
    android:versionCode="1"
    android:versionName="1.0" >

    <uses-sdk
        android:minSdkVersion="8"
        android:targetSdkVersion="17" />

    <application
        android:icon="@drawable/ic_launcher"
        android:label="@string/app_name"
        android:theme="@style/AppTheme"
        android:allowBackup="true" >
        <activity
            android:name=".TipCalculatorActivity"
            android:label="@string/title_activity_tip_calculator"
            android:screenOrientation="portrait">
            <intent-filter>
                <action android:name="android.intent.action.MAIN" />

                <category android:name="android.intent.category.LAUNCHER" />
            </intent-filter>
        </activity>
    </application>

</manifest>
```

Description

- The *Android manifest* (the AndroidManifest.xml file) specifies some essential
  information about an app that the Android system must have before it can run the
  app, including the first activity to launch when the app is started.

Figure 1-10   The Android manifest file

Perspective

The goal of this chapter has been to introduce you to some concepts and terms that you should understand before getting started with Android programming. However, this chapter is just an overview of Android programming. In the next few chapters, you'll learn the hands-on skills you need to develop a simple Android app like one shown in this chapter.

Terms

Android	Android Virtual Device (AVD)
application	Android debug bridge (ADB)
app	soft keyboard
Open Handset Alliance	activity
Android Open Source Project (AOSP)	element
Software Development Kit (SDK)	parent element
Android stack	child element
Dalvik virtual machine (VM)	attribute
Java virtual machine (JVM)	layout
Integrated Development Environment (IDE)	relative layout
	padding
project	density-independent pixel (dp)
building	control
Java source code	widget
Java bytecodes	text view
Dalvik executable files	label
Android package	scale-independent pixel (sp)
Android manifest	editable text view
signed app	text box
emulator	localization

Summary

- *Android* is a Linux-based operating system designed primarily for touch-screen mobile devices such as smartphones and tablet computers. It was first released in 2008. Android code is open-source.

- Android developers write *applications*, or *apps*.

- Android system architecture, known as the *Android stack*, consists of four layers: Linux, native libraries, the application framework, and Android apps.

- An Android app is typically developed using an *IDE (Integrated Development Environment)* like Eclipse, using Java as the programming language.

- Android apps can be run on a physical Android device or on an *emulator*, also called an *Android Virtual Device (AVD)*.

- An Android project must be built before it is run, compiling the *Java source code* (.java files) into *Java bytecodes* (.class files) and then into *Dalvik executable files* (.dex files).

- All of the files for an Android project are put into an *Android package* (.apk file), which includes a binary version of the AndroidManifest.xml file.

- To run an app on an emulator or device, it must be digitally *signed* with a certificate.

- The *Android debug bridge (ADB)* lets your IDE communicate with an emulator or a physical Android device.

2

How to use Eclipse
for Android development

Eclipse is an *Integrated Development Environment* (*IDE*) that you can use to develop Android apps. Eclipse is open-source, available for free, and runs on all modern operating systems. Although it's possible to use other IDEs, Eclipse has support from most of the Android development community, including many of the developers at Google who helped develop Android. As a result, most Android developers use Eclipse.

How to work with existing projects

If you read chapter 1, you're ready to start working with existing Eclipse projects. In particular, you're ready to learn how to open and run any of the applications for this book. You can download these applications from our web site as described in the appendix.

An introduction to Eclipse projects

When you first start Eclipse as described in the appendix, you may notice that it displays a splash screen that says, "Android Developer Tools". In addition, the title bar for the application uses a green icon instead of the traditional blue Eclipse icon and says ADT instead of Eclipse. That's because the appendix shows how to install a version of Eclipse that has been customized to include the Android Developer Tools (ADT).

Figure 2-1 shows the Eclipse *workbench*, which is where you work in Eclipse. The different parts of the workbench are known as *views*. In this figure, for example, the workbench shows the Package Explorer view, the Outline view, and the Console view.

In this figure, the Package Explorer only contains one *project*, which is a directory that contains all the files for an app. This lone project is for the Tip Calculator app. However, it's common to import several projects into the workbench. Then, you can use the Package Explorer to view all of the projects that you have imported.

Within a project, the src directory contains the Java source code. Within the src directory, the .java files can be organized into *packages*. In this figure, the project has a single package named com.murach.tipcalculator. Within this package, the project has a single .java file named TipCalculatorActivity.java. In this figure, this file is open in the Java *code editor*. You'll learn more about working with this code editor later in this chapter.

Within a project, the res/layout directory contains the XML file or files that define the layout of the user interface. In this figure, the project has a single XML layout file named activity_tip_calculator.xml. You'll learn more about working with this XML file later in this chapter.

Within a project, the res/values directory contains the XML file that defines the display text for the user interface. These strings are stored in a file named strings.xml. Again, you'll learn more about working with this XML file later in this chapter.

The Eclipse workbench with a project

Description

- When you first start Eclipse, it may display a Welcome page. If so, you can click the Workbench icon to go to the *workbench*, which is where you develop apps. The workbench contains multiple *views* such as the Package Explorer, Outline, and Console views.

- The workbench provides *code editors* for working with Java and XML files.

- An Eclipse *project* consists of a top-level directory that contains the subdirectories and files for an application.

- The src (source) directory contains the Java packages that store the source code (.java files) for the project. Within the src directory, the .java files can be organized into one or more *packages,* where each package corresponds with a subdirectory.

- The res (resources) directory contains other resources for the project, including the XML files that define the user interface.

- The directories, files, and libraries that make up a project are listed in the Package Explorer. If this view isn't visible, you can display it by selecting the Window→Show View→Package Explorer item.

- You can expand and collapse the nodes of the Package Explorer by clicking on the triangles to the left of each node.

Figure 2-1   An introduction to Eclipse

How to set the workspace

Figure 2-2 shows how to set the *workspace* for an Eclipse session. The workspace is the directory that's used by Eclipse to store the subdirectories and files it needs to work.

If you download and install the source code for this book as described in the appendixes, the workspace for this source code is available here:

```
\murach\android\workspace
```

As a result, the easiest way to get started with the source code that's available for this book is to use the Workspace Launcher dialog box to specify this workspace. Then, you can use this workspace when working on projects from this book, and you can use another workspace when working on other projects.

The Workspace Launcher dialog box

How to switch the Eclipse workspace

1. Start Eclipse.
2. Select the File→Switch Workspace→Other item from the menu bar. This should display the Workspace Launcher dialog box.
3. Click the Browse button and use the resulting dialog box to select the workspace directory. When you do, Eclipse will restart and load the new workspace.

Description

- In Eclipse, a *workspace* is a directory that stores information about how to configure the workbench and which projects to display. In addition, the workspace typically contains the directories and files for the projects, though it's possible to store a project outside of the workspace.

- Once you have set a workspace, it will appear in the Switch Workspace menu, and you can select it to switch to it. As a result, it's easy to switch between existing workspaces.

Figure 2-2   How to set the workspace

How to import a project into the workspace

Figure 2-3 shows how to import an existing project into a workspace. To do that, select the Import item from the File menu. In the first dialog box, expand the node for the General category and select the Existing Projects into Workspace option. In the second dialog box, click the Browse button and select the directory that contains the project or projects that you want to import. When you do, all of the possible projects will be displayed in the Projects list box so you can select the ones you want to import. In this figure, for example, the project for the Tip Calculator app presented in chapter 3 is selected. As a result, clicking the Finish button imports this project into the current workspace.

Often, when you import an existing project, the directory that you specify in the Import dialog box is in a subdirectory of the current workspace. In that case, you don't need to use the "Copy projects into workspace" check box since the project is already in the workspace. However, if you specify a directory other than the directory for the current workspace, you may want to select the "Copy projects into workspace" check box. That way, the directories and files for the projects that you select will be copied into the current workspace, and the changes you make to these projects won't change the directories and files in the original location. On the other hand, if you want your changes to these projects to affect the directories and files in the original location, don't select the "Copy projects into workspace" check box.

The dialog box for importing an existing project

How to import a project

1. Select File→Import from the menu bar.
2. In the first dialog box, select the General→Existing Projects into Workspace option and click on the Next button.
3. In the second dialog box, click the Browse button and browse to the directory that contains the project you want to import. This should display the project in the Projects pane and select it.
4. Click the Finish button to import the selected project or projects.

Description

*   To import all projects in the book_apps directory, navigate to the book_apps in step 3. This should select all projects in the book_apps directory.
*   If you get an error that says, "Unable to resolve target", you can fix the problem by right-clicking on the project name in the Package Explorer and selecting the Android Tools→Fix Project Properties item.

Figure 2-3   How to import an existing project

How to remove a project from the workspace

To keep your workspace organized, you may want to remove a project from the workspace as shown in figure 2-4. For example, if you want to remove the project named ch03_TipCalculator that's shown in figure 2-1, you can right-click on the project in the Package Explorer and select the Delete item.

When you do that, you'll get a Delete Resources dialog box like the one shown in this figure. Then, you can remove the project from the workspace without deleting the files for the project by making sure that the "Delete project contents on disk" option is not selected. Or, you can remove the project from the workspace and also delete all the files associated with the project by selecting that option.

The dialog box for removing a project from the workspace

Description

- To remove a project from the workspace, right-click on the project in the Package Explorer, select the Delete item, make sure the "Delete project contents" check box is *not* selected, and click the OK button.

- To delete a project and its directories and files, right-click on the project in the Package Explorer, select the Delete item, select the "Delete project contents" check box, and click the OK button. Then, if necessary, respond to the Delete Launch Configurations dialog box.

Figure 2-4   How to remove a project from the workspace

How to work with the user interface

Figure 2-5 shows how to work with the user interface. To start you can open the XML file for the user interface by expanding the res/layout directory for the project and double-clicking on the XML file for the layout. Then, if you want to view that user interface in the Graphical Layout editor, you can click on the Graphical Layout tab. In this figure, for example, the user interface for the Tip Calculator app is shown in the Graphical Layout editor.

In this figure, the workbench shows the Package Explorer view, the Palette view, and the Outline view. In addition, the Properties view is shown at the bottom of the Outline view. Finally, the bottom of the workbench provides tabs that let you access more views. If a view you want to use isn't displayed, you can display it by using the Window→Show View menu to select the view that you want.

Once you've opened a user interface in the Graphical Layout editor, you can add widgets to the layout by clicking on a category in the Palette view to display the widgets for that category. Then, you can add a widget by dragging it from the Palette to the layout, and you can align a widget by dragging it to where you want it to be displayed. As you do that, the Graphical Layout editor typically displays guidelines that can help you align the widget.

When you add widgets, it's sometimes hard to find the widget you want in the Palette. In this figure, for example, it's easy to see that the TextView and Button widgets are available from the Form Widgets category. However, where is the EditText widget? As it turns out, several variations of an EditText widget are available from the Text Fields category, and they all have different names. Each variation is designed for accepting different types of text input, such as email addresses, phone numbers, dates, times, and numbers. To add an EditText widget that allows the user to enter a decimal number, you can use the widget named Number (Decimal). This creates an EditText widget and automatically sets its attributes so it only allows the user to enter decimal numbers.

To work with an individual widget, you need to select it by clicking on it in the Graphical Layout editor or in the Outline view. Then, you can use the Properties view to set its properties. To do that, scroll through the properties for the widget and select or enter a value for any property that you want to modify. In this figure, for example, the label for the bill amount is selected, and the properties that control its alignment, are shown in the Properties view.

The Graphical Layout editor

Description

- To open the XML file for an app's user interface, expand the res\layout directory for the project, and double-click on the file to open it.

- To display a user interface in the Graphical Layout editor, click on the Graphical Layout tab.

- To display a view such as the Outline view, use the Window→Show View menu.

- To add widgets to the layout, click on the widget category in the Palette view to display the widget, and drag the widget onto the layout.

- The TextView and Button widgets are in the Form Widgets category of the Palette.

- All of the widgets in the Text Fields category are variations on the EditText widget. For example, the Number (Decimal) widget is an EditText widget with its properties set to make it easy for the user to enter decimal numbers.

- To align widgets, drag the widget within the layout.

- To select a widget, click it in the Graphical Layout editor or in the Outline view.

- To set the properties of a widget, select it and use the Properties view.

Figure 2-5　How to work with a user interface (part 1 of 2)

When you develop a user interface, you often need to view its XML in an XML editor. To do that, you can open the XML file for the user interface and click on the tab that displays the name of the XML file. When you view the XML for a user interface, Eclipse makes it easier to for you to recognize the syntax by using different colors for elements and attributes. In addition, Eclipse provides standard File and Edit menus and keystroke shortcuts that let you save and edit the source code. For example, you can press Ctrl+S to save your source code, and you can use standard commands to cut, copy, and paste code.

Although you can use the XML editor to create a user interface, it's usually easier to use the Graphical Layout editor. Then, you can use the XML editor to fine-tune the user interface if necessary. As a result, you don't usually need to use the XML editor to add widgets.

However, it often makes sense to use the XML editor to modify the XML that's generated by the Graphical Layout editor. To do that, you can set the properties of a widget by adding, editing, or deleting the XML attributes for the widget. As you do this, you can take advantage of the code completion feature to enter property names and values. This feature is described later in this chapter.

If you want to include comments in the XML for your user interface, you can use the code editor to add them. To do that, use the same syntax as an HTML comment (start with <!-- and end with -->). In this figure, for example, the code includes a comment that identifies the starting point for the elements that define the two widgets for the bill amount row.

The XML editor

The syntax for an XML comment

```
<!-- your comment goes here -->
```

Description

- Once the XML file for a user interface is open, you can display it in the XML editor by clicking on the tab that displays the name of the XML file.

- In the XML editor, you can add widgets to the layout. To do that, enter the XML code for the widget.

- In the XML editor, you can set the properties of a widget by adding or editing the XML attributes for the widget.

- In the XML editor, you can code comments. To do that, use the same syntax as an HTML comment.

- When you use the XML editor, you can use the code completion feature described later in this chapter to help you enter the names and values of the properties for a widget.

Figure 2-5   How to work with a user interface (part 2 of 2)

How to work with other XML resources

Figure 2-6 shows how to use Eclipse to work with other XML resources besides the user interface. These resources can define display text (strings.xml), styles (styles.xml), colors (colors.xml), and so on. In this figure, for example, the strings.xml file that's stored in the res/values directory is displayed in the XML editor. At this point, you can use the XML editor to work with this file. Or, if you want to use the Resource editor to work with this resource, you can click on the Resources tab to display the Resource editor for this file. In most cases, though, it's easier to use the XML editor.

In general, the techniques for working with XML resources are similar to the techniques for working with the user interface. As a result, once you understand how to work with the XML for a user interface, you shouldn't have much trouble working with other XML resources.

The strings.xml file

Description

- You can use Eclipse to work with the XML files that define display text (strings.xml), styles (styles.xml), colors (colors.xml), and so on.

- To open the XML file for a resource, expand the appropriate subdirectory of the res directory and double-click on the XML file.

- To switch to the XML editor, click on the tab that contains the name of the XML file.

- To switch to the Resources editor, click on the Resources tab.

Figure 2-6   How to work with other XML resources

How to work with the Java code

To work with the Java source code for an app, you can use the Package Explorer to expand the src directory and expand the package that contains the .java file. Then, you can double-click on that file to open it in the Eclipse code editor. In figure 2-7, for example, the .java file for the Tip Calculator activity is open in the code editor. This file is stored in the com.murach.tipcalculator package of the src directory.

Once you open Java code in the editor, Eclipse makes it easier for you to recognize the syntax by using different colors for different parts of the syntax, such as Java keywords, instance variables, and comments. In addition, the Outline view lists the instance variables and methods available from the current Java class, and you can jump to an instance variable or method by clicking on it.

You can also rename or delete a .java file from the Package Explorer. To do that, just right-click on the file and select the appropriate item from the resulting menu. If you rename a file, Eclipse automatically changes both the name of the .java file and the name of the class. Since the name of the .java file must match the name of the class, this is usually what you want.

The Java code for the Tip Calculator app

Description

- To display Java code in the Eclipse text editor, use the Package Explorer to expand the src directory, expand the package that contains the Java code, and double-click on the .java file that you want.

- To jump to an instance variable or method, you can click on the instance variable or method in Outline view.

- To rename a .java file, right-click on the file, select the Refactor→Rename item, and enter a new name for the file. This automatically renames the class that corresponds with the file.

- To delete a .java file, right-click on the file, select the Delete item, and confirm the deletion.

Figure 2-7   How to work with Java code

How to set the run configuration

For an Android project, the *run configuration* specifies what Eclipse should do when you run the Android app. This includes which emulator or physical device to use.

Figure 2-8 shows how to set the run configuration for an Android project. To start, you can view the run configuration for a project by right-clicking on the project in the Package Explorer and selecting the Run As→Run Configurations item from the menus. Then, to change the run configuration for the selected project, you can click on the tabs and set the options you want. In most cases, you only need to use the Target tab to set the target device for the project.

In this figure, the Target tab has been set to the "Always prompt" option. That way, when you run a project, Eclipse always displays a dialog box like the one shown in the next figure that prompts you to select the emulator or physical device that you want to use.

If you want Eclipse to automatically pick a device, you can select the "Automatically pick compatible device" option. Then, if a compatible physical device is connected to the computer, Eclipse will run the app on that device. Otherwise, Eclipse will run the app on a compatible emulator. If that emulator isn't already running, Eclipse will start it.

The Run Configurations dialog box

Description

- To view the *run configuration* for an app, right-click on the project in the Package Explorer, and select the Run As→Run Configurations item from the menus. Then, click on the tabs and set the options for the run configuration. Click the Apply button when you're done.

- On the Android tab, you can specify the activity that's started. In most cases, you can leave this tab at its default values.

- On the Target tab, you can specify the device that you want to use. When testing Android apps, you often need to change the selected option on this tab.

- On the Common tab, you can specify some other options for the run configuration. In most cases, you can leave this tab at its default values.

Figure 2-8   How to set the run configuration

How to run an app on a physical device

Figure 2-9 shows how to run an app on a connected device. To do that, make sure a compatible physical device is connected to your computer. Typically, you use a USB cable to connect your device to your computer. Then, to run an app, you can right-click on the project in the Package Explorer and select the Run As→Android Application item from the menus.

When you're working with a project, you typically set it as the current project by selecting it or one of its files in the Package Explorer. Most of the time, you do this without even realizing it. For example, when you open a file within a project, you typically select that file. Then, you can run the project for that file by pressing Ctrl+F11 or by clicking on the Run button that's available on the toolbar.

Depending on your run configuration, running a project runs the app on the connected device, or it displays an Android Device Chooser dialog box like the one shown in this figure. If you get this dialog box, you can select the "Choose" option, select the device, and click the OK button. In this figure, for example, the "Choose" option would allow you to pick the physical device named "asus-nexus_7".

Once your app is running on the physical device, you can test that app using any of the hardware that's available on that device. For example, you can rotate the device to test how it handles screen orientation changes. Similarly, if the device has a physical keyboard, you can test the physical keyboard to see if it works properly for your app.

By default, Eclipse automatically compiles an app before running it. Since this saves a step in the development process, this is usually what you want. Sometimes, though, Eclipse gets confused and doesn't compile a class that's needed by your project. For example, Eclipse sometimes doesn't compile the R class that contains the compiled resources needed by your project. In that case, you can usually fix this issue by selecting the Project→Clean item from the menu bar. This cleans the project and rebuilds it from scratch.

How to run an app on an emulator

This figure also shows how to run an app on an emulator. To do that, you use the same technique for running an app on a physical device. However, before you run the app, you must disconnect the physical device from your computer. Or, you must set the run configuration for the project so it allows you to choose the device. Then, when you get the dialog box shown in this figure, you can select an emulator that's already running by selecting the "Choose" option and selecting the emulator. Or, you can start a new emulator, by selecting the "Launch" option and selecting the emulator.

Sometimes, an emulator may be too big to fit on your screen. Other times, the size of the emulator that's displayed on your computer may be larger than the size of the device that you want to emulate. In those cases, you can make the emulator smaller by selecting the "Launch" option, selecting the emulator,

The Android Device Chooser dialog box

Description

- To run any app, right-click on the project in the Package Explorer and select the Run As→Android Application item. This runs the app on the default device, or it displays the Android Device Chooser dialog box to let you pick the device.

- To set the current project, click on the project or one of its files in the Package Explorer.

- To run the current project, press Ctrl+F11 or click on the Run button in the toolbar.

- To run on a physical device or on an emulator that's already running, select the "Choose" option, select the device or emulator, and click the OK button.

- To start an emulator, select the "Launch" option, select the emulator, and click the OK button.

- If an emulator is too big for your screen, or if you want to emulate the actual size of a device, select the "Launch" option, select the emulator, click the Start button, and use the resulting dialog box to set the DPI and screen size.

- If the Android Device Chooser dialog box doesn't appear, you can modify your run configurations for the project as shown in the previous figure.

- By default, a project is compiled automatically before it is run, which is usually what you want.

- To clean and rebuild a project, select the Project→Clean item from the menu bar and respond to the resulting dialog box.

Figure 2-9   How to run an app (part 1 of 2)

clicking the Start button, and using the resulting dialog box to set the DPI (dots per inch) and screen size.

Once your app is running in an emulator, you can use the emulator to test your app. To do that, you can use your mouse to interact with the touchscreen to test the app. In this figure, for example, I clicked in the bill amount text box and used the soft keyboard that was displayed to enter an amount. I also clicked on the Increase (+) button several times to increase the tip to 20%. In addition, if the keyboard and D-pad are enabled you can use your keyboard or the D-pad on the emulator to test how your app responds to the keyboards and D-pads that may be available from physical devices.

When you're done testing an app in an emulator, you can leave the emulator open. That way, the next time you run the app, Eclipse won't need to boot the emulator, which takes about as much time as it does to boot a smartphone. As a result, Eclipse will be able to run your app much more quickly the next time.

The Tip Calculator app running in an emulator

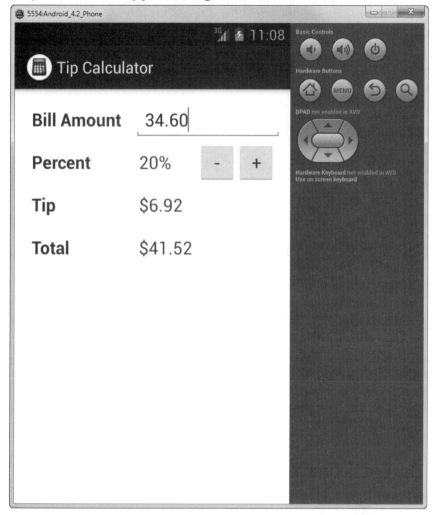

Description

- To test an app on a device, you can use any of the hardware that's available on that device.

- To test an app in an emulator, you can use your mouse to interact with the touch-screen that's shown on the left side of the emulator. Or, if the keyboard and D-pad are enabled you can use your keyboard or the D-pad that is shown on the right side of the emulator to emulate the hardware that's available from some Android devices.

- When you're done testing an app in an emulator, you can leave the emulator open. That way, the next time you start the app, it will start more quickly.

Figure 2-9   How to run an app (part 2 of 2)

How to work with a new project

Now that you know how to work with existing Android projects in Eclipse, you're ready to learn how to develop new Eclipse projects. Of course, many of the skills for working with existing projects also apply to working with new projects and vice versa.

How to create a new project

Figure 2-10 shows how to create a new Eclipse project. To do that, select the File→New→Project item from the menu bar. In the resulting dialog box, expand the Android category, select the Android Application Project item, and click the Next button. This should display a New Android Application dialog box like the one in this figure.

In the New Android Application dialog box, you can enter a name for the application as well as a name for the project and package. In this figure, for example, the name of the app is "Tester", the name of the project is "Tester", and the name of the package is com.murach.tester.

Since you almost always want Eclipse to generate the code for the first activity in your app, you should make sure that the "Create activity" check box is selected as you step through these dialog boxes. Then, when the New Android Application dialog box asks you what type of activity you want to create, select the BlankActivity option. This is what you usually want, especially when you're getting started. Then, enter a name for the activity. For example, you can enter a name of "TestActivity" for the activity for this project. After that, you can click the Finish button. When you do, Eclipse creates a directory that corresponds with the project name, and it creates some additional files that it uses to configure the project.

If you want to create a custom launcher icon for the app, you can select the "Create custom launcher icon" check box. Then, when you click the Next button, the New Android Application dialog box will prompt you to create the icon. In most cases, you can do this by selecting the Image option and browsing to the image that you want to use for the icon. If you don't already have an image, you can often download a suitable image from the Internet. Or, if you prefer, you can skip the custom icon by deselecting the "Create custom launcher icon" check box.

The New Android Application dialog box

How to create a new project

1. Select the File→New→Project item from the menu bar.

2. Select the Android→Android Application Project item and click the Next button. This should display the New Android Application dialog box.

3. Enter a name for the application, project, and package and click the Next button.

4. Make sure the "Create activity" option is selected and click the Next button until you reach the step that allows you to select the type for the activity.

5. Select the BlankActivity option and click the Next button.

6. Enter a name for the activity and click the Finish button.

Note

- If you want to create a custom launcher icon for the app, make sure to select the "Create custom launcher icon" option in step 4. Then, the New Android Application dialog box will prompt you to create the icon. In most cases, you can do this by selecting the Image option and browsing to the image that you want to use for the icon.

Figure 2-10   How to create a new project

How to work with the starting code

When you create a new project, Eclipse typically generates some code for you. Figure 2-11, for example, shows the code that was generated for an activity named TestActivity. Here, Eclipse generated the XML code that defines the user interface for the activity. In addition, it generated Java code that displays this user interface. If you want, you can delete or modify the generated code.

The first screen shows the XML for the user interface. This XML is stored in a file named activity_test.xml. Within this file, the XML defines a RelativeLayout element that's used to organize child elements. Then, within the RelativeLayout element, a single TextView element displays a string that says "Hello world!".

The second screen shows the Java code that displays the user interface. This code defines a class named TestActivity. Within this class, the method named onCreate displays the user interface. Then, the method named onCreateOptionsMenu displays a settings menu.

To practice using Eclipse, you can modify the generated code. For now, for example, you don't need to display a settings menu. As a result, you can delete the method named onCreateOptionsMenu.

Similarly, you can modify the XML code for the user interface so that the interface isn't centered horizontally and vertically. To do that, you can use the Graphical Layout editor to change the properties of the TextView element. Or, you can use the XML editor to edit or delete the attributes named layout_centerHorizontal and layout_centerVertical of the TextView element. If you delete these attributes, the TextView element uses its default values for these attributes, which causes the TextView element to be aligned in the top left corner of the screen.

The code editor with the starting XML code for an activity

The code editor with the starting Java code for an activity

Description

- By default, Eclipse generates some code when you create a project, but you can
  delete or modify this code.

Figure 2-11   How to work with the starting code

How to use the code completion feature

Figure 2-12 shows how to use the *code completion feature*, which is also known as the *Content Assist feature*, to help you enter code. This feature prevents you from making typing mistakes, and it helps you enter code more quickly and easily.

In this figure, for example, I am using the code completion feature to enter the Java statement that sets the text for the TextView control. To do that, I entered the name of the TextView control, entered "sett", and pressed Ctrl+Space. This displays a list of possible options. At this point, if I select the setText method from this list and press the Enter key, Eclipse will enter the name of the setText method into the code editor for me. In addition, it will automatically enter the opening and closing parentheses and placeholders for the arguments for this method.

The code completion feature can also make it easy for you to enter values for string variables. If you type a quotation mark to identify a string value, the code completion feature sometimes enters both opening and closing quotation marks and places the cursor between the two. At this point, you can enter the text for the string.

If you experiment with the code completion feature, you'll quickly see when it helps you enter code more quickly and when it makes sense to enter the code yourself. In addition, you'll see that it helps you understand the kinds of fields and methods that are available to the various classes and objects that you're working with. Similarly, it helps you understand the kinds of attributes that are available to the various XML elements, and the kinds of values that are available for the various attributes.

If the code completion feature doesn't work correctly on your system, it's probably because the Content Assist feature isn't configured correctly. To fix this issue, select the Window→Preferences item from the menu bar. Then, select the Java→Editor→Content Assist→Advanced node. Finally, in the first list box, select the Java Proposals item.

The code editor with a code completion list

Description

- You can use the *code completion feature*, which is also known as the *Content Assist feature*, to help you enter code. For Java code, this helps you enter the names of classes, objects, methods, fields, and other values. For XML, this helps you enter the names of elements, attributes, and values.

- To activate code completion, enter the letters you're looking for then press Ctrl+Space. Then, Eclipse displays a list of potential items that begin with those letters.

- To activate code completion for a method or field of a class or object, enter a period after a class or object name. Then, Eclipse displays a list of all the methods and fields for that class or object. You can limit these methods and fields by typing the first few letters of the method or field you want.

- To insert an item from a completion list, use the arrow keys to select the item and then press the Enter key. If the item requires parentheses, they're added automatically. If the item requires one or more arguments, default values are added for those arguments and the first argument is highlighted so you can enter its value. Then, you can press the Tab key and enter the values for any remaining arguments.

- If you enter the opening quote for a string value, Eclipse sometimes adds the closing quote and places the cursor between the two quotes so you can enter a value.

Figure 2-12   How to use the code completion feature

How to detect and correct errors and warnings

In Eclipse, an *error* is code that won't compile properly. Eclipse displays errors each time you save the source code. In figure 2-13, for example, Eclipse has displayed an error that indicates that the TextView class can't be resolved to a type. As you can see, this single error is marked in several places in the IDE with an icon that includes a red X. In the Package Explorer, this error is marked with icons for the project directory, src directory, package, and file that contains the source code. In addition, this error is marked in the code editor and is displayed in the Problems view.

To fix an error, you can jump to it by double-clicking on its message in the Problems view. This is especially helpful for projects that contain many classes or classes that have many lines of code.

Once you've got the error displayed in the code editor, you can position the cursor over the error icon to read a description of the error. This is the first step in fixing the error. If the error icon includes a light bulb as it does in this figure, you can click on the error icon to display a list of possible fixes. Then, you can select the fix that you want, and Eclipse will modify your code accordingly. This is known as the *Quick Fix feature*. In this figure, for example, the android.widget.TextView class needs to be imported. As a result, you can fix the problem by selecting the "Import" item from the resulting menu. This adds the necessary import statement to the code. Then, when you save the file, Eclipse removes the error icons from all of its views and editors.

If Eclipse doesn't suggest any fixes, you can usually determine the cause of the error by reading the error message. Then, you can enter the fix in the code editor yourself.

Eclipse may also display *warnings*. These icons identify lines of code that will compile but may cause problems, and they are marked with a yellow triangle that has an exclamation point in it. In general, warnings work like errors, and you can use similar skills for working with them.

However, there are times when you will want to ignore the warnings. If, for example, you have a good reason to use a line of code that Eclipse warns you about, you can ignore the warning. Then, if you want, you can remove the warning icons from the Eclipse views by clicking on the warning icon in the code editor and selecting the "Add @Suppress Warnings" item from the resulting menu. This adds a line of code known as an annotation that prevents the warning from being displayed in the Eclipse views.

The code editor with an error displayed and a possible solution

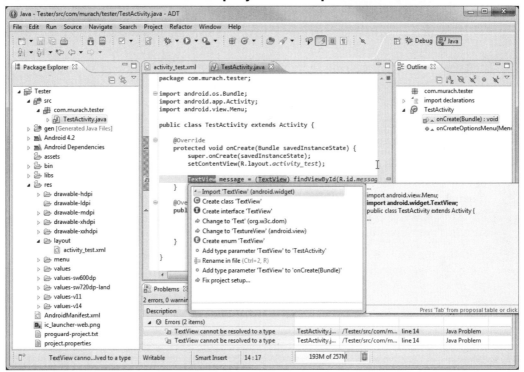

Description

- Eclipse marks *errors* in the Package Explorer view, Outline view, and code editors with an error icon that includes a red X.

- Eclipse marks *warnings* with a yellow triangle that has an exclamation point in it.

- To read a description of an error or warning, you can position the cursor over the error or warning icon or you can look in the Problems view.

- To fix an error or warning, you can often use the *Quick Fix feature*. To do that, click on the icon on the left side of the code editor to display a list of possible fixes. Then, select the fix that you want.

- You can also use the Quick Fix feature to suppress the warnings. To do that, click on the Warning icon and select the "Add @Suppress Warnings" item. Then, Eclipse will add a line of code that suppresses the warnings.

- Eclipse also lists all errors and warnings in the Problems view.

- To jump to an error or a warning in the code editor, you can double-click on it in the Problems view.

Figure 2-13   How to detect and correct errors and warnings

The Tester app

Now that you know how to use Eclipse to create and work with an app, you're ready to gain some hands-on skills by creating a simple app named Tester that displays some text. This app is a slightly modified version of the code that's generated by Eclipse for a new project.

The user interface

When you run the Tester app successfully, it displays "Success!" in the upper left corner of the app as shown in figure 2-14. This may not seem like much, but creating an app like this gives you a chance to practice modifying both the XML and Java source code for an app. In addition, it's a good test to make sure that your computer is set up and ready to begin with Android development.

The Tester app running in an emulator

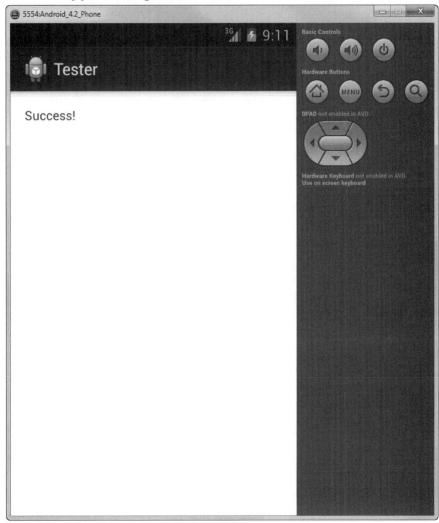

Description

- If it runs successfully, the Tester app displays "Success!" in the upper left corner of the screen.

Figure 2-14   The user interface for the Tester app

The XML for the user interface

Figure 2-15 shows the XML code for the user interface. Most of this code was generated by Eclipse when the project was created. However, I deleted the Android attributes that centered the TextView widget horizontally and vertically. As a result, this widget uses its default values for alignment, which aligns it in the top left corner of the screen. In addition, I added the four highlighted attributes to the TextView widget. The first of these attributes defines an ID for the widget. The next three attributes set the padding, text size, and text style for the widget.

The Java source code

Figure 2-15 also shows the Java source code for the app. Again, most of this code was generated by Eclipse when the project was created. However, I deleted the onCreateOptionMenu method that creates the options menu. As a result, this app does not display an options menu.

Within the onCreate method, I added the two highlighted statements. The first statement uses the ID of the TextView widget to get a reference to that widget. The second statement uses the setText method to display a message of "Success!" on the TextView widget.

The activity_test.xml file

```xml
<RelativeLayout xmlns:android="http://schemas.android.com/apk/res/android"
    xmlns:tools="http://schemas.android.com/tools"
    android:layout_width="match_parent"
    android:layout_height="match_parent" >

    <TextView
        android:id="@+id/messageTextView"
        android:layout_width="wrap_content"
        android:layout_height="wrap_content"
        android:padding="10dp"
        android:textSize="20sp"
        android:textStyle="bold"
        android:text="@string/hello_world"
        tools:context=".TestActivity" />

</RelativeLayout>
```

The TestActivity.java file

```java
package com.murach.tester;

import android.os.Bundle;
import android.app.Activity;
import android.view.Menu;
import android.widget.TextView;

public class TestActivity extends Activity {

    @Override
    public void onCreate(Bundle savedInstanceState) {
        super.onCreate(savedInstanceState);
        setContentView(R.layout.activity_test);

        TextView message = (TextView) findViewById(R.id.messageTextView);
        message.setText("Success!");
    }
}
```

Description

- The Tester app is a slightly modified version of the code that's generated by Eclipse for a new project. In this figure, all of the highlighted code has been added to the generated code.

Figure 2-15   The code for the Tester app

Perspective

In this chapter, you learned how to use Eclipse to create and run an Android app. With that as background, you're ready to learn how to write your own Android apps. But first, I recommend that you get some hands-on practice with Eclipse by doing the exercises at the end of this chapter.

Terms

Integrated Development Environment (IDE)	workspace
	run configuration
workbench	code completion
view	Content Assist feature
project	error
package	Quick Fix feature
code editor	warning

Summary

- Eclipse is an *Integrated Development Environment* (*IDE*) that's often used to develop Android apps. Eclipse is open-source, available for free, and runs on all modern operating systems.

- A *workspace* is a directory that's used by Eclipse to store the subdirectories and files for one or more apps.

- A *project* is a directory that contains all of the files for an app.

- In Eclipse, the *workbench* is where you can do most of your work. The different parts of the workbench are known as *views*.

- The Java files for a project are typically organized into *packages*. Each package has a corresponding directory in the file system.

- You can use a *code editor* to edit XML and Java files.

- To run an application, you use a *run configuration* that specifies the device that the app should run on.

- The *Content Assist feature*, which is also known as the *code completion feature*, helps you enter the code for XML and Java files.

- Eclipse displays an *error* icon to mark each line of code that won't compile properly. And Eclipse displays a *warning* icon to mark each line of code that will compile but may cause problems.

- You can often use the *Quick Fix feature* to fix errors and warnings automatically.

Before you do the exercises for this chapter

Before you do any of the exercises in this book, you need to install the JDK for Java SE, the Android SDK, Eclipse, and the ADT Plugin for Eclipse. In addition, you need to create an emulator, and you need to install the source code for this book. This may take several hours, so plan accordingly. See appendix A (PC) or B (Mac) for details.

Exercise 2-1   Import, run, and modify an existing app

Import and run the app

1. Start Eclipse. If you haven't already done it, set the workspace to:

   \murach\android\workspace

2. Within that workspace, import this project:

   \ex_starts\ch02_ex1_TipCalculator

3. Make sure the run configuration for the Tip Calculator project always prompts you to choose the emulator or device.

4. Run the Tip Calculator app in the emulator that you created in the appendix. When you do, Eclipse should start the emulator and run the Tip Calculator within it. This may take several minutes. To display the app in the emulator, you may need to unlock the emulator by dragging the lock icon to the right.

5. Test the Tip Calculator app by using the soft keyboard to enter a bill amount and by clicking on the Increase (+) and Decrease (-) buttons to modify the tip percent. When you're done testing, don't close the emulator!

6. If you have a physical Android device, connect it to your computer and repeat steps 4 and 5 for that device.

Modify the app

7. Open the .java file for the activity.

8. Scroll down to the onClick method and modify it so it increases or decreases the tip amount by 2% instead of 1%.

9. Run the Tip Calculator app in your emulator again. Note that the app displays much more quickly this time since the emulator doesn't need to be started.

10. Test the app to make sure it's working correctly.

11. Remove the Tip Calculator project from the Package Explorer, but *don't* delete the files for the project.

Exercise 2-2   Develop a new app

Create the app

1. Start Eclipse. If you haven't already done it, set the workspace for the project to this directory:

   `\murach\android\workspace`

2. Create a project for an Android application named "Tester". If you want, you can include a custom icon for this app. This app should be based on the BlankActivity template, and it should include an activity class named TestActivity that's stored in the com.murach.tester package.

3. Run the app in an emulator. This should display a message that says "Hello world!" in the center of the screen.

Modify the app

4. Open the XML file for the user interface. Use the Graphical Layout editor to modify the user interface so it looks like the user interface for the Tester app shown in figure 2-14.

5. Use the XML editor to view the code generated by the Graphical Layout editor. If necessary, modify this code so it works like the code shown in figure 2-15.

6. Open the .java file for the activity. Review the generated code. Add the two statements shown in figure 2-15 that set a message of "Success!". Make sure to use code completion to help you enter these statements.

7. Run the app in an emulator. This should display a message that says "Success!" at the top left corner of the screen.

Introduce and correct a compile-time error

8. In the code editor for the .java file, delete the import statement for the TextView class.

9. Press Ctrl+S to save the source code. Eclipse should display error icons in the code editor and the Package Explorer.

10. View the error message. It should say something like "TextView cannot be resolved to a type".

11. Use the Quick Fix feature to add the import statement for the TextView class.

12. Save the file again. Eclipse should remove all error icons.

13. Press Ctrl+F11 to run the app. Again, this should display a message that says "Success!" at the top left corner of the screen.

3

How to develop your first Android app

In this chapter, you'll learn how to develop the Tip Calculator app that you were introduced to in the first two chapters. That will get you off to a fast start and give you the perspective you need for learning rapidly. Since you should already have some Java experience, this chapter begins by focusing on the Eclipse skills that you need to develop the user interface. Then, it teaches the Java skills that you need to add functionality to that user interface.

How to develop the user interface

The first step in developing an Android app is to develop the user interface. The easiest way to do this is to use the Graphical Layout editor that's provided by Eclipse to develop most of the user interface. This generates most of the XML for the user interface. Then, if necessary, you can review and modify this code.

The Tip Calculator app

Figure 3-1 shows the user interface for the Tip Calculator that you'll learn how to develop in this chapter. In Android development, the components that make up the user interface are known as *widgets*. Widgets can also be referred to as *controls*. The Tip Calculator contains ten widgets: seven TextView widgets, one EditText widget, and two Button widgets.

A TextView widget can be referred to as a *text view* or a *label*. It especially makes sense to refer to a TextView as a label when that widget provides text that labels another widget. For example, the TextView widget that displays "Bill Amount" on the left side of the Tip Calculator labels the EditText widget on the right side of the user interface. Similarly, the TextView widget that displays "Percent" labels the TextView widget that displays the value for the tip percent.

An EditText widget can be referred to as an *editable text view* or a *text box*. When the user touches the editable text view for the Tip Calculator, the user interface automatically displays a soft keyboard that allows the user to enter a total for the bill. After the user enters a total for the bill and touches the Done key, the app automatically calculates the tip.

A Button widget can be referred to as a *button*. In the Tip Calculator, the user can click on the Increase (+) and Decrease (–) buttons to increase or decrease the default tip percent. This automatically recalculates the tip.

The user interface for the Tip Calculator

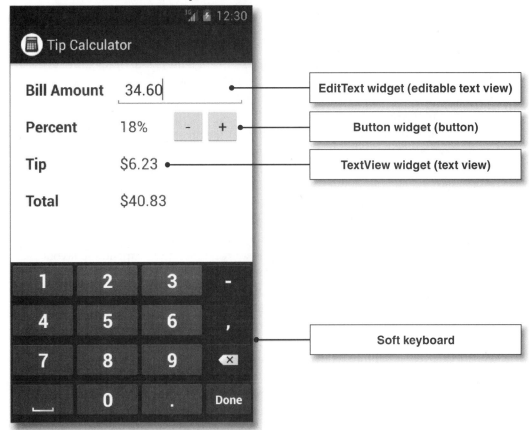

Description

- The user interface for the Tip Calculator contains ten *widgets*: seven TextView widgets, one EditText widget, and two Button widgets.

- A TextView widget can be referred to as a *text view* or a *label*, an EditText widget can be referred to an *editable text view* or a *text box*, and a Button widget can be referred to as a *button*.

- When the user clicks on the editable text view for the bill amount, the user interface automatically displays a soft keyboard that allows the user to enter a total for the bill. After the user enters a total for the bill and touches the Done key, the app automatically calculates the tip.

- The user can click on the Increase (+) and Decrease (–) buttons to increase or decrease the default tip percent. This automatically recalculates the tip.

- A widget can be referred to as a *control*.

Figure 3-1   The user interface for the Tip Calculator app

How to work with a layout

In Android development, a *layout* is a container that contains one or more child elements such as widgets and controls how they are displayed. A layout can be referred to as a *form*. Figure 3-2 shows the default layout of an *activity*, which defines a screen. An activity like this one is created automatically when you create a new Android project that's based on the BlankActivity template as described in the previous chapter.

Android provides several different types of layouts. In chapter 5, you'll learn more about working with different types of layouts. For now, you'll learn how to use the relative layout to create the interface for the Tip Calculator app. In this figure, the relative layout contains a single TextView widget that displays a message of "Hello world!".

Since a new project typically generates at least one layout file for you, you can usually begin by opening an existing layout. To do that, expand the project in the Package Explorer, expand the res (resources) directory, expand the layout directory, and double-click on the name of the XML file.

Once you've opened a layout, you can view it in the Graphical Layout editor by clicking on the Graphical Layout tab. Then, you can use the techniques described in the next few figures to add widgets to the layout, to align those widgets, and to set their properties. Then, if you want to review or edit the XML code that's generated for the layout, you can click on the tab that displays the filename of the layout.

If you need to change the type of layout, you can do that by right-clicking on the layout in the Outline window, selecting the Change Layout item, and selecting the type of layout. For example, if you're working with another type of layout such as a linear layout, you can use this technique to change it to a relative layout. Then, you can use the skills described in this chapter to create the user interface.

Similarly, when you need to add additional layouts to a project, you can do that by right-clicking on the project in the Package Explorer and selecting the New→Android XML File item. Then, you can enter the name of the file and select the type of layout. For now, select the relative layout.

Conversely, if you want to delete a layout from a project, you can do that by right-clicking on the layout in the Package Explorer, selecting the Delete command, and clicking the OK button.

The default layout for an activity in a new project

Description

- By default, a new Android application includes a layout file that uses a relative layout and a single TextView widget that displays "Hello world!".

- To open an existing layout, expand the project in the Package Explorer, expand the res (resources) directory, expand the layout directory, and double-click on the name of the XML file.

- To change the type of layout, right-click on the layout in the Outline view, select the Change Layout command, and select the type of layout.

- To create a new layout, right-click on project in the Package Explorer and select the New→Android XML File item. Then, enter the name of the file and select the type of layout. For now, select relative layout.

- To delete a layout, right-click on the layout in the Package Explorer, select the Delete command, and click the OK button.

- To view the layout in Graphical Layout editor, click on the Graphical Layout tab.

- To view the layout in XML view, click on the tab that displays the filename of the layout.

Figure 3-2   How to work with a layout

How to add widgets to a layout

Figure 3-3 shows how to add widgets to a layout. To do that, you just drag a widget from the Palette onto the layout. This figure, for example, shows the ten widgets for the Tip Calculator after they have been added to the layout but before all of their properties have been set correctly.

The Palette has several ways of displaying widgets. In this figure, I have displayed an icon and text for each widget. To do that, I right-clicked within the Palette and selected the "Show Icon and Text" item. However, if you prefer, you can display previews of the widgets by right-clicking within the Palette and selecting one of the "Preview" items.

As you add widgets to a layout, you often need to switch categories in the Palette so you can find the widget you want to add. To do that, you can click on the category for the widget. The Form Widgets category contains a variety of useful widgets including the TextView and Button widgets. The Text Fields category, on the other hand, only contains variations of the EditText widget. For this figure, I used the widget named Number (Decimal).

When you use the relative layout, it's a good practice to set the Id property for the widget as soon as you add it to the layout. That way, the other widgets can use that ID to align themselves relative to this widget. In this figure, for example, I set the Id property of the last TextView widget to @+id/totalTextView. Here, the prefix of @+id/ identifies the following text as the ID. As a result, the ID of the widget is totalTextView. This clearly indicates that this widget is a TextView widget for the total amount.

If you modify the Id property for a widget after you have aligned other widgets relative to that widget, Eclipse might not be able to align your widgets correctly. To fix that, you need to update all references to the Id property that you modified. An easy way to do that is to switch to the XML editor and use the Find/Replace feature.

By default, the Properties view is displayed below the Outline view. If it isn't there, you can display it by selecting the Window→Show View→Other item from the menus. Then in the resulting dialog box, you can expand the General category and select the Properties item.

In most cases, you can align a widget by dragging it to where you want it to be aligned. This sets properties of the widget in the Layout Parameters group such as the Align Baseline property, the Align Left property, and so on. In this figure, for example, the Properties view shows some of these properties for the selected TextView widget.

After you add widgets, you may need to delete one or more of them. To do that, click on the widget in the Outline view and press the Delete key. If you delete it in the Graphical Layout editor, it might not be deleted correctly.

As you add widgets to a layout, you may need to set its *theme*, which is a group of styles that define the look of the layout. To set the theme, select it from the toolbar that's displayed above the Graphical Layout editor. In this figure, the AppTheme has been selected, which is usually what you want when you're getting started. You'll learn more about working with themes in chapter 7.

A layout after some widgets have been added to it

Description

- To add a widget, click on the category for the widget in the Palette view and drag the widget from the Palette onto the layout. TextView and Button widgets are in the Form Widgets category, and EditText widgets are in the Text Fields category.

- To display the icon and text for a widget in the Palette, right-click within the Palette and select the "Show Icon and Text" item.

- When you use the relative layout, it's a good practice to set the Id property for the widget as soon as you add it to the layout. To do that, you can click on the widget and use the Properties view to set the Id property.

- By default, the Properties view is displayed below the Outline view. If it isn't there, you can display it by selecting the Window→Show View→Other item from the menus, expanding the General category, and selecting the Properties item.

- To align a widget, you can drag it to where you want it to be aligned. This sets properties of the widget in the Layout Parameters group such as the Align Baseline property, the Align Left property, and so on.

- To delete a widget, click on the widget in the Outline view and press the Delete key. If you delete it in the Graphical Layout editor, this might not work correctly.

- A *theme* is a group of styles that define the look of the layout. To set the theme for the layout, select the theme from the toolbar.

Figure 3-3   How add widgets to a layout

How to set the display text

The Android development environment encourages separating the display text from the rest of the user interface. The advantage of this approach is that it's easier to create international apps that work in multiple languages. The disadvantage is that it takes a little extra work to set the display text for the user interface.

However, figure 3-4 shows a technique that makes it easy to set the display text, and this extra step shouldn't slow down your development much once you get used to it. In this figure, for example, I set the Text property for the first TextView widget to a string named bill_amount_label that's stored in the strings. xml file. This is indicated by the value for the Text property which is set to:

```
@string/bill_amount_label
```

This string contains a value of "Bill Amount", and this value is displayed in the Graphical Layout editor.

A layout after the text has been set for the widgets

How to set the text property for a widget

1. In the Graphical Layout editor, select the widget.
2. In the Properties view, click the button (…) to the right of the Text property.
3. In the Resource Chooser dialog box, click the New String button to add a new string. Or, if the string is already available, select the string and click the OK button.
4. In the Create New Android String dialog box, enter a name and value for the string and click the OK button.

Description

- It's generally considered a best practice to store the display text for your app in a separate XML file. This makes it easy to internationalize your application.

Figure 3-4   How to set the display text for a widget (part 1 of 2)

After you click the button (…) to the right of the Text property, Eclipse displays the Resource Chooser dialog box. If the string you want to display is available from this dialog box, you can select it and click the OK button. This sets the display text to the selected string.

On the other hand, if you need to enter a new string, you can click on the New String button to display the Create New Android String dialog box. Then, you can enter the string you want to display and a name for the string. In this figure, for example, I entered a string of "Bill Amount" with a name of bill_amount_label. When you click the OK button, Eclipse displays the Resource Chooser dialog box with the string you just created. As a result, you can click the OK button to set the display text to your newly created string.

The Resource Chooser dialog box

The top of the Create New Android String dialog box

Figure 3-4   How to set the display text for a widget (part 2 of 2)

How to work with the strings.xml file

When you use the technique shown in the previous figure to set the display text for your app, Eclipse stores the text in a strings.xml file like the one that's shown in figure 3-5. If you want, you can open the strings.xml file and edit it just as you would edit any other XML file.

This figure shows the strings.xml file for the Tip Calculator app. Here, the first two strings are used in the manifest file to identify the name of the app and the name of the main activity of the app. Both of these strings are set to "Tip Calculator".

The third string is for a Settings item in the options menu. Since this app doesn't display an options menu, this string isn't displayed anywhere in the app. However, since this string is referred to by the unused options menu, deleting this string causes an error that prevents the project from compiling. As a result, I decided to leave this string and the unused menu in the project for now.

The last ten strings are used for the widgets. The strings that provide labels for widgets have names that end with "_label". Other strings provide default values for the bill amount, tip percent, tip amount, and total. These default values make it easier to use the Graphical Layout editor to set the properties for the widgets that display these values. In addition, they can make it easier to test an app. For example, when the app starts, it displays a value of 34.60 in the editable text view for the bill amount. As a result, you don't have to enter a bill amount to test the app.

However, before you put the app into production, you can modify the app so it sets these values to appropriate starting values for the app. For example, you probably want to supply an empty string for the bill amount. That way, when the app starts, the editable text view for the bill amount will be blank.

If you want to supply a strings.xml file for another country, you can create a values-*xx* directory within the res directory where *xx* is the ISO two-letter code for the country. For example, *fr* is the two-letter country code for France. Then, you can supply a strings.xml file within that directory that uses French instead of English. To find a list of these two-letter codes, you can search the Internet for "ISO two-letter country code".

The location of the strings.xml file

```
res/values/strings.xml
```

The strings.xml file for the Tip Calculator app

```xml
<resources>
    <string name="app_name">Tip Calculator</string>
    <string name="title_activity_tip_calculator">Tip Calculator</string>
    <string name="menu_settings">Settings</string>

    <string name="bill_amount_label">Bill Amount</string>
    <string name="bill_amount">34.60</string>

    <string name="tip_percent_label">Percent</string>
    <string name="tip_percent">15%</string>
    <string name="increase">+</string>
    <string name="decrease">-</string>

    <string name="tip_amount_label">Tip</string>
    <string name="tip_amount">$0.00</string>

    <string name="total_amount_label">Total</string>
    <string name="total_amount">$0.00</string>
</resources>
```

The location of a strings.xml file for France

```
res/values-fr/strings.xml
```

Description

- When you use the technique shown in the previous figure to set the display text for your app, Eclipse stores the text in the strings.xml file that's shown in the next figure.

- If you want, you can open the strings.xml file and edit it just as you would edit any XML file.

- To provide a strings.xml file for a country that uses another language, you can create a values-*xx* directory within the res directory to store the strings.xml file. Here, you can replace *xx* with the ISO two-letter country code.

Figure 3-5   How to work with the strings.xml file

How to set properties

Figure 3-6 shows how to use the Properties view in the Graphical Layout editor to set the properties of a widget. To do that, select the widget and use the Properties view to change the property. After you change a property from its default value, the Properties view displays the property in white. In this figure, for example, the Properties view shows that the Text Style property has been set to bold.

A layout after some properties of the widgets have been set

Description

- To set a property for a widget, select the widget and use the Properties view to change the property.

- The properties that have been changed from their default are displayed in white in the Properties view.

Figure 3-6   How to set properties

Common properties

Figure 3-7 shows the properties that are used by the Tip Calculator app. These properties are commonly used by most apps. As a result, you should review them and make sure you understand how they work.

For a layout, the Width and Height properties are typically set to a pre-defined value of "match_parent". This expands the layout so it takes up the entire parent minus padding. In most cases, this causes the layout to take up the entire screen. You may sometimes see a value of "fill_parent" instead of "match_parent". These values work the same, but "fill_parent" has been deprecated and replaced by "match_parent" as of Android 2.2 (API Level 8).

For the Tip Calculator app, the Padding property of the layout is set to a value of "10dp". As a result, the layout includes 10 *density-independent pixels* (*dp*) of space between the edge of the screen and the widgets contained by the layout. Although Android supports other units of measurement such as inches, it's a best practice to use density-independent pixels whenever possible.

When you use a relative layout, it's considered a best practice to specify the Id property for each widget. To do that, you begin the value of the Id property with a prefix of @+id/. This indicates that you are specifying an ID value for the widget. Then, you can specify the ID for the widget. In this figure, the Id property shows how to set the ID for a widget to billAmountLabel.

For a widget, the Width and Height properties are typically set to a predefined value of "wrap_content". This forces the width and height to be just big enough to contain the widget and its content. However, if you want, you can use density-independent pixels (dp) to specify the height or width of a widget.

For the Text property, you begin with a prefix of @string/. This indicates that you are specifying a value that's stored in the strings.xml file. In this figure, the Text property shows how to set the display text for the widget to the string in the strings.xml file that's named bill_amount_label.

For the Text Size property, it's generally considered a best practice to use *scale-independent pixels* (*sp*) whenever possible. In this figure, the Text Size property specifies a value of "20sp". This indicates that the text size should be 20 scale-independent pixels tall.

The last two properties are typically used with an EditText widget. Here, the Ems property sets the width of the EditText widget so it's wide enough to contain the specified number of the letter *m*.

The Input Type property sets the type of input that the EditText widget should accept. In this figure, for example, the Input Type property specifies a value of "numberDecimal". This indicates that the EditText widget should only accept characters for decimal numbers. In other words, it should only accept numeric characters and the decimal point character. Of course, the numberDecimal value is only one of the many values that specify the types of input that an EditText widget can accept. For example, an EditText widget can accept plain text, email addresses, telephone numbers, passwords, and so on. In chapter 5, you'll learn more about how to specify the type of input for an EditText widget.

Common properties for layouts

Property	Example settings
Width	`match_parent`
Height	`match_parent`
Padding	`10dp`

Common properties for widgets

Property	Example settings
Id	`@+id/billAmountLabel`
Width	`wrap_content`
Height	`wrap_content`
Align Parent Left	`true`
Align Parent Top	`true`
Padding	`10dp`
Text	`@string/bill_amount_label`
Text Size	`20sp`
Text Style	`bold`
Margin Left	`130dp`
Ems	`8`
Input Type	`numberDecimal`

Pre-defined values for setting height and width

Value	Description
`wrap_content`	Wraps the height or width so it is large enough to display the widget.
`match_parent`	Stretches the height or width to match the parent container.
`fill_parent`	Replaced by match_parent. Deprecated starting in API Level 8.

Common units of measurement

Name	Abbreviation	Typical uses
Density-independent pixels	dp or dip	margins, padding, etc.
Scale-independent pixels	sp	font sizes

Description

- For the Id property, you begin with a prefix of @+id/ to indicate that you are specifying an ID value for the widget.
- For the Text property, you begin with a prefix of @string/ to indicate that you are specifying a value that's stored in the strings.xml file.
- Although Android supports other units of measurement, such as inches, it's a best practice to use density-independent and scale-independent pixels whenever possible.

Figure 3-7   Common properties

To specify a density-independent pixel, it's possible to use an abbreviation of dip instead of dp. For example, you could use "10dip" instead of "10dp". In this book, I use dp for two reasons. First, it's shorter. Second, it's more consistent with the sp abbreviation that's used for scale-independent pixels.

The XML for the user interface

Figure 3-8 shows the XML for the user interface. This code is stored in the file named activity_tip_calculator.xml file.

The code in this figure is the same as the XML code presented in chapter 1. By now, though, you should understand how to use the Graphical Layout editor to generate this kind of code, you should understand how to use the XML editor to work with this code, and you should understand the details of how this code works. One main point to notice is that the layout attributes align these widgets relative to one another. These attributes begin with a prefix of "layout_".

To start, the RelativeLayout element defines a relative layout that contains the widgets for the user interface. Here, the first two attributes for the layout are set automatically, and you can usually ignore them. Then, the layout_width and layout_height attributes use a value of match_parent to specify that the layout should use the whole screen. Finally, the padding attribute determines that there should be 10dp of space between the edge of the layout and its child elements.

The first TextView element displays "Bill Amount" on the user interface. The first attribute of this widget defines an ID of billAmountLabel for the widget. The next three attributes for this widget set its width, height, and padding. Here, the width and height attributes use a value of wrap_content to specify that the widget should be just wide enough and tall enough to fit its content. The fifth attribute sets the text that's displayed on the widget to the value that's stored in the bill_amount string of the strings.xml file. The sixth attribute sets the size of the text to 20 scale-independent pixels (20sp). Finally, the last attribute sets the style of the text to bold.

The EditText element displays the widget that allows the user to enter a bill amount. By now, you should understand how most of the attributes of the EditText element work. Within a relative layout, the alignment attributes align the editable text view relative to other widgets. In this case, the alignment attributes align the editable text view to the right of the TextView element for the bill amount label with a margin of 5dp between the two elements. Also, the ems attribute specifies that the editable text view should be wide enough to fit 8 ems (8 of the letter *m*). And the inputType element specifies that the soft keyboard should only allow the user to enter decimal numbers.

The next two TextView elements define the widgets that display the tip percent. The Button elements define the buttons that allow the user to increase or decrease the tip percent. These widgets are square (45dp x 45dp). And the last four TextView elements define the widgets that display the tip and total amounts.

The XML for the user interface   Page 1

```xml
<RelativeLayout xmlns:android="http://schemas.android.com/apk/res/android"
    xmlns:tools="http://schemas.android.com/tools"
    android:layout_width="match_parent"
    android:layout_height="match_parent"
    android:padding="10dp" >

    <!-- The bill amount -->

    <TextView
        android:id="@+id/billAmountLabel"
        android:layout_width="wrap_content"
        android:layout_height="wrap_content"
        android:padding="10dp"
        android:text="@string/bill_amount_label"
        android:textSize="20sp"
        android:textStyle="bold" />

    <EditText
        android:id="@+id/billAmountEditText"
        android:layout_width="wrap_content"
        android:layout_height="wrap_content"
        android:layout_alignBaseline="@+id/billAmountLabel"
        android:layout_marginLeft="5dp"
        android:layout_toRightOf="@+id/billAmountLabel"
        android:ems="8"
        android:inputType="numberDecimal"
        android:text="@string/bill_amount"
        android:textSize="20sp" >

        <requestFocus />
    </EditText>

    <!-- The tip percent -->

    <TextView
        android:id="@+id/percentLabel"
        android:layout_width="wrap_content"
        android:layout_height="wrap_content"
        android:layout_alignLeft="@+id/billAmountLabel"
        android:layout_below="@+id/billAmountLabel"
        android:padding="10dp"
        android:text="@string/tip_percent_label"
        android:textSize="20sp"
        android:textStyle="bold" />

    <TextView
        android:id="@+id/percentTextView"
        android:layout_width="wrap_content"
        android:layout_height="wrap_content"
        android:layout_alignBaseline="@+id/percentLabel"
        android:layout_alignLeft="@+id/billAmountEditText"
        android:padding="5dp"
        android:text="@string/tip_percent"
        android:textSize="20sp" />
```

Figure 3-8   The XML for the user interface (part 1 of 3)

The XML for the user interface   **Page 2**

```
<Button
    android:id="@+id/percentDownButton"
    android:layout_width="45dp"
    android:layout_height="45dp"
    android:layout_alignBaseline="@+id/percentTextView"
    android:layout_marginLeft="25dp"
    android:layout_toRightOf="@+id/percentTextView"
    android:text="@string/decrease"
    android:textSize="20sp" />

<Button
    android:id="@+id/percentUpButton"
    android:layout_width="45dp"
    android:layout_height="45dp"
    android:layout_alignBaseline="@+id/percentDownButton"
    android:layout_toRightOf="@+id/percentDownButton"
    android:text="@string/increase"
    android:textSize="20sp" />

<!-- the tip amount -->

<TextView
    android:id="@+id/tipLabel"
    android:layout_width="wrap_content"
    android:layout_height="wrap_content"
    android:layout_alignLeft="@+id/percentLabel"
    android:layout_below="@+id/percentLabel"
    android:padding="10dp"
    android:text="@string/tip_amount_label"
    android:textSize="20sp"
    android:textStyle="bold" />

<TextView
    android:id="@+id/tipTextView"
    android:layout_width="wrap_content"
    android:layout_height="wrap_content"
    android:layout_alignBaseline="@+id/tipLabel"
    android:layout_alignLeft="@id/billAmountEditText"
    android:padding="5dp"
    android:text="@string/tip_amount"
    android:textSize="20sp" />
```

Figure 3-8   The XML for the user interface (part 2 of 3)

The XML for the user interface   **Page 3**

```xml
<!-- the total -->

<TextView
    android:id="@+id/totalLabel"
    android:layout_width="wrap_content"
    android:layout_height="wrap_content"
    android:layout_alignLeft="@+id/tipLabel"
    android:layout_below="@+id/tipLabel"
    android:padding="10dp"
    android:text="@string/total_amount_label"
    android:textSize="20sp"
    android:textStyle="bold" />

<TextView
    android:id="@+id/totalTextView"
    android:layout_width="wrap_content"
    android:layout_height="wrap_content"
    android:layout_alignBaseline="@+id/totalLabel"
    android:layout_alignLeft="@+id/tipTextView"
    android:padding="5dp"
    android:text="@string/total_amount"
    android:textSize="20sp" />

</RelativeLayout>
```

Description

- The activity_tip_calculator.xml file contains the XML that defines the user interface.

Figure 3-8   The XML for the user interface (part 3 of 3)

How to write the Java code

Once you have the user interface working the way you want, you can write the Java code for the app. This is the code that provides the functionality for the app.

How to work with an activity

Figure 3-9 shows how to work with an activity. When you create an Android project based on the BlankActivity template, Eclipse automatically creates a class like the one shown in this figure. This class inherits the Activity class and overrides the onCreate method of that class. Android runs this method when the activity is started.

Within the onCreate method, the first statement passes the Bundle parameter to the onCreate method of the superclass (the Activity class). Then, the second statement displays the user interface for the activity. To do that, it calls the setContentView method of the superclass. Within the parentheses for this method, the code specifies the resource for the layout. To do that, this code uses the R (resources) class.

The R class is created automatically when you build your project, and it provides a way to access the compiled resources for your project. In the onCreate method, for example, this code:

```
R.layout.activity_tip_calculator
```

refers to a compiled version of this resource:

```
res\layout\activity_tip_calculator.xml
```

By default, the activity class includes an onCreateOptionsMenu method that displays an options menu. If your app doesn't need an options menu, you can delete this method. If your app needs menus, you can learn more about working with them in chapter 7.

If you're creating an app like Tip Calculator that only uses a single activity, you can usually modify the generated class to get it to work the way you want. However, if you need to add an additional screen to your app, you need to add a new activity. To do that, you can right-click on the project name in the Package Explorer, select the New→Other item. Then, expand the Android category, select the Android Activity item, and respond to the resulting dialog box. This adds an XML file for the user interface as well as a Java file that defines a class that inherits the Activity class that's similar to the one shown in this figure.

The default Java code for an activity

```
package com.murach.tipcalculator;

import android.os.Bundle;
import android.app.Activity;
import android.view.Menu;

public class TipCalculatorActivity extends Activity {

    @Override
    public void onCreate(Bundle savedInstanceState) {
        super.onCreate(savedInstanceState);
        setContentView(R.layout.activity_tip_calculator);
    }

    @Override
    public boolean onCreateOptionsMenu(Menu menu) {
        getMenuInflater().inflate(R.menu.activity_tip_calculator, menu);
        return true;
    }
}
```

Description

- When you create an Android project that's based on the BlankActivity template, it automatically creates a class for the first activity in your app. This class displays the user interface for the activity.
- To create a new activity, you can right-click on the project name in the Package Explorer, select the New→Other item. Then, expand the Android category, select the Android Activity item, and respond to the resulting dialog box.
- Android runs the onCreate method when the activity is started.
- The onCreate method usually passes the Bundle parameter to the onCreate method of the superclass.
- The onCreate method usually uses the setContentView method of the Activity class to specify the resource for the layout.
- The R class provides a way to access the compiled resources for your project. This class is created automatically when you build your project.
- Android runs the onCreateOptionsMenu method when it's ready to display the options menu.
- To remove the options menu from an app, delete the onCreateOptionsMenu method.

Figure 3-9   How to work with an activity

How to get references to widgets

The previous figure shows the starting Java code for displaying a user interface. Now, figure 3-10 shows how to get references to the widgets on the user interface. This is the first step to adding functionality to a user interface.

To make it easy to work with widgets, you typically start by adding import statements that import the classes for the widgets. Most widgets are stored in the android.widgets package. In this figure, for example, the TextView and EditText classes are both stored in this package.

After you import the classes for the widgets, you typically declare the widgets as private instance variables of the activity. In this figure, for example, three widgets have been declared as private instance variables. Here, the first widget is an EditText widget, and the other two are TextView widgets.

In the onCreate method, you can use the findViewById method to get references to the widgets that are declared in the layout. To do that, you can use the R (resources) class to access compiled resources. In this figure, for example, the first findViewById method specifies an argument of:

`R.id.billAmountEditText`

This gets a reference to a compiled resource with an ID of billAmountEditText. If you look at the XML file for the user interface, you'll find that this ID corresponds with the ID of the EditText widget on the user interface. Similarly, the other two IDs specified in this figure correspond with the IDs of two of the TextView widgets.

The findViewById method returns a View object that can be cast to the correct type for each widget. In this figure, for instance, the first statement casts the View object for the bill amount to the EditText type. Then, the next two statements cast the View object that's returned to the TextView types. This works because the View class is the superclass for all widgets.

An activity that gets references to the widgets

```
package com.murach.tipcalculator;

import android.os.Bundle;
import android.widget.EditText;    // 1. Import the classes for the widgets
import android.widget.TextView;
import android.app.Activity;

public class TipCalculatorActivity extends Activity {

    // 2. Declare variables for the widgets
    private EditText billAmountEditText;
    private TextView tipTextView;
    private TextView totalTextView;

    @Override
    public void onCreate(Bundle savedInstanceState) {
        super.onCreate(savedInstanceState);
        setContentView(R.layout.activity_tip_calculator);

        // 3. Get references to the widgets
        billAmountEditText = (EditText) findViewById(R.id.billAmountEditText);
        tipTextView = (TextView) findViewById(R.id.tipTextView);
        totalTextView = (TextView) findViewById(R.id.totalTextView);
    }
}
```

Description

- You can add import statements to import the classes for the widgets from the android.widgets package.

- You can declare widgets as private instance variables of the activity.

- You can use the findViewById method to get references to the widgets that are declared in the layout.

- The findViewById method accepts an argument for the ID of the widget. The R class provides a way to get the ID for any of the compiled resources for your project.

- The findViewById method returns a View object for the widget with the specified ID. This View object can be cast to the correct type of the widget.

Figure 3-10   How to get references to the widgets

How to handle the EditorAction event

Now that you have references to some of the widgets on the user interface, you're ready to learn how to code an event handler. An *event handler* is a special type of method that's executed when an *event* occurs. Typically, an event occurs when a user interacts with widgets. In figure 3-11, for example, you'll learn how to handle the EditorAction event. This event typically occurs when a user interacts with the EditText widget to display a soft keyboard and presses an *action key* such as the Done key.

In general, there are three steps to handling an event. First, you need to import the interface that defines a *listener*, which is an object that listens for an event. In this figure, the sixth import statement imports the OnEditorActionListener interface.

Second, you need to create a listener. To do that, you can implement the listener interface. In this figure, the class declares that it implements the OnEditorActionListener interface, and then it implements the method that's defined by the interface: the onEditorAction method. This method contains the code that's executed when the listener detects that an *action event* occurred on the widget.

Third, you must connect, or *wire*, the listener to a widget. In this figure, the last statement in the onCreate method wires the EditText widget for the bill amount to the onEditorAction listener that's defined by the current class. To do that, this code calls the setOnEditorActionListener method from the EditText widget for the bill amount. Then, it uses the keyword named *this* to pass the object for the current class to that method.

When you're implementing interfaces, Eclipse can automatically generate the method or methods defined by the interface for you, which is usually what you want. To do that, you can code the implements keyword followed by the name of the interface. Then, you can click on the error icon that Eclipse displays to the left of this code and select the "Add unimplemented methods" item. When you do, Eclipse generates the method or methods for you. If necessary, you can modify the names of the arguments. In this figure, for example, I changed the default names of the arguments to make them more readable.

Within the onEditorAction method, you can handle the events that occur on specific action keys such as the Done key by comparing the second parameter (the ID of the action key) with the constants of the EditorInfo class. In this figure, this method begins by checking whether the Done button was clicked or whether an unspecified action occured such as an Enter key being pressed on an older device. If either condition is true, this code sets the tip to $10 and the amount to $110. If not, the tip and amount are left at their default values. Either way, this method returns a false value, which causes the soft keyboard to be hidden, which is usually what you want.

If you review the onEditorAction method, you'll notice that the first and third parameters aren't used in this code. That's because the first parameter is used to determine which widget triggered the event. For this code, there is only one widget that has been wired to the event, so that widget is the only one

An activity that handles the EditorAction event

```
package com.murach.tipcalculator;

import android.os.Bundle;
import android.view.KeyEvent;
import android.view.inputmethod.EditorInfo;
import android.widget.EditText;
import android.widget.TextView;
import android.widget.TextView.OnEditorActionListener;   // 1. Import listener
import android.app.Activity;

public class TipCalculatorActivity extends Activity
implements OnEditorActionListener {      // 2a. Implement the listener

    private EditText billAmountEditText;
    private TextView tipTextView;
    private TextView totalTextView;

    @Override
    public void onCreate(Bundle savedInstanceState) {
        super.onCreate(savedInstanceState);
        setContentView(R.layout.activity_tip_calculator);

        billAmountEditText = (EditText) findViewById(R.id.billAmountEditText);
        tipTextView = (TextView) findViewById(R.id.tipTextView);
        totalTextView = (TextView) findViewById(R.id.totalTextView);

        // 3. Set the listener
        billAmountEditText.setOnEditorActionListener(this);
    }

    // 2b. Implement the listener
    @Override
    public boolean onEditorAction(TextView v, int actionId, KeyEvent event) {

        if (actionId == EditorInfo.IME_ACTION_DONE ||
            actionId == EditorInfo.IME_ACTION_UNSPECIFIED
        {
            tipTextView.setText("$10.00");
            totalTextView.setText("$110.00");
        }
        return false;
    }
}
```

Description

- An *event handler* is a special type of method that's executed when an *event* occurs. Typically, an event occurs when a user interacts with widgets.

- A *listener* is an object that listens for an event. To create a listener, import the interface for the listener and implement that interface. After you create a listener, connect, or *wire*, the listener to an event that's available from a widget.

- The EditorAction event typically occurs when the user uses a soft keyboard to enter text into an editable text view.

Figure 3-11   How to handle the EditorAction event (part 1 of 2) that could possibly trigger the event. The third parameter, on the other hand, is used to determine whether the Enter key triggered the event. For now, this app assumes that the device is a touchscreen device that doesn't have a physical Enter key. As a result, this app doesn't handle the Enter key.

Part 2 of figure 3-11 shows a few of the constants that are available from the EditorInfo class. All of these constants begin with IME, which stands for input method editor. An *input method editor* (*IME*) is a user control such as a soft keyboard that enables users to input data, often text.

The first constant shown in this figure is used in part 1 of this figure to check whether the Done key was pressed. The next three constants can be used to check for the Go, Next, and Search keys. Although these keys aren't available from the soft keyboard for the Tip Calculator shown in this chapter, they are available from other types of soft keyboards. For example, the Go key is available from the soft keyboard for entering a URL. Finally, the last constant can be used to check for an unspecified action such as an Enter key that's available from an older device.

As you review this code, you should be aware that there are several possible ways to define a class that implements the interface for a listener. In this figure, I decided to use the class for the activity to implement this interface. That way, that class defines the activity, and it acts as the listener for the EditorAction event. However, in some cases, it makes sense to define a separate class for the listener. You'll learn more about how this works in chapter 6.

A few constants from the EditorInfo class

Constant	The action key performs...
IME_ACTION_DONE	A "done" operation, typically closing the soft keyboard.
IME_ACTION_GO	A "go" operation, typically taking the user to the target specified by the user.
IME_ACTION_NEXT	A "next" operation, typically taking the user to the next field that needs text input.
IME_ACTION_SEARCH	A "search" operation, typically searching for the text specified by the user.
IME_ACTION_UNSPECIFIED	No specified operation, typically an action key on an older device such as an "enter" key.

Description

- After you declare the interface for the listener, you can automatically generate the declaration for the method by clicking on the error icon and selecting the "Add unimplemented methods" item.

- An *input method editor* (*IME*) is a user control such as a soft keyboard that enables users to input data, often text.

- To determine which widget triggered the event, you can use the first argument of the onEditorAction method.

- To determine which *action key* triggered the event, you can use the second argument of the onEditorAction method. To do that, compare the second argument with the constants of the EditorInfo class.

- To determine whether the action was trigged by pressing the Enter key, you can use the third argument of the onEditorAction method.

- To hide the soft keyboard, you can return a false value from the onEditorAction method.

- To keep the soft keyboard displayed, you can return a true value from the onEditorAction method.

Figure 3-11   How to handle the EditorAction event (part 2 of 2)

How to get and set the text for widgets

Figure 3-12 shows how to get and set the text that's displayed on a widget. Typically, you need to get the text from widgets such as editable text views that allow the user to enter text. Conversely, you often need to set the text on widgets that display text such as labels. However, these techniques work for most types of widgets.

You can use the getText method to get the text that's displayed on a widget. To do that, you call the getText method from the widget. This returns an Editable object. In most cases, you want to call the toString method from the Editable object to convert it to a String object. In this figure, for instance, the first example gets the text from the editable text view for the bill amount and converts that text to a string.

You can use the setText method to set the text that's displayed on a widget. To do that, you call the setText method from the widget and you pass a string argument that contains the display text. In this figure, for instance, the second example sets the display text for the TextView widget to "Test 1".

The calculateAndDisplay method shown in this figure uses both the getText and setText methods. To start, the first statement gets the display text from the editable text view for the bill amount and converts that text to a string. Then, this code uses the parseFloat method of the Float class to convert that string to a float value. Since the editable text view for the bill amount only allows the user to enter an empty string or a string for a decimal number, this code doesn't include a try statement that handles the exception that occurs if the user enters non-numeric data. However, this code does use an if statement to convert an empty string to a zero.

After getting a float value for the bill amount, this code calculates the tip and total amount for the bill. Here, the code assumes that the tip is a float value of 15%, and it uses the float type for all of these variables. To specify the literal value for the tip percent, this code appends an f to the number to indicate that it is a float value.

Before this code can display the tip and total amounts, it must convert them to string values. To do that, this code uses the NumberFormat class. First, it calls the static getCurrencyInstance method of the NumberFormat class to return a NumberFormat object for currency formatting. Then, it uses the format method of the NumberFormat object to convert numeric values to strings that have currency formatting, and it displays these strings on the TextView widgets for the tip and total amounts.

Although the code in this figure shows how to work with float values, you can use a similar technique to work with double and integer values. For example, you can use the parseInt method of the Integer class to convert a string to an int value.

Although the code in this figure shows how to work with currency formatting, you can use a similar technique to work with percent formatting. In particular, you can use the static getPercentInstance method of the NumberFormat class to return a NumberFormat object that you can use to convert numbers to strings that have percent formatting.

Two methods for working with widgets

Method	Description
`getText()`	Returns an Editable object that represents the text that the widget is displaying. You can use the toString method to convert the Editable object to a string.
`setText(string)`	Sets the text that's displayed on the widget.

Examples

How to get text
```
String billAmountString = billAmountEditText.getText().toString();
```

How to set text
```
tipTextView.setText("Test 1");
```

A method that gets input, calculates amounts, and displays output
```
public void calculateAndDisplay() {

    // get the bill amount
    billAmountString = billAmountEditText.getText().toString();
    float billAmount;
    if (billAmountString.equals("")) {
        billAmount = 0;
    }
    else {
        billAmount = Float.parseFloat(billAmountString);
    }

    // calculate tip and total
    float tipPercent = .15f;
    float tipAmount = billAmount * tipPercent;
    float totalAmount = billAmount + tipAmount;

    // display the results with formatting
    NumberFormat currency = NumberFormat.getCurrencyInstance();
    tipTextView.setText(currency.format(tipAmount));
    totalTextView.setText(currency.format(totalAmount));
}
```

Description

- For most widgets, you can use the getText and setText methods to get and set the text that's displayed on the widget.
- If necessary, you can use the Integer, Double, and Float classes to convert the strings that are returned by widgets to int, double, and float values.
- If necessary, you can use the NumberFormat class to convert numeric values to string values that have currency or percent formatting so they can be displayed on widgets.

Figure 3-12   How to get and set the text for widgets

How to handle the Click event

Figure 3-13 shows how to handle the Click event. This event is available from most widgets, but it's almost always handled for buttons, which are designed to be clicked. The steps for handling the Click event are the same as the steps for handling the EditorAction event that you saw earlier in this chapter.

First, you import the interface for the listener. Often, Eclipse does this for you automatically when you declare the interface. As a result, you can often skip this step.

Second, you implement the interface by declaring it and by implementing any methods declared by the interface. In this figure, the class for the Tip Calculator activity declares that it implements the OnClickListener interface, and it includes the onClick method that's defined by this interface.

Third, you wire the listener to a widget. In this figure, the two statements wire the event handler shown in this figure to both of the buttons on the Tip Calculator app. To do that, this code calls the setOnClickListener method from each button and passes the object for the current class to that method.

Since the onClick method is wired to two buttons, the code in the onClick method includes a switch statement that checks which button was clicked and executes the appropriate code. To do that, the switch statement gets the ID of the widget by calling the getId method from the View object that's the first and only argument of the onClick method. Then, it compares this value to the IDs of the Button widgets. To get these IDs, this code uses the R class to access information about the compiled resources for the project. For instance, to get the ID for button named percentDownButton, it uses this code:

```
R.id.percentDownButton
```

Earlier in this chapter, the onEditorAction method didn't need to include code like this because that event handler was only wired to a single widget.

Since a class can implement multiple interfaces, a class can handle multiple events. In this figure, for instance, step 2a shows that the class implements both the OnClickListener and OnEditorActionListener interfaces.

How to handle the Click event

Step 1: Import the interface for the listener

```
import android.view.View.OnClickListener;
```

Step 2a: Implement the interface for the listener

```
public class TipCalculatorActivity extends Activity
implements OnEditorActionListener, OnClickListener {
```

Step 2b: Implement the interface for the listener

```
@Override
public void onClick(View v) {
    switch (v.getId()) {
    case R.id.percentDownButton:
        tipPercent = tipPercent - .01f;
        calculateAndDisplay();
        break;
    case R.id.percentUpButton:
        tipPercent = tipPercent + .01f;
        calculateAndDisplay();
        break;
    }
}
```

Step 3: Set the listeners

```
percentUpButton.setOnClickListener(this);
percentDownButton.setOnClickListener(this);
```

Description

- The steps for handling the Click event are the same as the steps for handling the EditorAction event. However, since the Click event is wired to two buttons, the code in the onClick method includes a switch statement that checks which button was clicked and executes the appropriate code.

- To specify a literal value for the float type, you can append an f or an F to the number.

Figure 3-13   How to handle the Click event

The lifecycle of an activity

The Activity class defines methods that are called by the Android operating system at different points in the lifecycle of an activity. This is shown by figure 3-14.

When you create an activity, you create a class that inherits the Activity class. Then, to get your app to work the way you want, you override the lifecycle methods to control the lifecycle of the activity. For example, you have already seen how you can override the onCreate method to display and prepare the user interface for an activity. However, you often need to override other methods to make sure that your app doesn't lose its progress if the user leaves the app and returns to it later. Similarly, you often need to make sure that your app doesn't lose its progress when the user switches between landscape and portrait orientation.

When an app starts for the first time, Android calls the onCreate, onStart, and onResume methods. As this happens, the activity quickly passes through the *created state* and the *started state*. Of these three methods, the onCreate method is typically used to create the user interface and prepare it for display. In the Tip Calculator activity, for example, this method displays the user interface, gets references to important widgets, and wires the event handlers.

After the onResume method executes, the app is in the *running state*, which is also known as the *active state* or *resumed state*. In this state, the app is visible and has the focus. Unlike the created and started states, an app may remain in the running state for a long time.

If the running activity loses the focus, becomes partially covered by another activity, or if the device goes to sleep, Android calls the onPause method. This causes the activity to go into the *paused state*. Android attempts to maintain the state of all activities in the paused state. As a result, if the activity regains the focus or the device wakes up, Android can call the onResume method and return to the running state without losing any of the user's progress.

If the user gets a phone call or navigates to another activity, Android calls the onPause method followed by the onStop method. This causes the activity to go into the *stopped state*. In the stopped state, Android attempts to maintain the state of the activity for as long as possible. However, an activity in the stopped state has a lower priority than an activity in the running or paused states. As a result, Android often kills activities that are in the stopped state.

Most lifecycle methods have a corresponding method. If you implement one method, it often makes sense to implement the other. For example, if you use the onPause method to save data when the activity goes into the paused state, you typically want to use the onResume method to restore the data when the activity returns to the running state. Similarly, if you use the onStop method, you typically want to use the onStart method, and possibly the onRestart method.

Although this diagram shows the most commonly used lifecycle methods for an activity, it doesn't show them all. For example, in chapter 8, you'll learn more about working with the lifecycle methods for displaying menus. Or, you can learn more about other lifecycle methods by looking up the Activity class in the documentation for the Android API.

The lifecycle of an activity

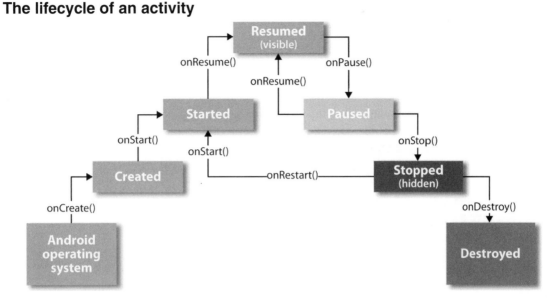

Three common states

State	Description
Resumed	This state is also known as the *active state* or the *running state*. A running activity is visible and has the focus. Android only destroys a resumed activity in extreme situations, such as if the activity tries to use more memory than is available on the device.
Paused	A paused activity has lost the focus and may be partially hidden by other activities, or the device has gone to sleep. A paused activity typically maintains all states. Android only destroys a paused activity if necessary to keep the activity that's in the running state stable and responsive.
Stopped	The activity has lost the focus and is completely hidden by another activity. Android tries to retain the state for a stopped activity for as long as possible. However, stopped activities have a lower priority than the other two states. As a result, Android destroys activities in the stopped state whenever it's necessary to allow activities with higher priorities to run.

Description

- When you create an activity, Android calls the methods of the Activity class at different points in its lifecycle. To get your app to work the way you want, you can override these methods.

- An activity passes through the *created state* and the *started state* quickly, but it can spend a long time in the *resumed state*, the *paused state*, and the *stopped state*.

- Although this diagram shows the most commonly used lifecycle methods for an activity, it doesn't show them all. To learn more about other lifecycle methods available to an activity, you can look up the Activity class in the documentation for the Android API.

Figure 3-14   The lifecycle of an activity

How to save and restore values

Android doesn't always save the data of an activity when the activity moves from the running state into the paused or stopped states. These states occur when the user navigates to another activity or when the user changes the orientation of the screen.

For the Tip Calculator app, this means that the activity won't always save the bill amount and the tip percent. As a result, if you only implement the onCreate method, the tip percent will be reset to its default value of 15% every time the user changes the orientation of the screen. Worse, the bill amount and the tip percent will be reset to their default values every time the user navigates to another activity and returns to the current activity. Since you usually want to save a user's progress, you typically want to save and restore these values as described in figure 3-15.

To start, you need a way to permanently save values. Fortunately, the SharedPreferences class provided by the Android API makes this easy to do. First, you import the SharedPreferences class and its Editor class. Then, you declare a SharedPreferences object as an instance variable, and you use the onCreate method to create an instance of this object. In this figure, for example, the onCreate method uses the getSharedPreferences method to get a SharedPreferences object named savedValues. This object saves values to a file named SavedValues, and this file is only available to this activity.

Once you have created a SharedPreferences object, you can override the onPause method and use the SharedPreferences object to save values. To do that, you can call the edit method of the SharedPreferences object to get an Editor object. Then, you can use the methods of the Editor object to store the bill amount string and the tip percent, which are both instance variables. Next, this code calls the commit method from the Editor object to save the values in the editor object to disk. Finally, the last statement calls the onPause method of the superclass, which is necessary for the superclass to work correctly.

To restore values that you have saved, you can override the onResume method. Within this method, the first statement calls the onResume method of the superclass, which is required for the superclass to work correctly. Then, the next two statements use the get methods of the SharedPreferences object to restore the instance variables for the bill amount string and tip percent.

When you work with the Editor and SharedPreferences objects, you can use them to put and get values for strings and most primitive types. In this figure, for example, the instance variable for the bill amount variable is a string value, and the instance variable for the tip percent is a float value. However, the Editor and SharedPreferences objects don't allow you to work directly with double values. So, if you want to use a double value, you must first cast it to a float value.

To test the onPause and onRestore methods in an emulator, you can run the Tip Calculator activity and navigate to an activity in another app. Then, you can navigate back to the Tip Calculator activity. Or, if you want, you can change the orientation of the Tip Calculator activity in your emulator by pressing Ctrl+F11 or Numpad 7.

How to import the SharedPreferences class and its Editor class

```
import android.content.SharedPreferences;
import android.content.SharedPreferences.Editor;
```

How to set up the instance variables

```
// define SharedPreferences object
private SharedPreferences savedValues;

@Override
public void onCreate(Bundle savedInstanceState) {
    // other code goes here

    // get SharedPreferences object
    savedValues = getSharedPreferences("SavedValues", MODE_PRIVATE);
}
```

How to use the onPause method to save values

```
@Override
public void onPause() {
    // save the instance variables
    Editor editor = savedValues.edit();
    editor.putString("billAmountString", billAmountString);
    editor.putFloat("tipPercent", tipPercent);
    editor.commit();

    super.onPause();
}
```

How to use the onResume method to restore values

```
@Override
public void onResume() {
    super.onResume();

    // get the instance variables
    billAmountString = savedValues.getString("billAmountString", "");
    tipPercent = savedValues.getFloat("tipPercent", 0.15f);
}
```

Description

- Android doesn't always save the data of an activity when you change orientation or when you navigate away from an activity to another activity or app.
- To save data for an activity, you can override the onPause method.
- To restore data for an activity, you can override the onResume method.
- You can use a SharedPreferences object and its Editor class to save and restore data.
- To change orientation in an emulator, you can press Ctrl+F11 or Numpad 7.

Figure 3-15   How to save and restore values

How to use the documentation for the Android API

One of the most difficult aspects of learning Android is mastering the hundreds of classes and methods that your apps will require. To do that, you frequently need to study the documentation for the Android API. If you've used the documentation for the Java API, you'll find that the Android API works similarly.

Figure 3-16 summarizes some of the basic techniques for navigating through the documentation for the Android API. This shows one screen of the documentation available from the Activity class, which has many screens. To view the documentation for the Activity class, you click on the package name (android. app) in the upper left frame and then on the class name (Activity) in the lower left frame.

If you scroll through the documentation for this class, you'll get an idea of the scale of the documentation that you're dealing with. After a few pages of overview information, you come to a summary of the constants, fields, and constructors available from the class. After that, you come to a summary of the dozens of public and protected methods that the class offers. That in turn is followed by more detail about these constants, fields, constructors, and methods.

At this point in your Android education, this is far more information than you can handle. That's why one of the goals of this book is to introduce you to the dozens of classes and methods that you'll use in most of the Android apps that you develop. Once you've learned those, the documentation for the Android API will make more sense to you, and you'll be able to use that documentation to research classes and methods that aren't presented in this book.

It's never too early to start using the documentation, though. So by all means use the documentation to get more information about the methods that are presented in this book and to research the other methods that are offered by the classes that are presented in this book. After you learn how to use the Activity class, for example, take some time to do some research on that class.

The documentation for the Activity class

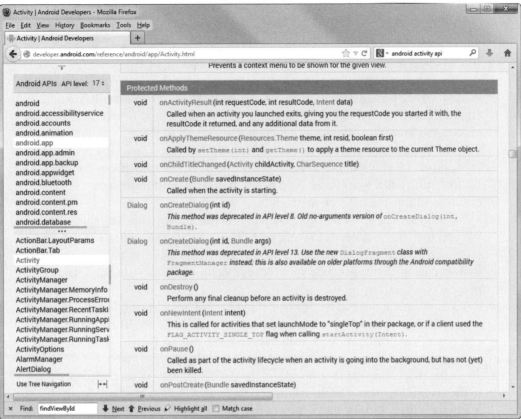

Description

- The Android API contains hundreds of classes and methods that you can use in your apps.

- You can use a browser to view the documentation for the Android API by going to this address:

  http://developer.android.com/reference/

- You can select the name of the package in the top left frame to display information about the package and the classes it contains. Then, you can select a class in the lower left frame to display the documentation for that class in the right frame.

- Once you display the documentation for a class, you can scroll through it or click on a hyperlink to get more information.

- To make it easier to access the documentation for the Android API, you should bookmark the index page. Then, you can easily redisplay this page whenever you need it.

Figure 3-16   How to use the documentation for the Android API

The Java code for the app

Figure 3-17 shows the Java code for the Tip Calculator app. The code in this figure is similar to the Java code presented in chapter 1. By now, though, you should understand the details of how this code works.

To start, the package statement stores the class for the Tip Calculator activity in the package named com.murach.tipcalculator. Then, it imports all Java and Android classes and interfaces that are needed for this class. If you review these classes and interfaces, you should be familiar with all of them.

The TipCalculatorActivity class inherits the Activity class that's provided by Android. In addition, this class implements the listener interfaces for the EditorAction and Click events. As a result, this class must implement the onEditorAction and onClick methods.

Within the class, the first six statements define the instance variables for the widgets that the class needs to work with. Then, the seventh statement defines the SharedPreferences object that's used to save and restore values, and the next two statements define instance variables for the bill amount string and the tip percent.

The onCreate method begins by calling the onCreate method of the super-class, which is necessary for the superclass to work correctly. Then, it uses the setContentView method that's available from the superclass to display the user interface that's defined in the XML file for the activity.

After displaying the user interface, the onCreate method gets references to the six widgets that it declared earlier. To do that, it calls the findViewById method for each widget. Then, it casts the View object that's returned to the appropriate type for the widget.

After getting references to the widgets, this code sets the listeners. First, it sets the current class as the listener for the EditorAction event on the editable text view for the bill amount. Then, this code sets the current class as the listener for the Click event on both of the buttons.

The Java code   Page 1

```java
package com.murach.tipcalculator;

import java.text.NumberFormat;

import android.os.Bundle;
import android.view.KeyEvent;
import android.view.View;
import android.view.View.OnClickListener;
import android.view.inputmethod.EditorInfo;
import android.widget.Button;
import android.widget.EditText;
import android.widget.TextView;
import android.widget.TextView.OnEditorActionListener;
import android.app.Activity;
import android.content.SharedPreferences;
import android.content.SharedPreferences.Editor;

public class TipCalculatorActivity extends Activity
implements OnEditorActionListener, OnClickListener {

    // define variables for the widgets
    private EditText billAmountEditText;
    private TextView percentTextView;
    private Button   percentUpButton;
    private Button   percentDownButton;
    private TextView tipTextView;
    private TextView totalTextView;

    // define SharedPreferences object
    private SharedPreferences savedValues;

    // define an instance variable for the tip percent
    private String billAmountString = "";
    private float tipPercent = 15f;

    @Override
    public void onCreate(Bundle savedInstanceState) {
        super.onCreate(savedInstanceState);
        setContentView(R.layout.activity_tip_calculator);

        // get references to the widgets
        billAmountEditText = (EditText) findViewById(R.id.billAmountEditText);
        percentTextView = (TextView) findViewById(R.id.percentTextView);
        percentUpButton = (Button) findViewById(R.id.percentUpButton);
        percentDownButton = (Button) findViewById(R.id.percentDownButton);
        tipTextView = (TextView) findViewById(R.id.tipTextView);
        totalTextView = (TextView) findViewById(R.id.totalTextView);

        // set the listeners
        billAmountEditText.setOnEditorActionListener(this);
        percentUpButton.setOnClickListener(this);
        percentDownButton.setOnClickListener(this);
```

Figure 3-17   The Java code (part 1 of 3)

The onPause method saves both of this activity's instance variables: the bill amount string and the tip percent. These instance variables need to be saved for the app to work correctly when orientation changes and when the user navigates away from and back to this activity.

The onResume method restores both of the activity's instance variables. Then, it sets the bill amount string as the display text on the bill amount EditText widget. This is necessary because Android sets the display text to its default value if the user navigates away from and back to this activity. Finally, this code calls the calculateAndDisplay method.

The calculateAndDisplay method calculates the tip and total amounts for the bill and displays all current data on the user interface. First, this method gets the bill amount from the EditText widget and converts this string to a float value. Then, it calculates the tip and total amounts and stores them as float values. Finally, it formats these float values and displays them on their corresponding widgets.

The Java code   **Page 2**

```java
        // get SharedPreferences object
        savedValues = getSharedPreferences("SavedValues", MODE_PRIVATE);
    }

    @Override
    public void onPause() {
        // save the instance variables
        Editor editor = savedValues.edit();
        editor.putString("billAmountString", billAmountString);
        editor.putFloat("tipPercent", tipPercent);
        editor.commit();

        super.onPause();
    }

    @Override
    public void onResume() {
        super.onResume();

        // get the instance variables
        billAmountString = savedValues.getString("billAmountString", "");
        tipPercent = savedValues.getFloat("tipPercent", 0.15f);

        // set the bill amount on its widget
        billAmountEditText.setText(billAmountString);

        // calculate and display
        calculateAndDisplay();
    }

    public void calculateAndDisplay() {

        // get the bill amount
        billAmountString = billAmountEditText.getText().toString();
        float billAmount;
        if (billAmountString.equals("")) {
            billAmount = 0;
        }
        else {
            billAmount = Float.parseFloat(billAmountString);
        }

        // calculate tip and total
        float tipAmount = billAmount * tipPercent;
        float totalAmount = billAmount + tipAmount;

        // display the other results with formatting
        NumberFormat currency = NumberFormat.getCurrencyInstance();
        tipTextView.setText(currency.format(tipAmount));
        totalTextView.setText(currency.format(totalAmount));

        NumberFormat percent = NumberFormat.getPercentInstance();
        percentTextView.setText(percent.format(tipPercent));
    }
```

Figure 3-17   The Java code (part 2 of 3)

The onEditorAction method is executed whenever the user presses an action key on a soft keyboard such as the Done key. This method begins by using an if statement to check whether the action key is the Done key. If so, it calls the calculateAndDisplay method to perform the calculation and display the results on the user interface.

The onClick method is executed whenever the user clicks on either of the buttons. Within this method, a switch statement checks which button was clicked. Then, if the Decrease (-) button is clicked, this code decreases the tip percent by 1 percent and calls the calculateAndDisplay method to perform the calculation and display the results on the user interface. Conversely, if the Increase (+) button is clicked, this code increases the tip percent by 1 percent and calls the calculateAndDisplay method to perform the calculation and display the results on the user interface.

The Java code

```java
@Override
public boolean onEditorAction(TextView v, int actionId, KeyEvent event) {
    if (actionId == EditorInfo.IME_ACTION_DONE)
    {
        calculateAndDisplay();
    }
    return false;
}

@Override
public void onClick(View v) {
    switch (v.getId()) {
    case R.id.percentDownButton:
        tipPercent = tipPercent - .01f;
        calculateAndDisplay();
        break;
    case R.id.percentUpButton:
        tipPercent = tipPercent + .01f;
        calculateAndDisplay();
        break;
    }
}
}
```

Figure 3-17   The Java code (part 3 of 3)

Perspective

The goal of this chapter has been to get you off to a fast start with Android development. Now, if you understand the Tip Calculator app, you should also be able to develop simple Android apps of your own. Keep in mind, though, that this chapter is just an introduction to the Android essentials that will be expanded upon in section 2 of this book.

Terms

widgets	event
controls	action key
text view	listener
label	action event
editable text view	wire
text box	input method editor
button	created state
layout	started state
form	running state
activity	active state
density-independent pixels	resumed state
scale-independent pixels	paused state
theme	stopped state
event handler	

Summary

- In Android development, the components that make up the user interface are known as *widgets*.
- In Android, a *layout* consists of one or more child elements and controls how those child elements are displayed.
- In Eclipse, you can add widgets to a layout by dragging them from the Palette onto the layout in the Graphical Layout editor.
- It's generally considered a best practice to store the display text for your app in a separate XML file. This makes it easy to internationalize the text for your application.
- Although Android supports other units of measurement, such as inches, it's a best practice to use *density-independent* and *scale-independent pixels* whenever possible.
- An *event handler* is a special type of method that's executed when an *event* occurs.
- A *listener* is an object that listens for an event.
- An EditorAction event typically occurs when the user presses an action key, such as the Done key, on a soft keyboard.

- A Click event typically occurs when the user clicks a widget such as a button.

- You can use the SharedPreferences class to permanently save values in your app.

- To save values, you can override the onPause method.

- To restore values you have saved, you can override the onResume method.

Before you do the exercises for this chapter

If you haven't already done so, you should install the software and the source code for this book as described in the appendix.

Exercise 3-1   Create the Invoice Total app

In this exercise, you'll create the Invoice Total app. When you're done, a test run should look something like this:

Create the project

1. Create a project for an Android app named Invoice Total and store it in a project named ch03_ex1_Invoice in this directory:

   `\murach\android\workspace`

   This project should be stored in a package named com.murach.invoice.

2. Optionally, use the New Android Application dialog box to create a custom launcher icon. You should be able to download possible icons by searching the Internet.

3. Use the New Android Application dialog box to create a blank activity with a Java class for the activity named InvoiceTotalActivity and a layout file for the activity named activity_invoice_total.

Create the user interface

4. If necessary, open the XML file for the user interface that's in the res\layout directory and switch to the Graphical Layout editor.

5. Use the Outline view to make sure the layout is a relative layout. If the Task List view is open, I recommend closing it.

6. Use the Outline view to delete the TextView widget that displays the "Hello world!" message.

7. Add the seven TextView widgets and one EditText widget to the layout. Set the Id and Text properties of each widget immediately after you add the widget. When you're done, the user interface should have the widgets and text shown above. However, these widgets may look different since you haven't set their properties yet.

8. Set the Text Size property for all eight widgets to 20sp.

9. Set the Text Style property for the four widgets in the left column to bold.

10. Switch to the XML editor and review the code. Note how the XML attributes match the properties that you have set with the Graphical Layout editor.

11. Open the strings.xml file that's in the res/values directory. If necessary, switch to the XML editor view. Then, review the code.

12. Delete the string named hello_world.

13. Test the user interface by running the app on a physical device or an emulator. The app should allow you to enter a subtotal, but that won't change the discount percent, discount amount, or total.

Write the Java code

14. Open the InvoiceTotalActivity.java file that's in the src directory of the project.

15. Delete all methods except the onCreate method.

16. Use the onCreate method to get references to the EditText widget and the three TextView widgets that display data.

17. Create a method named calculateAndDisplay. This method should get the subtotal value. Then, it should calculate the discount amount and total. It should give a 20% discount if the subtotal is greater than or equal to 200, a 10% discount if the subtotal is greater than or equal to 100, and no discount if the subtotal is less than 100.

18. Add code to the end of the calculateAndDisplay method that displays the results of the calculation on the widgets for the discount percent, discount amount, and total.

19. Override the onResume method and use it to call the calculateAndDisplay method.

20. Handle the EditorAction event for the EditText widget so that it executes the calculateAndDisplay method when the Done key is pressed.

21. Test the app. It should display the starting values that are coded in the strings.xml file.

22. Modify the strings.xml file so it doesn't display a starting value for the subtotal.

23. Test the app again. This time, it shouldn't display a starting value for the subtotal. Enter some values for the subtotal and make sure it works correctly.

24. Change the orientation of the activity. On an emulator, you can do that by pressing Ctrl+F11 or Numpad 7. The activity should retain all of its data.

25. Press the Back key to navigate away from the app. Then, navigate back to the app. In an emulator, you can do this by clicking on the Settings icon, selecting the Apps tab, and clicking on the Invoice Total app. The activity should lose all of its data.

26. Override the onPause method so it saves the string for the subtotal. Then, modify the onResume method so it gets the string for the subtotal. To get these methods to work correctly, you need to set up instance variables for the subtotal string and for a SharedPreferences object that you can use to save and get this string.

27. Test the app again. This time, the app should always remember the last subtotal that you entered even if you navigate away from the app and return to it. In addition, the app should always calculate and display the correct values for the discount percent, discount amount, and total.

Exercise 3-2   Use the documentation for the Android API

This exercise steps you through the documentation for the Android API. This should give you an idea of how extensive the Android API is.

1. Open a browser and display the documentation for the Android API.

2. Click the android.app package in the upper left frame and the Activity class in the lower left frame to display the documentation for the Activity class. Then, scroll through this documentation to get an idea of its scope.

3. Skim through the overview information for the Activity class.

4. Scroll through the public methods. These methods include the findViewById and setContentView methods that you learned how to use in this chapter.

5. Scroll through the protected methods. These methods include the onCreate, onPause, and onResume methods that you learned how to override in this chapter.

6. Go to the documentation for the TextView class, which is in the android.widget package. Review the attributes for this class as well as the public methods for this class. The public methods include the getText and setText methods you learned about in this chapter.

7. Go to the documentation for the SharedPreferences interface, which is in the android.content package. Then, review public methods. These methods provide a way to get strings as well as values of the boolean, int, long, and float types. However, they don't provide a way to get values of the double type.

How to test and debug an Android app

As you develop an Android app, you need to test it to make sure that it performs as expected. Then, if there are any problems, you need to debug your app to correct any problems. This chapter shows how to do both.

Basic skills for testing and debugging

When you *test* an app, you run it to make sure that it works correctly. As you test the app, you try every possible combination of input data and user actions to be certain that the app works in every case. In other words, the goal of testing is to find errors (*bugs*) and make an app fail.

When you *debug* an app, you fix the bugs that you discover during testing. Each time you fix a bug, you test again to make sure that the change you made didn't affect any other aspect of the app.

Typical test phases

When you test an app, you typically do so in phases, like the four that are summarized in figure 4-1.

In the first phase, you should test the user interface. To start, you can use the Graphical Layout editor to check the widgets to make sure they display properly. As you do this, you should test portrait and landscape orientation, all target screen sizes, and all target platforms. You'll learn how to do that in the next figure. Then, you should run the app on a device or emulator to make sure that all the widgets work correctly. For instance, you can click on an EditText widget to make sure the soft keyboard displays correctly.

In the second phase, you should test the app with valid data. To start, you can run the app and enter data that you would expect a user to enter. Then, you should enter valid data that tests the limits of the app. You should change the orientation of the app to make sure it works correctly when the user switches between portrait and landscape orientation. And you should test other changes in the lifecycle of each activity. For instance, you should navigate away from an activity and return to it.

In the third phase, you should try to make the app fail by testing every combination of invalid data and user action that you can think of. For instance, if possible, you should try entering letters where numbers are expected.

In the first three phases, you can test the app on a single device or emulator. If possible, this device should be the primary target device. For example, if you primarily want your app to run on a current touchscreen smartphone, you can begin by testing on that device or an emulator that emulates that device.

In the fourth phase, the app should be working correctly on the primary target device or emulator. However, in this phase, you test the app on all other target devices or on emulators for those devices. This may include devices that have physical keyboards, D-pads, or other custom hardware.

If possible, you should use physical devices since this is the only true way to test your app. However, whenever necessary, you can use an emulator instead of a physical device. As you test on various target devices, you will usually find some bugs that you'll need to fix. This is one of the most difficult aspects of Android programming.

The Tip Calculator with a logic error

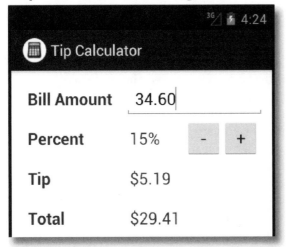

Four test phases

- Check the user interface to make sure that it works correctly.
    - Check in portrait and landscape orientation.
    - Check on all target screen sizes.
    - Check on all target platforms.
- Test the app with valid input data to make sure the results are correct.
    - Test changing the orientation.
    - Test other changes in the lifecycle of the activity.
- Test the app with invalid data or unexpected user actions. Try everything you can think of to make the application fail.
- Test on all target devices. This may include devices that have a physical keyboard, D-pad, or other kinds of hardware. Use a physical device if possible. Otherwise, use an emulator.

Description

- The goal of *testing* is to find all errors (*bugs*) in the app.
- The goal of *debugging* is to fix all of the bugs that you find during testing.

Figure 4-1   An introduction to testing and debugging

How to check the layout

Figure 4-2 shows how to use the Graphical Layout editor to check the layout. To start, you can use the buttons on the toolbar to zoom in or out. Or, you can emulate the real size of the device by clicking on the Emulate Real Size button. These skills can help you view the widgets on the user interface and make sure that they are being displayed correctly.

If you are going to allow the app to switch orientations, you need to make sure the layout works for both portrait and landscape orientation. To do that, you can click on the Flip Orientation button in the toolbar to switch between portrait and landscape orientation. Then, you can check to make sure the layout displays correctly in both orientations. If it doesn't, you can tweak it until it does, or you can create a separate layout for each orientation as described in the next chapter.

Once the layout displays correctly in both orientations, you should test it for different screen sizes. To do that, you can use the drop-down list in the toolbar to select a device with a different screen size. In this figure, the Nexus One smartphone is selected. This device has a 3.7 inch screen. However, if you want you can select a tablet with a 7 or 10 inch screen.

Once you are satisfied with the layout in different screen sizes, you should test it on different platforms. To do that, you can select any platform that's installed on your system from the drop-down list in the toolbar. In this figure, for example, four platforms are installed on the user's system. When you do this, it may take your system a minute or so to render the user interface, so be patient! When the device is displayed on the new platform, the widgets may look different, but the layout should still display correctly.

Finally, if you need to change the theme, you can do that by selecting a different theme from the drop-down list in the toolbar. In this figure, the theme named AppTheme has been selected, which is usually what you want. However, if necessary, you can use the drop-down list to select another theme.

The Graphical Layout editor for the Android 2.2 (API 8) platform

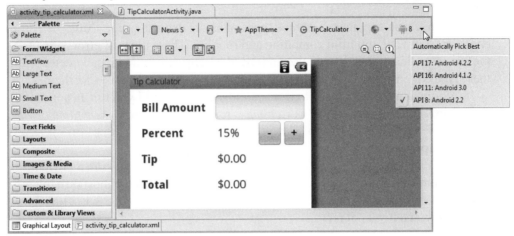

Description

- To zoom in or out, use the buttons on the toolbar to change the zoom level. To emulate the real size of the device, click on the Emulate Real Size button.

- To view on a different screen size, select a device with a different screen size from the drop-down list in the toolbar.

- To change the orientation, click on the Flip Orientation button in the toolbar.

- To change platforms, select a different platform from the drop-down list in the toolbar. This drop-down list only displays the platforms that are installed on your system.

- To change the theme, select a different theme from the drop-down list in the toolbar.

Figure 4-2   How to check the layout

The three types of errors

As you test an app, there are three types of errors that can occur. *Syntax errors* prevent your app from compiling and running. These types of errors are the easiest to find and fix. If you use an IDE like Eclipse, it automatically detects these types of errors as you type and gives you suggestions for how to fix them.

Unfortunately, some types of errors can't be detected until you run an app. These types of errors are known as *runtime errors*, and they throw *exceptions* that stop the execution of an app. This is often referred to as "crashing" or "blowing up."

Even if an app runs without crashing, it may contain *logic errors* that prevent the app from working correctly. These types of errors are often the most difficult to find and correct. For example, the Tip Calculator app in figure 4-1 has a logic error. Can you tell what it is?

How to handle runtime errors

As you test an app, you may encounter runtime errors that cause it to crash. When that happens, the emulator or device displays an error message like the one in figure 4-3, and Eclipse prints some information about the error to its LogCat view. This information includes the name of the exception. In this figure, for example, a NumberFormatException caused the app to crash.

By examining the LogCat view closely, you can often figure out the cause of the bug, which is the first step in fixing the bug. In this figure, for example, the LogCat data indicates that there was a problem parsing a float value. As a result, you can begin by looking at the code that parses the float value to see if you can figure out what's causing the bug. If that doesn't work, you can use the debugging skills that are presented in the rest of this chapter to figure out what's causing the bug. Once you figure out what's causing the bug, fixing the bug is often the easy part.

The three types of errors that can occur

- *Syntax errors* violate the rules for how the code must be written. These errors are caught by the IDE or the compiler before you run the app.

- *Runtime errors* don't violate the syntax rules, but they throw *exceptions* that stop the execution of the application.

- *Logic errors* are statements that don't cause syntax or runtime errors, but produce the wrong results.

The error that's displayed when an app crashes

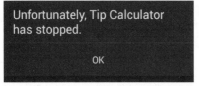

The LogCat view after a crash

```
Java - ch03_TipCalculator/src/com/murach/tipcalculator/TipCalculatorActivity.java - ADT

File  Edit  Run  Source  Refactor  Navigate  Search  Project  Window  Help

                                                                          Debug   Java

Package Explor          activity_tip_calculator.xml     TipCalculatorActivity.java        Outline

  ch03_TipCalculator         88                                                              android.util.Log
    Android 4.1.2            89      // get the bill amount                                  android.view.KeyEvent
    Android Dependencies     90      billAmountString = billAmountEditText.getText().toString()  android.view.View
    src                      91      float billAmount= Float.parseFloat(billAmountString);   android.view.View.OnC
      com.murach.tipcalcu     92                                                              android.view.inputmeth
        TipCalculatorActiv   93      // calculate tip and total                              android.widget.Button
    gen [Generated Java Files 94     float tipAmount = billAmount * tipPercent;
    assets                   95      float totalAmount = billAmount + tipAmount;
    bin                      96
    libs
    res
    AndroidManifest.xml
    ic_launcher-web.png    Problems  @ Javadoc  Declaration  LogCat    Console
    proguard-project.txt
    project.properties     Saved Filters    -      Search for messages. Accepts Java regexes. Prefix with pid:, app:, tag: or text: to  verbose

                           All messages (no filters)
                           com.murach.tipcalculat   Tag                 Text
                           com.murach.tipcalculat   AndroidRuntime      FATAL EXCEPTION: main
                                                    AndroidRuntime      java.lang.NumberFormatException: Invalid float: ""
                                                    AndroidRuntime      at java.lang.StringToReal.invalidReal(StringToReal.jav
                                                    AndroidRuntime      at java.lang.StringToReal.parseFloat(StringToReal.java
                                                    AndroidRuntime      at java.lang.Float.parseFloat(Float.java:300)
                                                    AndroidRuntime      at com.murach.tipcalculator.TipCalculatorActivity.calc
                                                                        lay(TipCalculatorActivity.java:91)
                                                    AndroidRuntime      at com.murach.tipcalculator.TipCalculatorActivity.onEd
                                                                        ipCalculatorActivity.java:116)
                                                    AndroidRuntime      at android widget TextView onEditorAction(TextView jav

  115M of 240M                               Writable      Smart Insert     89 : 1
```

Description

- When an app encounters an error that causes it to end, the emulator or device displays an error message that indicates that the app has stopped, and Eclipse displays some information about the error in the LogCat view.

- When an error occurs, the LogCat view includes the name of the exception that caused the error.

Figure 4-3   An introduction to errors and how to handle runtime errors

How to trace code execution

When you *trace* the execution of an app, you add statements to your code that display messages or variable values at key points in the code. This is typically done to help you find the cause of a bug.

How to use LogCat logging

The most common way to trace code execution is to use the LogCat logging as shown in figure 4-4. To do that, you begin by importing the Log class that's in the android.util package. Then, you use one of the methods of the Log class to send data to the LogCat view.

When you use LogCat for debugging, you typically use the d method to send a debug message to the log. However, if necessary, you can use other methods to send other types of messages such as an error, warning, or informational message. Each type of message displays in a different color in the LogCat view, which makes it easy to distinguish between the types of errors.

All of the methods in this figure have two parameters. The first parameter is for a tag that's used to identify the message in the LogCat view. Often, the name of the activity is used as the tag for a message. As a result, it's a common practice to define a constant named TAG for each activity that includes the name of the activity. Then, you can use this constant as the first argument for all logging statements in the activity, and you can code any string as the second argument.

In this figure, the first Log statement prints a message that indicates that the calculateAndDisplay method is starting. Then, the second statement prints the value of the variable named billAmount. Printing the values of variables like this can often help you find a bug. Then, when you fix the problem, you can remove the logging statements.

When you use this technique, you usually start by adding just a few logging statements to the code. Then, if that doesn't help you find and fix the bug, you can add more. Once you find and fix the bug, you typically remove these logging statements from your code.

When you run the app, Eclipse displays the messages in the LogCat view. Sometimes, Eclipse automatically creates and selects a filter for the current app. That way, the LogCat view shows messages for the current app, not for other apps, which is usually what you want. In this figure, for example, Eclipse has created and selected a filter for the Tip Calculator app. As a result, the LogCat view only shows messages for that app.

Other times, Eclipse does not automatically create and select a filter for the current app. In that case, the LogCat view mixes the messages for your app with the messages for all other apps, which isn't usually what you want. To fix this, you can create a filter for your app by clicking the Add button in the LogCat view. To create a filter like the one shown in this figure, for example, you can add a filter named com.murach.tipcalculator that only displays messages for the application named com.murach.tipcalculator.

Eclipse with the LogCat view displayed

A few methods of the Log class

Method	Message type	Display color
d(tag, message)	Debug	Blue
e(tag, message)	Error	Red
w(tag, message)	Warn	Orange
i(tag, message)	Info	Green

How to use the Log class

Import the Log class
```
import android.util.Log;
```

Declare a constant for the tag parameter
```
private static final String TAG = "TipCalculatorActivity";
```

Send messages to the log
```
Log.d(TAG, "calculateAndDisplay method started");
Log.d(TAG, "billAmount: " + billAmount);
```

Description

- A simple way to *trace* the execution of an app is to use methods of the Log class to send messages to the log at key points in the code. In Eclipse, these messages are shown in the LogCat view.

Figure 4-4   How to use LogCat to trace code execution

How to use toasts

Another way to trace code execution is to use *toasts*, which are messages that are briefly displayed on the user interface. In figure 4-5, for example, the bottom of the Tip Calculator app is displaying a toast that says, "onCreate method". Since toasts are only displayed briefly, they're only useful for certain situations. If you experiment with them, you'll quickly discover when they can be helpful.

The first step in displaying a toast is to import the Toast class. This class is stored in the android.widget package.

The second step is to create a Toast object. To do that, you call the static makeText method of the Toast class and pass this object three parameters. The first parameter is the context, and it determines where the toast is displayed. In most cases, you can use the this keyword to identify the current activity as the context. The second parameter is the message. You can specify any string you want for this parameter, including strings that display the value of a variable as shown in the previous figure. The third parameter is the length of time that the toast is displayed. You can use the LENGTH_SHORT and LENGTH_LONG constants of the Toast class for this parameter.

The third step is to call the show method from the Toast object that you created. There are two ways to do this. If you have created a variable for the object, you can call the method from that variable as shown in the second to last example. However, it's more common to call the show method directly from the makeText method as shown in the last example. This technique is referred to as *method chaining*.

Method chaining is commonly used in Android programming, and it has a couple advantages over using multiple statements. First, it only requires a single statement. Second, you don't have to provide extra object names that aren't used in other parts of your app. This results in code that's shorter and still easy to read.

When you use toasts for debugging, you typically only use a toast or two. Then, when you're done debugging, you need to remove these statements.

Although it's common to use toasts for debugging, you can also use a toast in an app to display a message to the user. In an email app, for example, you may want to use a toast to display a message that says, "Message deleted" whenever a user deletes a message. As you progress through this book, you'll see some examples of this.

A toast displayed in an emulator

Two methods of the Toast class

Method	Description
makeText(context, message, length)	Returns a Toast object with a message that can be displayed in the specified context for the specified length.
show()	Displays the toast.

Two constants of the Toast class

Constant	Description
LENGTH_SHORT	A short length of time.
LENGTH_LONG	A long length of time.

How to display a toast

Import the Toast class
```
import android.widget.Toast;
```

Use two statements
```
Toast t = Toast.makeText(this, "onCreate method", Toast.LENGTH_SHORT);
t.show();
```

Use a single statement (method chaining)
```
Toast.makeText(this, "onCreate method ", Toast.LENGTH_SHORT).show();
```

Figure 4-5   How to use toasts to trace code execution

How to use the debugger

When you use LogCat logging or toasts for debugging, you have to add statements to your code. Then, when you're done debugging, you need to remove these statements. This creates extra work for you. Fortunately, Eclipse includes a powerful tool known as a *debugger* that you can use to debug an app without having to add or remove statements.

However, the debugger is a complex tool, and you may find it easier to use LogCat logging or toasts for some types of bugs. As a result, you'll have to choose the debugging technique that you prefer for any given situation.

How to set and remove breakpoints

The first step in debugging an app is to find the cause of the bug. To do that, it's often helpful to view the values of the variables at different points in the execution of the app.

The easiest way to view the variables while an app is running is to set a *breakpoint* as shown in figure 4-6. Then, when you run the app with the debugger, execution stops just prior to the statement at the breakpoint, and you can view the variables that are in scope at that point in the app.

To set a breakpoint, you double-click to the left of the line of code on the vertical bar that's on the left side of the code editor. Then, the breakpoint is marked by a blue circle to the left of the line of code. Note, however, that you can only set a breakpoint on a line of code that can be executed, not on a declaration, comment, brace, or parenthesis.

When debugging, it's important to set the breakpoint before the line in the app that's causing the bug. Often, you can figure out where to set that breakpoint just by knowing which statement caused the crash. However, sometimes you need to experiment a little before finding a good location for a breakpoint.

After you set one or more breakpoints, you need to run the app with the debugger. To do that, you can use the Debug button that's available from the toolbar (just to the left of the Run button), or you can use the Debug command that's available from various menus. For example, you can right-click on the project in the Package Explorer and select the Debug As→Android Application item.

When you run an app with the debugger, Eclipse switches from the Java perspective to the Debug perspective. But first, it may prompt you with a dialog box like the one shown in this figure. Then, you can click Yes to confirm the switch. If you don't want to receive this message in the future, you can select the "Remember my decision" check box.

After you run an app with the debugger, the breakpoints remain where you set them. If you want to remove a breakpoint, you can do that by double-clicking on the blue circle for the breakpoint. Or, you can select the Run→Remove All Breakpoints item.

The Java perspective with a breakpoint

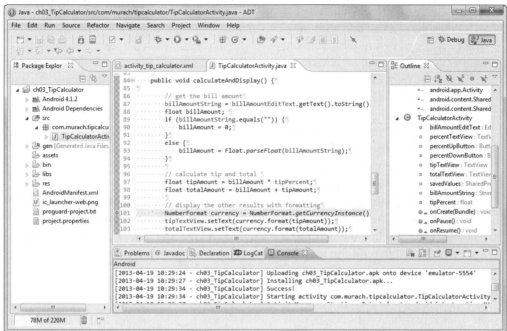

The Confirm Perspective Switch dialog box

Description

- A *breakpoint* is indicated by a small blue circle icon that's placed to the left of the line of code. When an application is run in debug mode, it stops just before executing the statement at the breakpoint.

- To set a breakpoint, double-click on the vertical bar to the left of a line of code in the code editor. To remove a breakpoint, double-click on it.

- Once you set a breakpoint, you can use the Debug button on the toolbar to begin debugging. This works much like the Run button described in chapter 2, except that it lets you debug the application. When the application encounters the breakpoint, Eclipse switches into the Debug perspective as shown in the next figure.

Figure 4-6   How to set and remove breakpoints

How to step through code

When you run an app with the debugger and it encounters a breakpoint, execution stops before the statement at the breakpoint. Then, a blue arrow marks the next statement that's going to be executed. In addition, Eclipse switches to the Debug perspective, and opens several new views, including the Debug, Variables, and Breakpoints views shown in figure 4-7.

The Debug perspective also displays some toolbar buttons to the right of the tab for the Debug view. Then, you can use the Step Into button to *step through* the statements in the app, one statement at a time. This lets you observe exactly how and when the variable values change as the app executes, and that can help you determine the cause of a bug.

Once you've stepped through the code that you're interested in, you can click the Resume button to continue execution until the next breakpoint is reached. You can click the Terminate button to end the app. Or, you can click the Disconnect button to disconnect the debugger.

How to inspect variables

When you set breakpoints and step through code, the Variables view automatically displays the variables that are in scope. In this figure, the execution point is in the calculateAndDisplay method of the TipCalculatorActivity class. Here, the billAmount variable is a local variable that's declared to store the amount of the bill. In addition, even though they aren't shown in this figure, the instance variables of this class are in scope. To view these variables, you can expand the variable named *this*, which refers to the current object. In this case, that object is the object created from the TipCalculatorActivity class.

For numeric variables and strings, the value of the variable is shown in the Variables view. However, you can also view the values for an object by expanding the variable that refers to the object. In this figure, for example, you can expand the variable named *this* by clicking on the plus sign to its left. Then, you can view the values of its variables.

How to inspect the stack trace

In the Debug perspective, the Debug view shows the *stack trace*, which is a list of methods in the reverse order in which they were called. You can click on any of these methods to display the method and highlight the line of code that called the next method. This opens a new code editor if necessary.

In this figure, for example, clicking on the onEditorAction method jumps to the onEditorAction method and displays the line of code that called the calculateAndDisplay method. Since the onEditorAction method is in the same class as the calculateAndDisplay method, this doesn't open a new code editor. However, if they were in different classes, it would.

The Debug perspective

Description

- When program execution stops, the arrow in the code editor marks the line that will be executed next.
- To step through code one statement at a time, click the Step Into button.
- To execute the code until the next breakpoint is reached, select the Resume button.
- To end the application's execution, select the Terminate button.
- To disconnect the debugger, select the Disconnect button.
- The Variables view shows the values of the variables that are in scope for the current method. This includes static variables, instance variables, and local variables. If a variable refers to an object, you can view the values for that object by expanding the object and drilling down through its variables.
- The Debug view shows the *stack trace*, which is a list of methods in the reverse order in which they were called. You can click on any of these methods to display its code in the code editor.
- To switch back to the Java perspective, click on the Java button in the upper right corner of Eclipse.

Figure 4-7   How to work with the Debug perspective

How to configure step filters

When you're stepping through code, you typically only want the debugger to step through your code, not the code of the core Java and Android librar-ies. However, some versions of Eclipse step through the source code libraries. This usually isn't what you want and often results in the debugger displaying a "Source not found" message instead of displaying the source code.

To fix this problem, you can turn on Step Filtering and add all packages that you want to skip to the filter as shown in figure 4-8. Unfortunately, this process requires a bit of trial and error. To start, you click the Select All button to select all packages that are already in the filter. Then, you can use the Add Filter button to add filters for the rest of the packages shown in this figure. As you debug, you may find other packages that you may want to add to your step filter, and you can use the Add Filter button to do that.

The Step Filtering preferences

Common packages to add to step filtering

```
android.*
com.android.*
com.google.*
dalvik.*
libcore.*
org.apache.*
```

Description

- If Eclipse attempts to step through core libraries that you don't want to step through, you can turn on Step Filtering and add all packages that you want to skip to the filter.

- To display the Step Filtering settings, select the Window→Preferences item from the menus. Then, select the Java→Debug→Step Filtering categories.

- To select all packages and classes in the dialog box, click the Select All button.

- To add a filter for a package, click the Add Filter button, type the name of the package with wildcard modifier, and click the OK button.

Figure 4-8   How to configure step filters

How to configure your emulators

Even after you've tested and debugged your app for your primary target device, you need to test and debug your app for the rest of your target devices. If possible, you should test and debug your app on physical devices. However, since it's usually not possible to use physical devices to test all target devices, you'll probably need to use emulators to emulate some target devices.

In the appendix, you learned how to create a generic emulator for an Android 4.2 phone. Now, you'll learn how to create a generic emulator for an Android 2.2 phone. This provides a way to test an old Android phone. If your app works on both of these emulators, it has a good chance of working on most devices.

In addition, you'll now learn how to create a generic emulator for a tablet. This provides a way to test your app on a device that has a large screen.

How to add an emulator for an old phone

Figure 4-9 shows how to add an emulator for an old phone. To start, part 1 shows how to create a device definition for the phone. This lets you specify the screen size in inches and the resolution in pixels. It lets you specify the amount of RAM for the device. It lets you specify whether the device provides hardware buttons such as the Home, Back, and Menu buttons. It lets you specify whether the device inludes a hard keyboard. And it lets you specify whether the device includes a DPad or trackball.

In this figure, the device definition is for a device that has a 4 inch screen that has a resolution of 480 by 800 pixels. In addition, this device has hardware that supports a keyboard, a DPad, and other buttons. It makes sense that this emulator would provide this support since hard keyboards, DPads, and other buttons were available on many older devices.

The device definition for a phone with a hard keyboard and DPad

How to create a new device definition

1. Start Eclipse.
2. In the toolbar, click the button for the Android Virtual Device Manager. This should start the Android Virtual Device Manager.
3. Click on the Device Definitions tab.
4. Select the New Device button. This should display the dialog box shown above.
5. Enter a name for the device. Then, set its screen size and resolution.
6. Specify a RAM amount for the device. For most systems a RAM setting of 512 MiB is adequate. However, you can increase or decrease this amount if you encounter problems.
7. Use the Buttons drop-down list to control whether the device supports the Home, Back, and Menu hardware buttons.
8. Use the Input category to control whether the device hardware supports a keyboard, DPad, or trackball.

Description

- If the built-in device definitions aren't adequate for you, you can use the Create New Device dialog box to create your own device definition.

Figure 4-9   How to add an emulator for an old phone (part 1 of 2)

Part 2 shows how to create an emulator that's based on the device definition that was created in part 1. To start, you use the Device drop-down list to select a custom or built-in device definition. Then, you use the Target drop-down list to select the version of Android for the device. If necessary, you use the CPU/API drop-down list to select the type of processor for the device. Finally, you use the "Hardware keyboard present" check box to specify whether you want to be able to use your keyboard to enter text.

In this figure, the emulator uses the custom device definition that was created in part 1. It runs Android 2.2. It uses an ARM processor, which is the most common type of processor for smartphones and tablets. And it allows you to use your computer's keyboard to enter text.

In this figure, no value is specified for the SD card size. As a result, this emulator does not support an SD card. Since the apps presented in this book don't use the SD card, that's fine. However, if your apps need to use the SD card, you can specify an SD card size. Since the SD card size you specify takes up the specified space on your computer, it's generally a good practice to only specify a size that's large enough for testing your application.

In this figure, the Snapshot and Use Host GPU check boxes are not selected. Although these options can help your emulator start faster, they can also cause problems. For example, if you use the Snapshot option and then edit the settings for the emulator, it won't start correctly.

The emulators presented so far in this book are for generic devices (an Android 4.2 phone and an Android 2.2 phone). If you wanted to emulate a particular device more closely, you can begin by searching the Internet for the hardware and software specs of the device. Then, you can create a device definition that matches the hardware specs for the device, and you can create an AVD based on that definition that specifies the software specs.

An emulator that supports a hard keyboard and DPad

How to create an emulator

9. From Android Virtual Device Manager, click on the Android Virtual Devices tab.

10. Select the New button. This should display the dialog box shown above.

11. Enter a name for the emulator. Then, select the device definition, target platform, and (if necessary) the CPU/ABI.

12. Use the "Hardware keyboard present" option to control whether the emulator allows you to use your computer's keyboard to enter text.

13. When you have the options set the way you want, click on the OK button.

Description

*   You can use the Create New Android Virtual Device (AVD) dialog box to create an emulator that's based on a built-in or custom device definition. In addition, you can specify the version of Android that's running on this device.

Figure 4-9   How to add an emulator for an old phone (part 2 of 2)

How to work with an emulator for an old phone

Figure 4-10 shows the emulator created in the last figure with the Tip Calculator app running on it. Since this emulator uses Android 2.2 (API 8), the widgets look a little different than they did in previous chapters.

In addition, all of the hardware buttons on the right side of the emulator are enabled. As a result, you can use the Home, Menu, Back, and Search buttons to emulate the hardware buttons on the device. You can use the DPad buttons to emulate the DPad on the device. And you can use your computer's keyboard to emulate the physical keyboard on the device.

If you use this emulator to test the Tip Calculator app, you'll find that it still displays a soft keyboard for EditText widgets. Although this soft keyboard looks different than the soft keyboard for newer devices, it still works the same.

Unfortunately, the first time you use this keyboard, it might display some Japanese characters, and it might not include a decimal point or a Done key. As a result, you can't use it to enter a decimal number. To fix this issue, you can long-click on the EditText widget, select the "Input method" item, and select the "Android keyboard" item. After you do that, the soft keyboard should not display Japanese characters, and it should display a decimal key and a Done key.

The soft keyboard on an emulator for an old phone

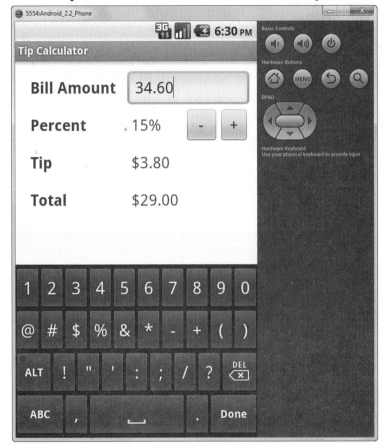

Description

- On this emulator, the hardware buttons are enabled. As a result, you can use this emulator to test the Home, Menu, Back, and Search keys.
- On this emulator, the DPad is enabled. As a result, you can use this emulator to test the DPad.
- On this emulator, the hardware keyboard is enabled. As a result, you can use your computer's keyboard to test the hard keyboard.
- On this emulator, the soft keyboard for an EditText widget might display Japanese characters and might not include a decimal point or a Done key. To fix this issue, long-click on the EditText widget, select the "Input method" item, and select the "Android keyboard" item.

Figure 4-10   How to work with an emulator for an old phone

How to add an emulator for a tablet

Figure 4-11 shows how to create an emulator for a tablet. The steps for doing this are the same as creating an emulator for a phone. However, you use the Device drop-down list to select a built-in or custom device definition for a larger screen. In addition, tablets typically use Android 3.0 (API 11) or later.

In this figure, the emulator uses the built-in device definition for a device that has a 7 inch screen with 1024 by 600 pixels. This emulator runs Android 4.2. It uses an ARM processor. And it does not allow you to use your computer's keyboard to enter text. Instead, you must use the soft keyboard that's provided by the emulator.

An emulator for a tablet

AVD Name:	Android_4.2_Tablet|
Device:	7.0" WSVGA (Tablet) (1024 × 600: mdpi) ▼
Target:	Android 4.2.2 - API Level 17 ▼
CPU/ABI:	ARM (armeabi-v7a) ▼
Keyboard:	☐ Hardware keyboard present
Skin:	☑ Display a skin with hardware controls
Front Camera:	None ▼
Back Camera:	None ▼
Memory Options:	RAM: 512 VM Heap: 16
Internal Storage:	200 MiB ▼

Create new Android Virtual Device (AVD)

SD Card:
- ◉ Size: [   ]   MiB ▼
- ○ File: [   ]   Browse...

Emulation Options: ☐ Snapshot   ☐ Use Host GPU

☐ Override the existing AVD with the same name

[OK]   [Cancel]

Description

- To create an emulator for a tablet, use the Device drop-down list to select a built-in or custom device definition for a larger screen.
- Tablets typically use Android 3.0 (API 11) or later.
- Tablets typically use an ARM processor.

Figure 4-11   How to add an emulator for a tablet

Perspective

Before you release an app, it should be thoroughly tested and debugged. Now that you've completed this chapter, you should be able to thoroughly test your apps. And if you find bugs during testing, you should be able to use the skills presented in this chapter to fix those bugs.

As your apps grow more complex, you may need testing and debugging skills that go beyond the skills covered in this chapter. For example, you may want to experiment with some of the other debugging views and toolbar buttons. If you do, you can probably learn a few new debugging skills.

Terms

test	trace
bugs	toasts
debug	method chaining
syntax errors	debugger
runtime errors	breakpoint
exceptions	step through
logic errors	stack trace

Summary

- To *test* an app, you run it to make sure that it works properly no matter what combinations of valid or invalid data you enter.

- When you *debug* an app, you find and fix all of the errors (*bugs*) that you find when you test the app.

- *Syntax errors* violate the rules for how Java statements must be written. These errors are caught by Eclipse or the Java compiler before you can run the app.

- *Runtime errors* occur after you run an app. These types of errors throw *exceptions* that stop the execution of the app.

- *Logic errors* don't cause the app to crash, but they prevent it from working correctly.

- A simple way to *trace* the execution of an app is to insert LogCat logging statements at key points in the code.

- Another way to trace code execution is to use *toasts*, which are messages that are briefly displayed on the user interface.

- Android often allows a coding technique known as *method chaining*. This technique allows you to call one method directly from another method.

- Eclipse includes a powerful tool known as a *debugger* that can help you find and fix errors.

- You can set a *breakpoint* on a line of code to stop code execution just before that line of code. Then, you can *step through* the code and view the values of the variables as the code executes.

- A *stack trace* is a list of methods in the reverse order in which they were called.

Exercise 4-1   Test and debug the Tip Calculator app

This exercise guides you through the process of using Eclipse to test and debug an app.

Test the app with invalid data

1. Start Eclipse and open the project named ch04_ex1_TipCalculator that's in this directory:

    `\murach\android\workspace\ex_starts`

2. Run this project and test the app with a valid subtotal like 100. This should work correctly.

3. Test the app with an invalid subtotal by leaving the bill amount blank and pressing the Done key. This should cause the app to crash with a run-time error, and it should display an error message in the LogCat view.

4. Study the error message that's displayed in red. You can focus on the first few lines of this message. These lines give information about the exception that caused the app to crash. Based on this information, you should be able to figure out that the app crashed because an empty string isn't a valid float value.

5. Fix the bug by using a try/catch statement to handle the exception. The catch clause for the exception should set the billAmount variable to zero.

Use LogCat logging

6. At the end of the onCreate method, add a logging statement that prints "onCreate executed" to the LogCat view.

7. Add logging statements that print "onPause executed" and "onResume executed" to the ends of the onPause and onResume methods.

8. Run the app and view the logging statements as you change orientation. Note that this causes the onPause, onCreate, and onResume methods to be executed. This shows that the activity is destroyed and recreated when you change orientation.

9. Run the app and view the logging statements as you navigate away from and back to the app. Note that this only causes the onPause and onResume methods to be executed. This shows that the activity is paused when you navigate away from it and resumed when you navigate back to it.

Test on different emulators

10. Create an emulator for a phone that supports a hard keyboard.

11. Run the app in that emulator. Then, use your computer's keyboard to enter a bill amount and press the Enter key to submit this entry.

12. Check the soft keyboard for this emulator to make sure it works correctly. If it displays Japanese characters and doesn't provide a Done button, fix the issue as described in this chapter.

13. Create an emulator for a tablet.

14. Run the app in that emulator. It should work as before.

Use toasts

15. Add a toast to the onEditorAction method that displays the value of the actionID parameter.

16. Run the app, use the soft keyboard to enter a bill amount, and press the Done key. Make a note of the value of the actionID parameter.

17. Run the app in an emulator that supports a hard keyboard. Then, use the hard keyboard to enter a bill amount and press the Enter key to submit this entry. Make a note of the value of the actionID parameter.

Use the debugger

18. Check your preferences for step filtering and make sure that it is on and includes the Java and Android packages described in this chapter.

19. In the calculateAndDisplay method, set a breakpoint on this line of code:
    ```
    billAmountString = billAmountEditText.getText().toString();
    ```

20. Make sure the project for this exercise is selected in the Package Explorer. Then, click on the Debug button in the toolbar. This runs the project with the debugger on.

21. If necessary, confirm the switch to the Debug perspective.

22. When execution stops at the breakpoint, click the Variables tab to display the Variables view. Then, expand the variable named *this* and view the value for the instance variable named billAmountString. Next, collapse the variable named *this* so you can easily view other variables.

23. Click the Step Into button in the toolbar 4 times to step through the app one statement at a time. After each step, review the values in the Variables view to see how they have changed. Note how the app steps through the try/catch statement based on the value of the bill amount string.

24. Click the Resume button in the toolbar to continue the execution of the app.

25. Switch to the emulator, enter a new bill amount, and click the Done button.

26. Switch back to Eclipse and use the Variables view to view the value of the instance variable named billAmountString.

27. When you're done inspecting the variables, click the Terminate button to end the app. Then, click the Java button in the toolbar to switch back to the Java perspective. This should give you some idea of how useful the Eclipse debugging tools can be.

Section 2

Essential Android skills

Most of the chapters in this section expand upon the basic Android skills that you learned in section 1. In chapter 5, for example, you'll learn more about getting input from the user and displaying output. In chapter 6, you'll learn more about handling events. In chapter 7, you'll learn more about controlling the appearance of your app. And so on.

In addition, some of the chapters in this section present new skills. In chapter 8, for example, you'll learn how to work with menus and preferences. Then, in chapter 9, you'll learn how to work with fragments.

To illustrate these skills, this section adds new features to the Tip Calculator app that you learned about in section 1. This allows you to expand your skills while still working with an app that's familiar to you. When you're done with this section, you'll be ready to learn how to work with some more complicated apps such as the News Reader and Task List apps.

5

How to work with layouts and widgets

In chapter 3, you learned how to use a relative layout with three different types of widgets: text views, editable text views, and buttons. Now, this chapter shows you how to use some other types of layouts. In addition, this chapter shows you some new skills for working with widgets including how to use check boxes, radio buttons, spinners, and seek bars.

An introduction to layouts and widgets

This chapter begins by summarizing some of the layouts and widgets that are available from the Android API and by presenting the inheritance hierarchy for the classes for layouts and widgets.

A summary of layouts

Figure 5-1 begins by summarizing six layouts that are available from the Android API. Of these layouts, the relative layout described in chapter 3 is one of the most powerful since it provides a way to align widgets relative to one another. However, it is also one of the most difficult layouts to use.

That's why this chapter briefly presents three other layouts that can be useful. If you need to create a user interface that displays widgets in a vertical or horizontal line, you may want to use a linear layout. If you want to display widgets in rows and columns, you may want to use a table layout. Or, if you want to display widgets on top of each other in a stack, you may want to use the frame layout.

This chapter doesn't present the absolute layout because this layout was deprecated in Android 1.5 (API level 3). As a result, you shouldn't use this layout for new app development.

This chapter doesn't present the grid layout because this layout was introduced with Android 4.0 (API level 14). As a result, this layout doesn't work with older versions of Android unless you include its library as part of your app. Since you can usually get the same results as the grid layout with the relative layout or the table layout, it's often not worth the effort to use the grid layout, though the grid layout does provide some nice improvements over the relative and table layouts.

A summary of widgets

Figure 5-1 also summarizes some of the widgets that are available from the Android API. Keep in mind that this is just a summary, not a complete list. The Android API provides many other useful widgets.

You already learned about the TextView, EditText, and Button widgets in chapter 3. Now, you'll learn about the CheckBox, RadioButton, Spinner, SeekBar, ImageView, and ScrollView widgets.

Some of the widgets in this figure are often referred to using other terminology. For example, a TextView widget is often called a *label*. An EditText widget is often called an *editable text view*, a *text box*, or a *text field*. And a spinner is often called a *drop-down list*.

Once you understand how to use the widgets presented in this chapter, you should be able to learn about other widgets on your own. For example, once you learn how to use the Button and ImageView widgets, you should be able to use the ImageButton widget. Then, if you need help, you can look up the ImageButton class in the Android API documentation.

A summary of layouts

Layout	Description
RelativeLayout	Lays out widgets relative to one another. This layout is described in chapter 3.
LinearLayout	Lays out widgets in a vertical or horizontal line. This layout is described in figure 5-3.
TableLayout	Lays out widgets in a table. This layout is described in figure 5-4.
FrameLayout	Lays out widgets in a stack where one widget is displayed on top of the other. This layout is described in figure 5-5.
AbsoluteLayout	This layout was deprecated in Android 1.5 (API level 8) and is not covered in this book.
GridLayout	Lays out widgets in a grid. This layout was introduced with Android 4.0 (API level 14). As a result, it isn't compatible with older versions of Android by default. However, you can make it compatible with older versions of Android by including its library as part of your application. This layout is not covered in this book.

A summary of widgets

Widget	Description
TextView	Also known as a *label*, this widget displays text.
EditText	Short for *editable text view*, this widget is also known as a *text box* or *text field*. This widget allows the user to enter text such as names, email addresses, passwords, phone numbers, dates, times, and numbers.
Button	Performs an action when the user clicks it.
CheckBox	Allows the user to check or uncheck an option.
RadioButton	Allows the user to select a single option from a group that's defined with a RadioGroup widget.
Spinner	Also known as a *drop-down list*, this widget allows the user to select an option from a list.
ProgressBar	Displays a visual indicator of the progress of an operation.
SeekBar	Allows the user to select a value by dragging a thumb to the left or right.
RatingBar	Allows the user to rate something by selecting one or more stars.
ImageView	Displays an image.
ImageButton	Works like a regular button but displays an image instead of text.
DatePicker	Allows the user to select a date.
TimePicker	Allows the user to select a time.
CalendarView	Allows the user to select a date.
ScrollView	Automatically displays a vertical scroll bar if the layout doesn't fit on the screen vertically.

Figure 5-1   A summary of layouts and widgets

The View hierarchy

Figure 5-2 shows the inheritance hierarchy for the classes that define Android widgets and layouts. To start, the View class is the superclass for all widgets. As a result, any methods of this class are available to all widgets.

Under the View class, the ViewGroup class is the superclass for widgets that can contain other widgets. This includes layouts such as the RelativeLayout, LinearLayout, TableLayout, and FrameLayout classes. In addition, it includes the RadioGroup class that's used to group widgets created from the RadioButton class.

Under the View class, the TextView class is the superclass for widgets that display text such as the EditText and Button widgets. From the Button class, the CompoundButton class is the superclass for the CheckBox and RadioButton classes. As a result, any methods of the CompoundButton class are available to both CheckBox and RadioButton widgets.

The View class is also the superclass for other widgets that don't display text. For example, the View class is the superclass for the ProgressBar class, which is the superclass for the AbsSeekBar class, which is the superclass for the SeekBar and RatingBar classes. Similarly, the View class is the superclass for the ImageView class.

The View hierarchy

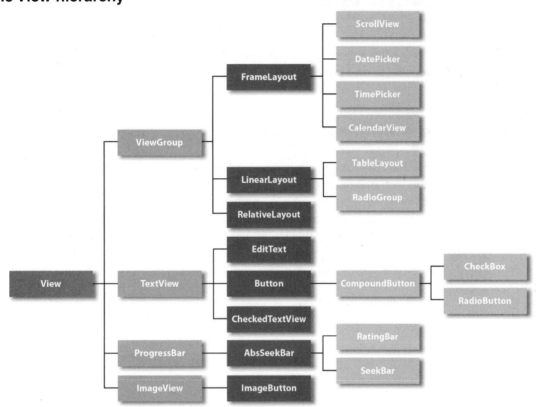

Description

- The View class is the superclass for all widgets.
- The ViewGroup class is the superclass for most widgets that can contain other widgets such as the RelativeLayout widget.
- The TextView class is the superclass for widgets that display text such as the EditText and Button widgets.

Figure 5-2   The View hierarchy

How to work with layouts

In chapter 3, you learned how to use a relative layout. Now, this chapter shows how to use three more layouts that are available from Android. In addition, it shows how to nest layouts and how to provide for separate landscape layouts.

How to use a linear layout

If you need to create a user interface that displays widgets in a vertical or horizontal line, you may want to use a linear layout as shown in figure 5-3. Here, the width and height attributes of the layout have been set to "match_parent". As a result, this layout stretches across the entire screen.

The orientation attribute of this layout has been set to "vertical". As a result, this layout displays the two buttons in a vertical line. In other words, this layout displays the two buttons in a column, one above the other.

A linear layout with vertical orientation and two buttons

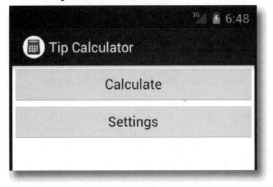

The XML for the linear layout

```xml
<?xml version="1.0" encoding="utf-8"?>
<LinearLayout xmlns:android="http://schemas.android.com/apk/res/android"
    android:layout_width="match_parent"
    android:layout_height="match_parent"
    android:orientation="vertical" >

    <Button
        android:id="@+id/calculateTipButton"
        android:layout_width="match_parent"
        android:layout_height="wrap_content"
        android:text="@string/calculate_tip" />

    <Button
        android:id="@+id/settingsButton"
        android:layout_width="match_parent"
        android:layout_height="wrap_content"
        android:text="@string/settings" />

</LinearLayout>
```

Description

- A linear layout displays a column or row of child widgets.

Figure 5-3   How to use a linear layout (part 1 of 2)

Part 2 of figure 5-3 shows some more examples for working with a linear layout. For all of these examples, the width of the buttons has been set to "wrap_content". That way, the buttons are only wide enough to display their text.

In the first example, the orientation attribute for the layout is set to a value of "horizontal". As a result, the buttons are displayed in a horizontal line. In other words, they are displayed in a row.

In the second example, the weight attribute for both buttons has been set to "1". As a result, both buttons are stretched so they take up the width of the layout, with each button taking half of the layout since both buttons have equal weight. If you wanted to have one button be wider than the other, you can experiment with the weight attributes of the buttons. For example, if you set the weight of the first button to 2 and the second button to 1, the first button takes 2/3 of the width, and the second button takes 1/3 of the width.

In the third example, the gravity attribute of both buttons has been set to "center". As a result, both buttons are centered horizontally in the layout, which has been set to vertical orientation.

Common attributes for working with linear layouts

Attribute	Description
orientation	Controls whether the linear layout uses vertical or horizontal orientation.
weight	Determines how much space Android allocates to a widget.
gravity	Aligns a widget with the top, bottom, center, left, or right of its layout.

A horizontal layout where the buttons have no weight

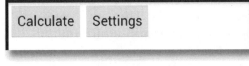

The orientation attribute for the layout
```
android:orientation="horizontal"
```

A horizontal layout where the buttons have equal weight

The weight attribute for both buttons
```
android:weight="1"
```

A vertical layout where the buttons are centered horizontally

The gravity attribute for both buttons
```
android:gravity="center"
```

Description

- The attributes of a linear layout and its child widgets control the appearance of the user interface.

Figure 5-3   How to use a linear layout (part 2 of 2)

How to use a table layout

If you need to create a user interface that displays controls in a grid, you may want to use the table layout as shown in figure 5-4. Here, the table layout displays the first two rows of the Tip Calculator app. To do that, this layout uses four columns. In the first row, the TextView widget is displayed in the first column, and the EditText widget is displayed in the next three columns. In the second row, one widget is displayed in each column.

To create a table layout, you use the TableRow element to define each row. Within each row, you can put as many widgets as you'd like. Then, if you need a widget to span multiple columns, you can use the layout_span attribute to specify the number of columns to span. In this figure, for example, the layout_span attribute of the EditText widget has been set to "3". As a result, this widget spans 3 columns.

A table layout has a couple of advantages over a relative layout. To start, when you use a table layout, you only have to provide the id attribute for the widgets that you need to use with your Java code. In this figure, for example, the widgets that just display text don't have id attributes. In addition, you don't have to provide layout attributes for positioning the widget relative to other widgets. Instead, you just use the layout_span attribute wherever necessary. As a result, you don't need to set as many attributes when you use a table layout.

However, a relative layout has a couple advantages over a table layout. To start, it gives you more control over the positioning and alignment of widgets. Second, a relative layout typically runs faster than a table layout.

A table layout with two rows and four columns

Common attributes for working with table layouts

Attribute	Description
layout_span	Specifies the number of columns that the widget should use.

The XML for the table layout

```xml
<?xml version="1.0" encoding="utf-8"?>
<TableLayout xmlns:android="http://schemas.android.com/apk/res/android"
    android:layout_width="match_parent"
    android:layout_height="match_parent" >

    <TableRow>
        <TextView
            android:text="@string/bill_amount_label"
            <!--    TextView widget attributes    -->    />
        <EditText
            android:id="@+id/billAmountEditText"
            android:text="@string/bill_amount"
            android:layout_span="3"
            <!--    EditText widget attributes    -->
            <requestFocus />
        </EditText>
    </TableRow>
    <TableRow>
        <TextView
            android:text="@string/tip_percent_label"
            <!--    TextView widget attributes    -->    />
        <TextView
            android:id="@+id/percentTextView"
            android:text="@string/tip_percent"
            <!--    TextView widget attributes    -->    />
        <Button
            android:id="@+id/percentDownButton"
            android:text="@string/decrease"
            <!--    Button widget attributes    -->    />
        <Button
            android:id="@+id/percentUpButton"
            android:text="@string/increase"
            <!--    Button widget attributes    -->    />
    </TableRow>

</TableLayout>
```

Description

- A table layout displays widgets in rows and columns.

Figure 5-4   How to use a table layout

How to use a frame layout

A frame layout is one of the simplest and most efficient types of layouts. You can use this type of layout to display one widget over another widget as shown in figure 5-5. Or, you can use a frame layout as a placeholder within another layout for displaying a single child layout or widget.

If a frame layout contains multiple child widgets, it stacks them on top of each other. This displays the first widget added to the layout on the bottom of the stack and the last widget added on the top. In this figure, for example, the first child widget is an ImageView widget that displays an image of a restaurant. Then, the second child widget is a TextView widget that displays white text. The result is that white text is displayed over an image.

The ImageView widget in this figure displays an image of a restaraunt. For now, don't worry about the details of how this widget works. These details are explained later in this chapter.

The TextView widget in this figure works like the TextView widgets presented in chapter 3. However, it uses the gravity attribute to center it horizontally. In addition, it uses the textColor attribute to set the text color to a hexadecimal color of "#ffffff", which is white. If you're familiar with HTML, you may already be familiar with hexadecimal colors, which are also known as *web colors*. If you aren't familiar with web colors, you can learn more by searching for "web colors wiki" on the Internet.

Interestingly, the DatePicker, TimePicker, and CalendarView classes inherit the FrameLayout class. That's because the DatePicker, TimePicker, and CalendarView widgets consist of different widgets stacked on each other.

A frame layout that displays an image behind some text

The XML for the frame layout

```xml
<?xml version="1.0" encoding="utf-8"?>
<FrameLayout xmlns:android="http://schemas.android.com/apk/res/android"
    android:layout_width="match_parent"
    android:layout_height="match_parent" >

    <ImageView
        android:layout_width="match_parent"
        android:layout_height="wrap_content"
        android:contentDescription="@string/photo"
        android:src="@drawable/restaurant" />

    <TextView
        android:layout_width="match_parent"
        android:layout_height="wrap_content"
        android:gravity="center"
        android:text="@string/welcome"
        android:textColor="#ffffff"
        android:textSize="30sp" />

</FrameLayout>
```

Description

- A frame layout is one of the simplest and most efficient types of layouts.
- A frame layout often displays only a single child layout or widget.
- If a frame layout contains multiple child widgets, it stacks them on top of each other displaying the first widget added to the layout on the bottom of the stack and the last widget added on the top.

Figure 5-5   How to use a frame layout

How to nest layouts

When working with linear layouts, it's common to *nest* one linear layout within another to align widgets in columns and rows as shown in figure 5-6. Here, two linear layouts with horizontal orientation are nested within a linear layout that has vertical orientation. This creates two rows, which can be filled with widgets. In this figure, the first row has a TextView and an EditText widget, and the second row has a Button widget.

If a layout only has a few levels of nested layouts, the nested layout probably won't degrade performance so much that it's noticeable to the user. However, nesting multiple levels can degrade performance significantly. As a result, if you are using multiple levels of nesting, and you notice that your app is taking too long to display, you should consider reducing the number of nested levels. One way to do that is to use a relative layout as described in chapter 3. When you use a relative layout, you typically don't need to nest any layouts, and that usually improves performance.

Interestingly, the table layout is a type of linear layout. If you look back to figure 5-2, you can see that the LinearLayout class is the superclass for the TableLayout class. As a result, the table layout uses nested layouts behind the scenes even though you don't explicitly specify nested layouts in the XML code for the layout.

Nested linear layouts

The XML for nested linear layouts

```xml
<?xml version="1.0" encoding="utf-8"?>
<LinearLayout xmlns:android="http://schemas.android.com/apk/res/android"
    android:layout_width="match_parent"
    android:layout_height="match_parent"
    android:orientation="vertical"
    android:padding="10dp" >

    <!-- the first row -->
    <LinearLayout
        android:layout_width="match_parent"
        android:layout_height="wrap_content"
        android:orientation="horizontal" >

        <!--  widgets go here -->

    </LinearLayout>

    <!-- the second row -->
    <LinearLayout
        android:layout_width="match_parent"
        android:layout_height="wrap_content"
        android:orientation="horizontal" >

        <!--  widgets go here -->

    </LinearLayout>

</LinearLayout>
```

Description

- You can *nest* one layout within another layout.
- Nesting multiple levels can degrade performance.

Figure 5-6   How to nest layouts

How to provide a landscape layout

By now, you should know that an Android activity needs to handle screen orientation changes. One way to do that is to prevent the activity from changing orientation. An easy way to do that is to edit the AndroidManifest.xml file for the app, and add a screenOrientation attribute to the activity that sets the screen orientation to portrait or landscape like this:

```
android:screenOrientation="portrait"
```

This forces the activity to display only in portrait orientation even if the user rotates the device to attempt to change to landscape orientation. This approach is shown in the Android manifest file from chapter 1.

Another way to handle screen orientation changes is to use the same layout for both portrait and landscape. This is the approach that's used for the Tip Calculator app in chapter 3.

However, there are times when you want to provide one layout for portrait orientation and another layout for landscape orientation. For example, you may want to rearrange the widgets on the activity to make better use of the space that's available in landscape orientation. To do that, you can create a directory named res/layout-land as shown in figure 5-7. Then, you can copy the layout for the portrait version of the layout into this directory. After that, you can modify the layout so it makes better use of the space that's available from landscape orientation. In this figure, for example, the two buttons from figure 5-3 have been rearranged to take advantage of landscape orientation.

The layout from figure 5-3 displayed in landscape orientation

The location of the XML files

Portrait
`res/layout/settings_activity.xml`

Landscape
`res/layout-land/settings_activity.xml`

The XML for landscape orientation

```
<?xml version="1.0" encoding="utf-8"?>
<LinearLayout xmlns:android="http://schemas.android.com/apk/res/android"
    android:layout_width="match_parent"
    android:layout_height="match_parent"
    android:orientation="horizontal" >

    <Button
        android:id="@+id/calculateTipButton"
        android:layout_width="match_parent"
        android:layout_height="wrap_content"
        android:layout_weight="1"
        android:text="@string/calculate_tip" />

    <Button
        android:id="@+id/settingsButton"
        android:layout_width="match_parent"
        android:layout_height="wrap_content"
        android:layout_weight="1"
        android:text="@string/settings" />

</LinearLayout>
```

Description

- To provide a separate XML file for landscape orientation, create the layout-land directory if necessary. Then, copy the XML file for the layout into this directory and modify it so it works correctly for landscape orientation.

Figure 5-7   How to provide a custom landscape layout

How to work with widgets

In chapter 3, you learned how to use three kinds of widgets: text views, editable text views, and buttons. Now, this chapter introduces several more widgets that are commonly used, and it presents some new skills for working with editable text views.

How to use editable text views

Figure 5-8 shows some new skills for working with editable text views. To start, this figure shows two editable text views. The first is for getting an email address from a user, and the second is for getting a password from a user. Here, the characters entered into the editable text view for the password are displayed as bullets, which is usually what you want for a password.

The XML for these two editable text views is mostly the same. However, for the first editable text view, the inputType attribute has been set to "textEmailAddress". For the second editable text view, the inputType attribute has been set to "textPassword". As a result, the first editable text view displays a soft keyboard like the one shown in this figure. This soft keyboard is optimized for entering an email address, and it only provides characters that are valid for an email address, which is usually what you want from an editable text view for an email address. Similarly, the second editable text view automatically displays bullets instead of characters, which is what you want from an editable text view for a password.

Two text views and two editable text views

Email:	joel@murach.com
Password:	••••••••

The soft keyboard for an editable text view for an email address

The XML for an editable text view for an email address

```
<EditText
    android:id="@+id/emailEditText"
    android:layout_width="wrap_content"
    android:layout_height="wrap_content"
    android:ems="10"
    android:inputType="textEmailAddress">
    <requestFocus />
</EditText>
```

The XML for an editable text view for a password

```
<EditText
    android:id="@+id/passwordEditText"
    android:layout_width="wrap_content"
    android:layout_height="wrap_content"
    android:ems="10"
    android:inputType="textPassword" />
```

Figure 5-8   How to use editable text views (part 1 of 2)

Part 2 of figure 5-8 begins by summarizing two commonly used attributes of an EditText widget. Of these, the inputType attribute is the most commonly used since you set this for almost every editable text view. However, you may occasionaly want to use the lines attribute to change the number of lines that are displayed by the editable text view. If you don't include this attribute, the editable text view only displays one line. The user can still use the Return key to enter more lines but will have to scroll up or down to view those lines. To allow the editable text view to display more than one line at a time, you can set the lines attribute to a value of 2 or higher. Then, the editable text view always displays the specified number of lines.

The common values of the inputType attribute show that Android provides for many types of editable text views. For example, you can change the input type so it's appropriate for a person's name, email address, password, number, decimal number, phone number, date, or time. In general, you should select an inputType attribute that's as specific as possible for the text that you want the user to enter. That way, Android can present a soft keyboard that's optimal. However, if you don't find an inputType attribute setting that's right for your app, you can always set this attribute to "text". This allows the user to enter any kind of text. Or, you can set this attribute to "textMultiline". This provides a soft keyboard that includes a Return key that you can use to start new lines.

When working with editable text views, the soft keyboard is sometimes displayed every time the user enters the activity. This typically happens because the activity gives the focus to the editable text view when the user enters the activity, which causes the soft keyboard to be displayed. In other words, every time you start an activity or change the orientation of an activity, Android displays the soft keyboard. Often, this isn't what you want. Instead, when the user enters the activity, you want the activity to display with no soft keyboard. That way, the user can view all of the widgets on the activity. Then, if the user moves the focus to the editable text view, you want to display the soft keyboard to the user.

To prevent the soft keyboard from being displayed automatically when the user enters an activity, you can edit the AndroidManifest.xml file. In this figure, for example, an attribute named windowSoftInputMode has been added to the activity element for the Tip Calculator activity. This attribute has been set to "stateUnchanged". As a result, Android only shows the keyboard if the keyboard was previously opened by the user. This is typically what you want. However, if you always want to hide the soft keyboard when entering an activity, you can set this attribute to "stateHidden".

In some rare cases, you may want to always show the soft keyboard when the user enters an activity. For example, if the widgets only fill the top half of the activity, you may always want to display the soft keyboard on the bottom half. In that case, you can change the windowSoftInputMode attribute to "stateVisible". That way, the user doesn't have to take any action to display the soft keyboard.

Two attributes of an EditText widget

Attribute	Description
inputType	The type of data for the editable text view. This typically determines the type of soft keyboard that's displayed for the user.
lines	The number of lines that are displayed by the editable text view. By default, this is set to 1.

Some common values for the inputType attribute

Value	Description
text	Any kind of text.
textPersonName	A person's name.
textEmailAddress	An email address.
textPassword	A password.
textMultiline	Multiple lines of text.
number	Any kind of number.
numberDecimal	A number that has a decimal.
phone	A phone number.
date	A date.
time	A time.

Some common values for the windowSoftInputMode attribute

Value	Description
stateHidden	Always hide the soft keyboard when entering the activity.
stateVisible	Always show the soft keyboard when entering the activity.
stateUnchanged	Only show the keyboard when entering the activity if it was previously opened by the user.

The activity element in the AndroidManifest.xml file

```
<activity
    android:name=".TipCalculatorActivity"
    android:label="@string/title_activity_tip_calculator"
    android:windowSoftInputMode="stateUnchanged" >
```

Description

- An *editable text view* lets the user enter text with a keyboard. A editable text view is also known as a *text field* or *text box*.

Figure 5-8   How to use editable text views (part 2 of 2)

How to use check boxes

Figure 5-9 shows show to use a *check box*, which is a widget that allows the user to check or uncheck an option. If you look at the XML for this widget, you should already be familiar with most of its attributes. The only new attribute is the checked attribute, which determines whether the box is checked or unchecked. In this figure, for example, the checked attribute has been set to "true", so the box is checked. However, you could set this attribute to "false" to uncheck the box.

When working with check boxes, you can use Java to work with a check box when the app runs. For example, it's common to need to set the checked attribute as shown in the first Java example. To do that, you can use the setChecked method to check or uncheck the box.

It's even more common to need to determine whether a check box is checked or unchecked. To do that, you can use the isChecked method. In this figure, for example, the second Java example uses an if statement to determine whether the box is checked. If so, you can execute some code. If not, you can execute some other code.

As you review this figure, note that the checked attribute and setChecked method both accomplish the same task. In Android, there's typically a corresponding set method for an attribute. As a result, you often have a choice as to whether you want to use XML or Java to work with a widget. Of course, you can only use XML to set attributes before the app is running. If you want to work with a widget dynamically, as the app is running, you need to use Java.

A check box

A common XML attribute for check boxes

Attribute	Description
checked	Checks or unchecks the box.

The XML code

```
<CheckBox
    android:id="@+id/rememberPercentCheckBox"
    android:layout_width="wrap_content"
    android:layout_height="wrap_content"
    android:checked="true"
    android:text="@string/remember_percent" />
```

Two common Java methods for check boxes

Method	Description
setChecked(boolean)	Checks or unchecks the box.
isChecked()	Returns a Boolean value that indicates whether the box is checked.

Java examples

Check or uncheck the box

```
rememberPercentCheckBox.setChecked(true);
```

Execute code if the box is checked

```
if (rememberPercentCheckBox.isChecked()) {
    // code to execute when the box is checked
}
else {
    // code to execute when the box is NOT checked
}
```

Description

- A *check box* allows the user to check or uncheck an option.

Figure 5-9   How to use check boxes

How to use radio buttons

Figure 5-10 shows how to use a *radio button*, which is a widget that lets the user select one option of several options. Selecting a radio button automatically deselects all other radio buttons in the same group.

When you add radio buttons, you typically begin by adding two or more radio buttons to a radio group. In this figure, for example, there are three radio buttons in a radio group.

When working with a radio group, you often need to set the orientation for the group. In this figure, for example, the first group of buttons has a vertical orientation, and the second group of buttons has a horizontal orientation.

The XML in this figure defines a radio group that displays three radio buttons in vertical orientation. Here, the orientation attribute of the RadioGroup element has been set to "vertical". Then, three RadioButton elements have been coded within the RadioGroup element. Here, the first radio button has a checked attribute that has been set to true. As a result, the first radio button is selected when the activity is first displayed.

The Java code for working with radio buttons is similar to the code for working with check boxes. For example, you can use the setChecked method to check a radio button, and you can use the isChecked method to determine which button is checked. The main difference is that only one radio button in a group can be checked. As a result, if you use the setChecked method to check one radio button, Android automatically unchecks all of the other radio buttons in that group.

Three radio buttons in a radio group with vertical orientation

Three radio buttons in a group with horizontal orientation

The XML code

```xml
<RadioGroup
    android:id="@+id/roundingRadioGroup"
    android:layout_width="wrap_content"
    android:layout_height="wrap_content"
    android:orientation="vertical" >

    <RadioButton
        android:id="@+id/noRoundingRadioButton"
        android:layout_width="wrap_content"
        android:layout_height="wrap_content"
        android:checked="true"
        android:text="@string/round_no" />

    <RadioButton
        android:id="@+id/roundTipRadioButton"
        android:layout_width="wrap_content"
        android:layout_height="wrap_content"
        android:text="@string/round_tip" />

    <RadioButton
        android:id="@+id/roundTotalRadioButton"
        android:layout_width="wrap_content"
        android:layout_height="wrap_content"
        android:text="@string/round_total" />

</RadioGroup>
```

Java examples

Check or uncheck the radio button

```java
roundTipRadioButton.setChecked(true);
```

Execute code if a radio button is checked

```java
if (roundTipRadioButton.isChecked()) {
    // code to execute when the button is checked
}
```

Description

- A *radio button* lets the user select one option of several options. Selecting a radio button automatically deselects all other radio buttons in the same group.

Figure 5-10   How to use radio buttons

How to use spinners

Figure 5-11 shows how to use a *spinner*, which is a widget that allows the user to select an item from a drop-down list. This type of widget is also known as a *drop-down list*. The top of this figure shows a spinner that allows the user to select how many ways to split a bill. The first item doesn't split the bill, the second item splits the bill 2 ways, the third item splits the bill 3 ways, and so on.

The XML for this spinner is simple. That's because most of the work for setting up a spinner is done with Java code as shown in part 2 of this figure.

An *array adapter* provides the list that a spinner should display. One easy way to provide a list is to store it in an array in the strings.xml file. In this figure, for example, the strings.xml file includes an element named split_array that provides the 4 items that are displayed in the spinner.

A spinner

The XML code

```
<Spinner
    android:id="@+id/splitSpinner"
    android:layout_width="wrap_content"
    android:layout_height="wrap_content" />
```

The array in the strings.xml file

```
<string-array name="split_array">
    <item>Split the bill? - No</item>
    <item>Split the bill? - 2 ways</item>
    <item>Split the bill? - 3 ways</item>
    <item>Split the bill? - 4 ways</item>
</string-array>
```

Description

- A *spinner*, also known as a *drop-down list*, allows the user to select an item from a list.
- An *array adapter* provides the list that a spinner should display.

Figure 5-11   How to use spinners (part 1 of 2)

In part 2 of figure 5-11, the first Java example sets up the spinner. To start, it gets a reference to the Spinner object defined by the XML. Then, it creates an array adapter for the specified array and layout. To do that, it specifies the current activity as the context, it specifies an array from the strings.xml file as the array, and it specifies this layout for the spinner:

```
android.R.layout.simple_spinner_item
```

This is a built-in Android resource that's commonly used to specify the layout for a spinner.

After creating the array adapter, this code sets the layout for the drop-down list of the spinner. To do that, it specifies this layout:

```
android.R.layout.simple_spinner_dropdown_item
```

This is another built-in Android resource that's commonly used to specify the layout for the items in the drop-down list of a spinner.

The second Java example selects the item at the specified position. Here, 0 is the position of the first item, 1 is the second item, and so on. As a result, this code selects the first item.

The third Java example gets the position of the selected item from the spinner. Again, 0 is the position of the first item, 1 is position of the second item, and so on.

The fourth Java example gets the selected text from the selected item. To do that, it uses the getSelectedItem method to return a generic object. Then, it casts the generic object to a String object.

Common methods of the ArrayAdapter class

Method	Description
`createFromResource(context, arrayID, layoutID)`	Creates an array adapter for the specified array and layout for the widget.
`setDropDownViewResource()`	Sets the layout for the items in the drop-down list.

Common methods for spinners

Method	Description
`setAdapter(arrayAdapter)`	Sets the array adapter for the spinner.
`setSelection(index)`	Selects an item from the spinner.
`getSelectedItemPosition()`	Returns an int value for the position of the selected item where 0 is the first item, 1 is the second item, and so on.
`getSelectedItem()`	Returns an Object type for the selected item.

Two Android resources for spinner layouts

```
android.R.layout.simple_spinner_item
android.R.layout.simple_spinner_dropdown_item
```

Java examples

Code that sets up the spinner

```java
// get a reference to the spinner
splitSpinner = (Spinner) findViewById(R.id.splitSpinner);

// create array adapter for specified array and layout
ArrayAdapter<CharSequence> adapter = ArrayAdapter.createFromResource(
    this, R.array.split_array, android.R.layout.simple_spinner_item);

// set the layout for the drop-down list
adapter.setDropDownViewResource(
    android.R.layout.simple_spinner_dropdown_item);

// set the adapter for the spinner
splitSpinner.setAdapter(adapter);
```

Code that selects an item

```java
splitSpinner.setSelection(0);  // select the first item
```

Code that gets the position of the selected item

```java
int position = splitSpinner.getSelectedItemPosition();
```

Code that gets the selected text from the selected item

```java
String selectedText = (String) splitSpinner.getSelectedItem();
```

Description

- You can use built-in Android resources to specify the layouts for the spinner and the items in the drop-down list of a spinner.

Figure 5-11   How to use spinners (part 2 of 2)

How to use seek bars

Figure 5-12 shows a *seek bar*, which is a widget that lets the user specify a value by dragging a *thumb* to the right or left. In this figure, the thumb of the seek bar is identified by a round circle. This thumb is set at a progress of 15 out of a total of 30, and the value for this progress is displayed in a label to the right of the seek bar.

By default, the max attribute of a seek bar is set to 100. As a result, a seek bar typically provides progress values from 0 to 100, which is usually what you want.

In this figure, however, the XML sets the max attribute for the seek bar to 30. As a result, the seek bar provides progress values from 0 to 30, which are more appropriate for a tip percent. Then, this XML sets the progress attribute to 15. As a result, when the seek bar is first displayed, the thumb is half way across the seek bar.

The first Java example sets the progress of the seek bar to a value of 20. To do that, it passes an int value of 20 to the setProgress method.

The second Java example gets the progress of the seek bar. To do that it uses the getProgress method to return an int value for the progress.

A seek bar and a text view

Two common XML attributes for seek bars

Attribute	Description
`max`	Sets the maximum value to the specified int value.
`progress`	Sets the value of the seek bar to the specified int value.

The XML code

```
<SeekBar
    android:id="@+id/percentSeekBar"
    android:layout_width="200dp"
    android:layout_height="wrap_content"
    android:max="30"
    android:progress="15" />

<TextView
    android:id="@+id/percentTextView"
    android:layout_width="0dp"
    android:layout_height="wrap_content"
    android:text="@string/percent" />
```

Two common Java methods for seek bars

Method	Description
`setProgress(int)`	Sets the value of the seek bar to the specified int value.
`getProgress()`	Gets the current value of the seek bar.

Java examples

How to set progress
```
percentSeekBar.setProgress(20);
```

How to get progress
```
int percent = percentSeekBar.getProgress();
```

Description

- A *seek bar* lets the user specify a value by dragging a *thumb* to the right or left.

Figure 5-12   How to use seek bars

How to display images

The ImageView widget in figure 5-13 displays an image of a restaraunt. Here, the contentDescription attribute points to a string in the strings.xml file. Android displays this text instead of the image if there's a problem displaying the image. Then, the src attribute points to an image in one of the drawable directories of the project.

An image that can be drawn on the screen is known as a *drawable resource*. Drawable resources are stored in one of the res/drawable directories of the project. When you work with image files, you typically store them in a directory that corresponds with the density of the screen. For example, you store images for extra high-density screens in this directory:

```
res/drawable-xhdpi
```

Extra high-density screens use approximately 320 dots per inch (dpi).

For this example, I put a JPG file for the image in the medium-density (mdpi) directory because medium-density is the baseline size for Android. Then, if there's no corresponding image for screens of other densities, Android does its best to scale the medium-density image for screens of other densities.

If you need more control over the size of your image on various screens, you can create image files for screens of different densities. For example, let's say you have a 300x300 pixel image that displays correctly on a high-density screen. Then, you can put the file for that image in this directory:

```
res/drawable-hdpi
```

For a medium-density screen, you can resize that image so its 200x200 pixels and put it in this directory:

```
res/drawable-mdpi
```

This works because, to be displayed at the same size, an image for a high-density (240dpi) screen should be 50% larger than the same image for a medium density (160dpi) screen.

To help automate the resizing of images, you can go to the Android Asset Studio web page, select the Generic Icon link, and use the resulting page to generate images of the correct size for the different density screens. Then, you can put those images in the correct directories.

Android supports PNG, JPG, and GIF images. It's generally considered a best practice to use PNG images, though JPG images are often acceptable. In general, it's considered a best practice to avoid using GIF images.

An image

The location of the image file

`res/drawable-mdpi/restaurant.jpg`

Two attributes of an ImageView widget

Value	Description
contentDescription	A string that describes the image. This string is only displayed if Android isn't able to display the image.
src	The location of the *drawable resource*, which is an image that can be drawn on the screen.

Four qualifiers for the drawable folder

Qualifier	Description
xhdpi	Extra high-density screen (approximately 320dpi).
hdpi	High-density screen (approximately 240dpi).
mdpi	Medium-density screen (approximately 160dpi).
ldpi	Low-density screen (approximately 120dpi).

The XML code

```
<ImageView
    android:layout_width="match_parent"
    android:layout_height="wrap_content"
    android:contentDescription="@string/photo"
    android:src="@drawable/restaurant" />
```

Description

- Android supports PNG, JPG, and GIF images. Its API documentation recommends PNG images, says that JPG images are acceptable, and discourages using GIF images.

Figure 5-13   How to display an image

How to show and hide widgets

Figure 5-14 shows how to use Java code to show or hide widgets. To start, this figure shows a layout with three rows. Here, the second row is the "Per Person" label and amount. The code in this figure shows this row when the bill is being split between multiple people, but hides this row when the bill is not being split.

In the Java code example, the first statement gets the position of the selected item from the Spinner widget. Then, the second statement adds an int value of 1 to the selected position and stores the result in a variable named split. That way, the split variable is equal to 1 if the bill is not being split, 2 if the bill is being split by two people, and so on. After that an if statemement checks whether the bill is being split. If not, this code uses the setVisible method to hide both widgets in the "Per Person" row. Otherwise, this code calculates the amount per person and uses the setVisible method to show both widgets in the "Per Person" row.

A layout with three rows

Split The Bill?   3 ways

Per Person $13.26

☑ Remember Tip Percent

The same layout with the per person amount hidden

Split The Bill?   No

☑ Remember Tip Percent

A method of the View class

Method	Description
setVisibility	Hides the widget when set to the GONE constant of the View class.
	Shows the widget when set to the VISIBLE constant of the View class.

Code that shows and hides the per person amount

```
int splitPosition = splitSpinner.getSelectedItemPosition();
int split = splitPosition + 1;
float perPersonAmount = 0;
if (split == 1) {   // no split - hide widgets
    perPersonLabel.setVisibility(View.GONE);
    perPersonTextView.setVisibility(View.GONE);
}
else { // split - show widgets
    perPersonAmount = totalAmount / split;
    perPersonLabel.setVisibility(View.VISIBLE);
    perPersonTextView.setVisibility(View.VISIBLE);
}
```

Description

- You can use Java code to dynamically show or hide widgets.

Figure 5-14   How to show and hide widgets

How to add scroll bars

Figure 5-15 shows how to add scroll bars to a layout. To do that, you typically add a layout within a ScrollView element. Then, if the layout doesn't fit on the screen, Android displays a *vertical scroll bar* so you can scroll through it.

In this figure, the XML begins by declaring a ScrollView element as the root element of a layout file. This ScrollView element contains a single TableLayout element. In turn, this layout can contain multiple TableRow elements, which can contain elements for other widgets. Then, if these widgets become too tall for the screen to display, the user can scroll up or down through these widgets.

In most cases, you only need to provide a vertical scroll bar for your layouts. However, you may occasionally need to provide a *horizontal scroll bar* that lets the user scroll right or left. To do that, you can use a HorizontalScrollView element. Since this element works similarly to the ScrollView element, you shouldn't have much trouble figuring out how to use it.

A scrollable layout

The XML

```
<ScrollView xmlns:android="http://schemas.android.com/apk/res/android"
    android:layout_width="wrap_content"
    android:layout_height="wrap_content">

  <TableLayout
    android:layout_width="match_parent"
    android:layout_height="wrap_content"
    android:padding="10dp" >

    <!-- All table rows and widgets go here -->

  </TableLayout>
</ScrollView>
```

Description

- A ScrollView widget can only have one child element, typically a layout that contains other elements.
- A ScrollView widget displays a *vertical scroll bar* that lets the user scroll up or down the child element.
- To display a *horizontal scroll bar* that lets the user scroll right or left across the child element, use a HorizontalScrollView element.

Figure 5-15   How to display scroll bars

Perspective

Now that you've finished this chapter, you should understand how to work with widgets on a layout. Although this chapter presented some of the most commonly used widgets, you'll learn how to use more widgets as you progress through this book. For example, in chapter 10, you'll learn how to work with the ListView widget.

In addition, once you learn how to use the widgets presented in this chapter, it should become easier to learn how to use other widgets. In general, you can use the same process to work with most widgets. To start, you can drag a widget from the Palette onto a layout. Then, you use the Graphical Layout editor to set the properties of the widget. Finally, you write Java code that uses the methods of the widget to work with it. To learn more about a widget, you can use the Android API documentation, or you can search the Internet for other information such as tutorials.

To get started with the widgets presented in this chapter, you can handle events like the Click event described in chapter 3. Then, you can use the methods shown in this chapter to work with the widgets when a Click event occurs. However, to provide a more responsive user interface, you often need to handle other events such as the ones described in the next chapter. That's why the next chapter starts by describing how to handle high-level events that occur when the user interacts with some of the widgets presented in this chapter.

Terms

label	radio button
editable text view	spinner
text box	array adapter
text field	seek bar
drop-down list	thumb
web colors	drawable resource
nest	vertical scroll bar
check box	horizontal scroll bar

Summary

- The View class is the superclass for all widgets.
- A linear layout displays a column or row of child widgets.
- The attributes of a linear layout and its child widgets control the appearance of the user interface.
- A table layout displays widgets in rows and columns.
- A frame layout is one of the simplest and most efficient types of layouts, and often displays only a single child layout or widget.
- You can *nest* one layout within another layout, but it can degrade performance.

- An *editable text view* lets the user enter text with a keyboard. An editable text view is also known as a *text box* or *text field*.
- A *check box* allows the user to check or uncheck an option.
- A *radio button* lets the user select one option of several options.
- A *spinner*, also known as a *drop-down list*, allows the user to select an item from a list.
- An *array adapter* provides the list that a spinner should display.
- A *seek bar* lets the user specify a value by dragging a *thumb* to the right or left.
- Android supports PNG, JPG, and GIF images.
- You can use Java code to dynamically show or hide widgets.
- A ScrollView widget displays a *vertical scroll bar* that lets the user scroll up or down the child element.
- To display a *horizontal scroll bar* that lets the user scroll right or left across the child element, use a HorizontalScrollView element.

Exercise 5-1   Modify the layout for the Tip Calculator app

In this exercise, you'll modify the Tip Calculator app so it uses a table layout instead of a relative layout.

Test the relative layout version of the app

1. Start Eclipse and import the project named ch05_ex1_TipCalculator that's in the ex_starts directory.
2. Test the Tip Calculator app. It should work as it did in previous chapters.
3. Open the XML file for the user interface that's in the res\layout directory and switch to the Graphical Layout editor to view the layout.

Use a table layout

4. Switch to the XML editor and change the RelativeLayout element to a TableLayout element.
5. Add TableRow elements to identify the rows of the table.
6. Review the XML for each widget and delete any old attributes that were needed for a relative layout but aren't needed for a table layout. Many of these will be marked as invalid by the XML editor.
7. Switch to the Graphical Layout editor to see if the widgets are being displayed correctly. Note that the buttons for setting the tip percent aren't displayed correctly.
8. Add the layout_span attribute to any widgets that should span multiple columns. For example, the EditText widget should span 3 columns. This should result in a grid that has four rows and four columns.

9. Switch to the Graphical Layout editor to view the widgets. They should be displayed correctly.

10. Test the new layout by running the app. The app should work correctly.

Exercise 5-2   Add radio buttons and a spinner

In this exercise, you'll modify the Tip Calculator app so that it uses radio buttons and a spinner. When you're done, the bottom of the layout should look like this:

Total	$37.60
Rounding	◉ None
	○ Tip
	○ Total
Split Bill?	2 ways ◢
Per Person	$18.80
Apply	

Test the app

1. Start Eclipse and import the project named ch05_ex2_TipCalculator that's in the ex_starts directory.

2. Test this app to see how it works. Note that the Apply button doesn't do anything yet.

Add radio buttons

3. Open the layout file and add a radio group that contains the three radio buttons shown above.

4. Open the .java file for the activity and modify the calculateAndDisplay method so it uses the selected radio button to determine whether to round the tip, total, or nothing. To round the tip, you can use code like this:

```
tipAmount = StrictMath.round(billAmount * tipPercent);
totalAmount = billAmount + tipAmount;
tipPercent = tipAmount / billAmount;
```

To round the total, you can use code like this:

```
float tipNotRounded = billAmount * tipPercent;
totalAmount = StrictMath.round(billAmount + tipNotRounded);
tipAmount = totalAmount - billAmount;
tipPercent = tipAmount / billAmount;
```

5. Test this change to make sure it works correctly. Note that the rounding is only displayed when you click the Apply button.

6. Modify the code so it automatically selects the None radio button when the user clicks on the increase (+) or decrease (-) buttons.

Add a spinner

7. Add a TextView widget and Spinner widget to the layout so the user can select the number of ways to split the bill as shown above.

8. Add TextView widgets to the layout that can display the amount per person when the bill is split.

9. Open the strings.xml file and add the array for the spinner.

10. Open the .java file for the activity and add the code that loads the spinner with the array.

11. Test this change to make sure it works correctly.

12. Modify the calculateAndDisplay method so it calculates and displays the correct amount per person when the bill is split. However, when the bill isn't split, this code should hide the TextView widgets that display the amount per person.

13. Test this change to make sure it works correctly.

6

How to handle events

In chapter 3, you learned one technique for handling events that occur on EditText and Button widgets. Now, this chapter expands on that knowledge to show you several techniques for handling events. In addition, it shows how to work with different types of events including events that occur on the CheckBox, RadioButton, RadioGroup, Spinner, and SeekBar widgets described in the previous chapter.

A summary of listeners

A *listener* is an object that listens for *events* that can occur on a widget. When an event occurs on a widget, the appropriate method of the listener is executed. Figure 6-1 summarizes some common listeners and divides them into two groups. Although this summary is far from complete, it shows how listeners work.

High-level events

High-level events only occur on specific types of widgets. For example, the EditorAction event only occurs on certain types of widgets such as an EditText widget. However, this event can't occur on other types of widgets such as a Button widget.

An interface that defines a listener for a high-level event is typically nested in the class for its corresponding widget. For example, the OnEditorActionListener interface is nested within the TextView class. As a result, this listener can work with TextView widgets or other widgets that inherit the TextView class such as the EditText widget.

Most listener interfaces define a single method for the event handler. For example, the OnEditorActionListener interface defines a single method named onEditorAction. However, some listener interfaces define multiple methods. For example, the OnItemSelectedListener interface defines two methods.

Low-level events

Low-level events occur on all types of widgets. The listeners for low-level events are nested within the View class. Since this class is the superclass for all widgets, these listeners can be wired to any type of widget. For example, the OnClickListener can be wired to any widget. This is necessary because a user can click on any type of widget.

Besides clicking on a widget and quickly releasing, it's common to click on a widget and hold the click for more than a second. This is known as a *long click*. You can use the OnLongClickListener to handle this type of event.

For phones that have hardware such as a keyboard or a D-Pad, it's common to use that hardware to enter text or to navigate through an app by moving the *focus* from one widget to another. Typically, Android handles these low-level events the way you want. However, if you need to modify Android's default behavior, you can use the OnKeyListener or OnFocusChangedListener to control how the hardware works with your app.

For some apps, such as games, you need more control over what happens when the user touches the screen. To get this control, you can implement the OnTouchListener.

Listeners for high-level events

Package/Class	Nested Interface	Methods
android.widget.TextView	OnEditorActionListener	onEditorAction
android.widget.CompoundButton	OnCheckedChangedListener	onCheckedChanged
android.widget.RadioGroup	OnCheckedChangedListener	onCheckedChanged
android.widget.AdapterView	OnItemSelectedListener	onItemSelected
		onNothingSelected
android.widget.SeekBar	OnSeekBarChangeListener	onProgressChanged
		onStartTrackingTouch
		onStopTrackingTouch

Listeners for low-level events

Package/Class	Nested Interface	Methods
android.view.View	OnClickListener	onClick
	OnLongClickListener	onLongClick
	OnKeyListener	onKey
	OnFocusChangeListener	onFocusChange
	OnTouchListener	onTouch

Description

- A *listener* listens for *events* that can occur on a widget. When an event occurs on a widget, the appropriate method of the listener is executed.
- *High-level events* are events that occur on a specific type of widget. An interface that defines a listener for a high-level event is typically nested within the class that defines the widget. These listeners can only be wired to an appropriate widget.
- *Low-level events* are events that occur on all types of widgets. An interface that defines a listener for a low-level event is typically nested within the android.view.View class. These listeners can be wired to any type of widget.

Figure 6-1   A summary of listeners

Four techniques for handling events

In chapter 3, you learned one technique for handling an event. Now, figure 6-2 reviews that technique. Then, it shows three other techniques for handling the same event.

Regardless of the technique you choose, you usually start by importing the class for the listener. That's why this is shown as step 1 for all four techniques.

How to use the current class as the listener

The first example shows how to handle the Click event for two buttons using the current class as the listener. By now, you should be familiar with this code since it's the same code that was presented in chapter 3. This code uses the current class (the TipCalculatorActivity class) to implement the onClick method defined by the OnClick listener interface. Then, it wires the listener to the two buttons. To do that, it uses the keyword named *this* to identify the current class as the listener.

How to use a named class as the listener

The second example shows how to handle the Click event for two buttons by creating a separate class named ButtonListener that handles the event. To do that, the ButtonListener class implements the listener interface. This class is stored in the same file as the class for the activity. Then, the code in the activity class creates a listener object from the listener class and assigns it to a variable named buttonListener. Finally, the activity class wires the listener to the two buttons.

Step 1: Import the interface for the listener

```
import android.view.View.OnClickListener;
```

Use the current class as the listener

Step 2a: Implement the interface for the listener

```
public class TipCalculatorActivity extends Activity
implements OnClickListener {
```

Step 2b: Implement the interface for the listener

```
@Override
public void onClick(View v) {
    switch (v.getId()) {
        case R.id.percentDownButton:
            tipPercent = tipPercent - .01f;
            calculateAndDisplay();
            break;
        case R.id.percentUpButton:
            tipPercent = tipPercent + .01f;
            calculateAndDisplay();
            break;
    }
}
```

Step 3: Set the listeners

```
percentUpButton.setOnClickListener(this);
percentDownButton.setOnClickListener(this);
```

Use a separate named class as the listener

Step 2: Code a separate class that implements the listener

```
class ButtonListener implements OnClickListener {
    @Override
    public void onClick(View v) {
        switch (v.getId()) {
            case R.id.percentDownButton:
                tipPercent = tipPercent - .01f;
                calculateAndDisplay();
                break;
            case R.id.percentUpButton:
                tipPercent = tipPercent + .01f;
                calculateAndDisplay();
                break;
        }
    }
}
```

Step 3: Create an instance of the listener

```
ButtonListener buttonListener = new ButtonListener();
```

Step 4: Set the listeners

```
percentUpButton.setOnClickListener(buttonListener);
percentDownButton.setOnClickListener(buttonListener);
```

Figure 6-2   Four techniques for handling events (part 1 of 2)

How to use an anonymous class as the listener

The third example shows how to handle the Click event for two buttons without specifying a name for the class that implements the listener. In other words, this code uses an *anonymous class* as the listener. Here, step 2 defines a private instance variable for an OnClickListener object named buttonListener. Then, it uses the new keyword to create the listener object, and it supplies all code for the class that defines the object within the braces ({ }). Next, step 3 uses the name of the instance variable to wire the listener to the two buttons.

How to use an inner anonymous class as the listener

The fourth example shows how to handle the Click event for two buttons without assigning the listener object to an instance variable. To do that, you can create the object for the listener within the parentheses for the method that wires the widget to the listener. This is known as an *inner anonymous class*.

An inner anonymous class is useful when you want to use a different event handler for each widget. In this figure, for example, the onClick method must be coded for both buttons. Here, the first onClick method is for the button that increases the tip percent, and the second onClick method is for the button that decreases the tip percent. When you use this technique, you know that the code for the onClick method will only be executed for a specific widget. As a result, you don't need to begin this method by using a switch statement to check which widget was clicked.

Conversely, the first three techniques described in this figure are useful when you want to use one event handler to handle an event for multiple widgets. In this figure, for example, the code for the first three examples uses a single onClick method to handle the Click event for two buttons.

When to use each technique

In most cases, you can use any of these four techniques for your app. So, which technique should you use? In most cases, that's largely a matter of personal preference. However, if you need to wire the exact same event handler to multiple widgets, you should reduce code duplication by using one of the first three techniques.

In general, I recommend using the technique that results in the most readable and maintainable code. I also recommend using the same technique throughout an app whenever possible for the sake of consistency. Throughout this book, I have used the first technique because I think it's the easiest to understand when you're getting started with Android. However, all four techniques are commonly used by professional Android programmers.

Use an anonymous class as the listener

Step 2: Create an instance variable for the listener

```
private OnClickListener buttonListener = new OnClickListener() {
    @Override
    public void onClick(View v) {
        switch (v.getId()) {
            case R.id.percentDownButton:
                tipPercent = tipPercent - .01f;
                calculateAndDisplay();
                break;
            case R.id.percentUpButton:
                tipPercent = tipPercent + .01f;
                calculateAndDisplay();
                break;
        }
    }
};
```

Step 3: Set the listeners

```
percentUpButton.setOnClickListener(buttonListener);
percentDownButton.setOnClickListener(buttonListener);
```

Use an inner anonymous class as the listener

Step 2: Set the listeners and implement the interfaces for the listeners

```
percentUpButton.setOnClickListener(new OnClickListener() {
    @Override
    public void onClick(View v) {
        tipPercent = tipPercent + .01f;
        calculateAndDisplay();
    }
});

percentDownButton.setOnClickListener(new OnClickListener() {
    @Override
    public void onClick(View v) {
        tipPercent = tipPercent - .01f;
        calculateAndDisplay();
    }
});
```

Description

- You can create an instance variable that creates an object from a class that implements the listener interface. Since this class doesn't have a name, it's known as an *anonymous class*.

- You can create an instance of a listener interface without assigning it to an instance variable. This is known as an *inner anonymous class*.

Figure 6-2   Four techniques for handling events (part 2 of 2)

How to handle high-level events

In chapter 3, you learned how to handle the EditorAction event, which is a high-level event. Now, you'll learn how to handle high-level events for a few other widgets including check boxes, radio buttons, radio groups, spinners, and seek bars.

How to handle events for check boxes and radio buttons

Figure 6-3 shows how to handle the CheckedChanged event that occurs when a check box or radio button is checked or unchecked. To do that, you implement the OnCheckedChangedListener that's nested in the CompoundButton class and wire it to a check box or radio button. In this figure, for example, the onCheckedChanged method is executed when the user checks or unchecks the Remember Tip Percent check box.

The first code example shows how to use both parameters of the onCheckedChanged method. Here, the first parameter is a CompoundButton object for the check box or radio button that was clicked. As a result, you can call the getId method from this object to get its ID. Then, you can use a switch statement to execute the appropriate code for each widget. In this figure, there's only one check box, so this switch statement isn't necessary. However, if you have multiple check boxes or radio buttons, this statement is usually necessary.

The second parameter of the onCheckedChanged method is a Boolean value that indicates whether the widget is checked. As a result, you can use an if statement to determine whether the widget is checked. Then, you can execute the appropriate code. In this figure, the code just sets the value of a Boolean variable named rememberTipPercent. However, if necessary, it could perform other actions too.

The second code example shows simplified code for the event handler. This event handler assumes that it is only wired to a single check box. As a result, it doesn't use a switch statement to determine which widget was checked. Similarly, this code doesn't use an if statement to execute different code depending on whether the check box is checked. Instead, it uses the second parameter to set the value of the variable named rememberTipPercent.

Although the code in this figure is for check boxes, you can use the similar skills for working with radio buttons whenever that's necessary. However, when you work with radio buttons, you often handle the CheckedChanged event that occurs on a radio group as described in the next figure instead of handling this event for individual buttons.

A check box

Remember Tip Percent

An event handler for a check box

```
@Override
public void onCheckedChanged(CompoundButton widget, boolean isChecked) {
    switch (widget.getId()) {
        case R.id.rememberPercentCheckBox:
            if (isChecked) {
                rememberTipPercent = true;
            }
            else {
                rememberTipPercent = false;
            }
            break;
    }
}
```

Another event handler for a check box

```
@Override
public void onCheckedChanged(CompoundButton widget, boolean isChecked) {
    rememberTipPercent = isChecked;
}
```

A method of the View class

Method	Description
getId()	Gets the ID of the widget on which the event occurred.

Description

- The CheckedChanged event of a check box or radio button occurs when a check box or radio button is checked or unchecked.

- The first parameter of the onCheckedChanged method is a CompoundButton object. This object can be cast to a CheckBox or RadioButton object.

- The second parameter of the onCheckedChanged method is a Boolean value that indicates whether the check box or radio button is checked.

Figure 6-3   How to handle events for check boxes and radio buttons

How to handle events for radio groups

Figure 6-4 shows how to handle the CheckedChanged event that occurs when a new radio button within a radio group is checked. To do that, you implement the OnCheckedChangedListener that's nested in the RadioGroup class and wire it to a radio group. In this figure, for example, the onCheckedChanged method is executed when the user selects a new radio button from the radio group.

The first code example shows how to use the second parameter of this onCheckedChanged method, which provides the ID of the new button that is checked. This code example uses the second parameter in a switch statement to execute the appropriate code depending on which radio button was clicked. Here, the code in the switch statement sets the value of a variable named rounding to a constant. After the switch statement, this code calls the calculateAndDisplay method. Although this method isn't shown here, it performs a calculation using the round variable and displays the results of the calculation on the user interface. As a result, the calculation is made whenever the user selects a new radio button from the group.

The second code example shows an onCheckedChanged method that doesn't use the second parameter. Instead, it calls another method named calculateAndDisplay. Then, that method can check which radio button is selected and execute the appropriate code. Later in this chapter, you'll see how that method works.

Three radio buttons in a group

An event handler for a radio group

```
@Override
public void onCheckedChanged(RadioGroup group, int checkedId) {
    switch (checkedId) {
        case R.id.noRoundingRadioButton:
            rounding = ROUND_NONE;
            break;
        case R.id.roundTipRadioButton:
            rounding = ROUND_TIP;
            break;
        case R.id.roundTotalRadioButton:
            rounding = ROUND_TOTAL;
            break;
    }
    calculateAndDisplay();
}
```

Another event handler for a radio group

```
@Override
public void onCheckedChanged(RadioGroup group, int checkedId) {
    calculateAndDisplay();
}
```

Description

- The CheckedChanged event of a radio group occurs when a new button within that group is checked.
- The first parameter of the onCheckedChanged method is the RadioGroup object for the radio group.
- The second parameter of the onCheckedChanged method is the ID of the radio button within the group that is checked.

Figure 6-4   How to handle events for radio groups

How to handle events for spinners

Figure 6-5 shows how to handle the events that occur when a spinner is first displayed and when a new item is selected from a spinner. To do that, you implement the OnItemSelectedListener that's nested in the AdapterView class and wire it to a spinner. In this figure, for example, the onItemSelected method is executed when the user selects a new item from the spinner.

Within the onItemSelected method, the third parameter provides the position of the selected item. In most cases, this is the only parameter that you need. However, if necessary, the first parameter provides the AdapterView object for the list of items, the second parameter provides a View object for the selected item, and the fourth parameter provides the ID of the selected item.

Within the onItemSelected method, the first statement adds 1 to the position of the selected item and stores this value in a variable named split. That way, the split variable stores a value of 1 for the first item, 2 for the second item, and so on. This is necessary because the first item in a spinner has a position of 0. Then, the second statement calls the calculateAndDisplay method to perform a calculation that uses the split variable and to display the results to the user.

Since the onNothingSelected method is defined by the interface for the listener, you must implement this method. However, the onNothingSelected method is only executed if the selection disappears. For example, the selection can disappear when the adapter for the spinner becomes empty. Since this rarely happens, you typically don't need to include code for this method.

A spinner

An event handler for a spinner

```
@Override
public void onItemSelected(AdapterView<?> parent, View v, int position,
        long id) {
    split = position + 1;
    calculateAndDisplay();
}

@Override
public void onNothingSelected(AdapterView<?> parent) {
    // You typically don't need to include any code here
}
```

Description

- The onItemSelected method is executed when the spinner is first displayed and whenever a new item is selected. However, it isn't executed when the user selects an item that was already selected.

- Within the onItemSelected method, the third parameter provides the position of the selected item.

- The onNothingSelected method is only executed when the selection disappears. Since this rarely happens, you typically don't need to include code for this method.

Figure 6-5   How to handle events for spinners

How to handle events for seek bars

Figure 6-6 shows how to handle the events that occur on a seek bar. To do that, you implement the OnSeekBarChangeListener that's nested in the SeekBar class and wire it to a seek bar.

The interface for this listener specifies three methods: onStartTrackingTouch, onProgressChanged, and onStopTrackingTouch. The onStartTrackingTouch method is executed when the user begins to change the progress of the seek bar, the onProgressChanged method is executed as the user changes the progress of the seek bar, and the onStopTrackingTouch method is executed when the user finishes changing the progress of the seek bar. Since all three methods are specified by the interface, you must implement all three of these methods. However, you only need to provide code for the methods that you intend to use. In this figure, for example, the onStartTrackingTouch method doesn't contain any code.

The onProgressChanged method contains a single statement. This statement uses the second parameter to get the progress value for the seek bar. Then, it creates a string by appending a percent sign (%) to the progress value, and it uses the setText method of a TextView widget named percentTextView to display this string. As a result, users get immediate feedback on the progress value as they change the value of the seek bar. This is necessary to provide a responsive user interface.

The onStopTrackingTouch method contains three statements. The first statement uses the getProgress method of the SeekBar parameter to get the final progress value for the seek bar. The second statement converts that int value to a float value by dividing it by 100 and uses the result to set the new tip percent. And the third statement calls the calculateAndDisplay method. As a result, the calculation is made and displayed when the user is done changing the value of the seek bar.

A seek bar and a label

An event handler for a seek bar

```
@Override
public void onProgressChanged(SeekBar seekBar, int progress,
        boolean fromUser) {
    percentTextView.setText(progress + "%");
}

@Override
public void onStartTrackingTouch(SeekBar seekBar) {
    // TODO Auto-generated method stub
}

@Override
public void onStopTrackingTouch(SeekBar seekBar) {
    int progress = seekBar.getProgress();
    tipPercent = (float) progress / 100;
    calculateAndDisplay();
}
```

Description

- The onStartTrackingTouch method is executed when the user begins to change the value of the seek bar.
- The onProgressChanged method is executed as the user changes the value of the seek bar.
- The onStopTrackingTouch method is executed when the user finishes changing the value of the seek bar.
- Within the onProgressChanged method, the second parameter provides the progress value for the seek bar.

Figure 6-6   How to handle events for seek bars

How to handle low-level events

In chapter 3, you learned how to handle the Click event, which is a low-level event. Now, you'll learn how to handle some other low-level events including the Key and Touch events.

How to handle Key events

Many Android devices are touchscreen devices that the user can interact with by touching the screen. However, many other Android devices include hardware components. This hardware may be a keyboard, a D-Pad, or a track-ball. Or, it may be other buttons such as Power, Home, or Back buttons. When a user presses a hardware button, a Key event occurs.

Figure 6-7 shows how to handle a Key event. To do that, you implement the OnKeyListener that's nested in the View class and wire that listener to any widget. For example, the listener in this figure is wired to an EditText widget. As a result, when the EditText widget has the focus and the user presses a hardware key, the onKey method in this figure is executed.

Within this method, the code begins by using a switch statement to determine which hardware key was pressed by the user. To do that, the switch statement checks the key code value that's stored in the method's second parameter and compares that key code to the constant values that are stored in the KeyEvent class.

If the user presses the Enter key or the Center key on a D-Pad, this code calls the calculateAndDisplay method to perform a calculation and display the results to the user. Then, the next two statements hide the soft keyboard if it is displayed. To do that, the first statement uses the getSystemService method to get an InputMethodManager object. Then, the second statement uses that object to hide the soft keyboard. For now, don't worry if you don't understand this code. You can learn more about working with system services in chapter 11.

After hiding the soft keyboard, this code returns a true value for the onKey method. This indicates that this method has consumed the event. As a result, the event is not passed on to the parents of the current widget. In other words, if the user presses the Center or Enter keys, the processing of this event stops and isn't passed on to the EditText widget or any of its parent widgets.

If the user presses the Left or Right keys on a D-Pad, this code begins by checking which widget the Key event occurred on. If the event occurred on the SeekBar widget, this code calls the calculateAndDisplay method. Then, the break statement exits the switch statement.

Outside the switch statement, this code returns a false value for the onKey method. This indicates that this method has not consumed the event. As a result, the event is passed on to the parents of the current widget. In other words, when the user presses most keys, such as the number or letter keys, the Key event is passed on to the EditText widget and its parents so it can be processed normally, which is usually what you want.

An event handler for the Key event

```
@Override
public boolean onKey(View view, int keyCode, KeyEvent event) {
    switch (keyCode) {
        case KeyEvent.KEYCODE_ENTER:
        case KeyEvent.KEYCODE_DPAD_CENTER:

            calculateAndDisplay();

            // hide the soft keyboard
            InputMethodManager imm = (InputMethodManager)
                    getSystemService(Context.INPUT_METHOD_SERVICE);
            imm.hideSoftInputFromWindow(
                    billAmountEditText.getWindowToken(), 0);

            // consume the event
            return true;
        case KeyEvent.KEYCODE_DPAD_RIGHT:
        case KeyEvent.KEYCODE_DPAD_LEFT:
            if (view.getId() == R.id.percentSeekBar) {
                calculateAndDisplay();
            }
            break;
    }
    // don't consume the event
    return false;
}
```

Some constants from the KeyEvent class

Constant	Description
KEYCODE_ENTER	The Enter key on a hard keyboard.
KEYCODE_DPAD_CENTER	The Center key on the D-Pad.
KEYCODE_DPAD_LEFT	The Left key on the D-Pad.
KEYCODE_DPAD_RIGHT	The Right key on the D-Pad.
KEYCODE_SPACE	The Space key on a hard keyboard.

Description

- The KeyEvent class contains constants for almost every possible hardware key on a device including the keys on a hard keyboard or a D-Pad.

- The onKey method returns a Boolean value that indicates whether this method has consumed the event. If this method returns a true value, the event is not passed on to the parents of the current widget.

- If you don't want to wire an event handler for a Key event to multiple child widgets, you can wire the Key event to a parent widget such as the root layout. Then, the event handler handles any Key events that aren't consumed by child widgets.

- You can use the getSystemService method to get an InputMethodManager object. Then, you can use that object to hide the soft keyboard. For more information about how system services work, see chapter 11.

Figure 6-7   How to handle Key events

If you don't want to wire an event handler for a Key event to multiple child widgets, you can wire the Key event to a parent widget such as the root layout. Then, the event handler handles any Key events that aren't consumed by child widgets.

Figure 6-7 only shows five constants from the KeyEvent class. However, the KeyEvent class contains constants for almost every possible hardware key on a device.

How to handle Touch events

When a user touches a widget on a touchscreen device, a Touch event occurs on the widget. Most of the time, you don't need to know how to handle this event. Instead, you can use another event such as a Click event to get your app to work the way you want.

Sometimes, though, you may need to handle a Touch event. To do that, you implement the OnTouchListener that's nested in the View class and wire that listener to any widget. In figure 6-8, for example, the onTouch method includes code that sends data to the LogCat view.

Within the onTouch method, the first statement declares four float variables for the X and Y values when the user presses down and lifts up. Then, the second statement gets an int value for the type of MotionEvent action that has occurred. Next, an if statement compares this int value to some constants of the MotionEvent class to determine what type of event occurred.

If the user presses down, the if statement sets the values of the downX and downY variables. Then, it sends these variables to the LogCat view. These X and Y values indicate the location of the down touch on the horizontal and vertical axis of the widget.

If the user lifts up, the if statement sets the values of the upX and upY variables. Then, it sends these variables to the LogCat view. Although this type of logging doesn't perform a useful task, it's often helpful when you're getting started with Touch events.

Like the onKey method, the onTouch method returns a Boolean value that indicates whether this method has consumed the event. In this figure, the onTouch method returns a value of true if the user presses down or lifts up to indicate that it has consumed the event. However, for all other touch actions, this method returns a false value to indicate that it has not consumed the event and that other elements can continue processing this event.

When the user touches the screen and moves the touch, it's common that the device can't process events quickly enough. In that case, Android stores the Touch events in a batch. Then, if necessary, you can use the "history" methods of the MotionEvent object to access these Touch events. To start, you can use the getHistorySize method to get the number of Touch events in the batch. Then, you can create a loop that loops through these events, and you can use the getHistoricalX and getHistoricalY methods to get the X and Y values for each event.

An event handler for a Touch event

```
@Override
public boolean onTouch(View v, MotionEvent event) {
    float downX, downY, upX, upY;
    int action = event.getAction();
    if (action == MotionEvent.ACTION_DOWN) {
        Log.d("MotionEvent", "ACTION_DOWN");
        downX = event.getX();
        downY = event.getY();
        Log.d("MotionEvent", "downX = " + downX);
        Log.d("MotionEvent", "downY = " + downY);
        return true;
    }
    else if (action == MotionEvent.ACTION_UP){
        Log.d("MotionEvent", "ACTION_UP");
        upX = event.getX();
        upY = event.getY();
        Log.d("MotionEvent", "upX = " + upX);
        Log.d("MotionEvent", "upY = " + upY);
        return true;
    }
    else {
        return false;
    }
}
```

Some constants of the MotionEvent class

Constant	Description
ACTION_DOWN	The start of the touch.
ACTION_MOVE	The touch has moved between the down and up actions.
ACTION_UP	The end of the touch.

Some methods of the MotionEvent class

Method	Description
getAction()	Gets an int value for the type of action.
getX()	Gets a float value for the location of the X axis of the touch.
getY()	Gets a float value for the location of the Y axis of the touch.
getHistorySize()	The number of historical events.
getHistoricalX(int i)	Gets a float value for the X axis at the specified position of the stored historical values.
getHistoricalY(int i)	Gets a float value for the Y axis at the specified position of the stored historical values.

Description

- The onTouch method returns a Boolean value that indicates whether this method has consumed the event.

Figure 6-8   How to handle Touch events

The Tip Calculator app

This chapter finishes by showing a new and improved Tip Calculator app that uses one seek bar, three radio buttons, a radio button group, and a spinner. The code for this app handles the events that occur on these widgets.

The user interface

Figure 6-9 shows the user interface for the new and improved Tip Calculator app. When you use this app, you can set a new tip percent by dragging the thumb on the seek bar to the right or left.

If you want, you can round the tip or total to the nearest dollar by selecting the appropriate radio button. Similarly, you can split the bill between multiple people by selecting the item for the number of people from the spinner. When you do that, the user interface displays the Per Person label and amount. Otherwise, it hides this label and amount.

The user interface

Description

- This version of the Tip Calculator includes a seek bar, a radio group, three radio buttons, and a spinner.
- To set a new tip percent, you can drag the thumb on the seek bar to the right or left.
- To round the tip or the total to the nearest dollar, you can select the appropriate radio button.
- To split the bill, you can select the appropriate item from the spinner.

Figure 6-9   The user interface

The Java code for the activity

Figure 6-10 shows the Java code for the new and improved Tip Calculator app. Since this code expands upon the Tip Calculator app described in chapter 3, I'll focus on the code that handles the events for the new widgets.

To start, this code imports the classes for all of the new widgets. In addition, it imports the five listeners needed to handle the events for this app. Then, the class declares that it implements all five of these listeners.

Within the class, the code defines instance variables for the widgets. Then, the onCreate method gets references to all necessary widgets and wires the five listeners to the appropriate widgets. Here, the listener for the Key event is wired to multiple widgets. In particular, it's wired to the EditText, SeekBar, and RadioGroup widget.

The onPause and onResume methods work much like they did in chapter 3. However, the onPause method contains new code that saves the instance variables that can be used to set the state of the widgets. Similarly, the onResume method contains new code that gets these variables and uses them to set the state of the widgets. Here, the event handlers for the RadioGroup and Spinner widgets are triggered by the code that sets the selected radio button and the position of the spinner. These event handlers both execute the calculateAndDisplay method.

The Java code for the activity

```java
package com.murach.tipcalculator;

import java.text.NumberFormat;

import android.os.Bundle;
import android.view.KeyEvent;
import android.view.View;
import android.view.View.OnKeyListener;
import android.view.inputmethod.EditorInfo;
import android.view.inputmethod.InputMethodManager;
import android.widget.AdapterView;
import android.widget.AdapterView.OnItemSelectedListener;
import android.widget.ArrayAdapter;
import android.widget.EditText;
import android.widget.RadioButton;
import android.widget.RadioGroup;
import android.widget.RadioGroup.OnCheckedChangeListener;
import android.widget.SeekBar;
import android.widget.SeekBar.OnSeekBarChangeListener;
import android.widget.Spinner;
import android.widget.TextView;
import android.widget.TextView.OnEditorActionListener;
import android.app.Activity;
import android.content.Context;
import android.content.SharedPreferences;
import android.content.SharedPreferences.Editor;

public class TipCalculatorActivity extends Activity
implements OnEditorActionListener, OnSeekBarChangeListener,
OnCheckedChangeListener, OnItemSelectedListener, OnKeyListener {

    // define variables for the widgets
    private EditText billAmountEditText;
    private TextView percentTextView;
    private SeekBar percentSeekBar;
    private TextView tipTextView;
    private TextView totalTextView;
    private RadioGroup roundingRadioGroup;
    private RadioButton roundNoneRadioButton;
    private RadioButton roundTipRadioButton;
    private RadioButton roundTotalRadioButton;
    private Spinner splitSpinner;
    private TextView perPersonLabel;
    private TextView perPersonTextView;

    // define the SharedPreferences object
    private SharedPreferences savedValues;

    // define rounding constants
    private final int ROUND_NONE = 0;
    private final int ROUND_TIP = 1;
    private final int ROUND_TOTAL = 2;
```

Figure 6-10   The Java code for the activity (part 1 of 6)

The Java code for the activity

```java
// define instance variables
private String billAmountString = "";
private float tipPercent = .15f;
private int rounding = ROUND_NONE;
private int split = 1;

@Override
public void onCreate(Bundle savedInstanceState) {
    super.onCreate(savedInstanceState);
    setContentView(R.layout.activity_tip_calculator);

    // get references to the widgets
    billAmountEditText = (EditText) findViewById(R.id.billAmountEditText);
    percentTextView = (TextView) findViewById(R.id.percentTextView);
    percentSeekBar = (SeekBar) findViewById(R.id.percentSeekBar);
    tipTextView = (TextView) findViewById(R.id.tipTextView);
    totalTextView = (TextView) findViewById(R.id.totalTextView);
    roundingRadioGroup = (RadioGroup)
            findViewById(R.id.roundingRadioGroup);
    roundNoneRadioButton = (RadioButton)
            findViewById(R.id.roundNoneRadioButton);
    roundTipRadioButton = (RadioButton)
            findViewById(R.id.roundTipRadioButton);
    roundTotalRadioButton = (RadioButton)
            findViewById(R.id.roundTotalRadioButton);
    splitSpinner = (Spinner) findViewById(R.id.splitSpinner);
    perPersonLabel = (TextView) findViewById(R.id.perPersonLabel);
    perPersonTextView = (TextView) findViewById(R.id.perPersonTextView);

    // set array adapter for spinner
    ArrayAdapter<CharSequence> adapter = ArrayAdapter.createFromResource(
        this, R.array.split_array, android.R.layout.simple_spinner_item);
    adapter.setDropDownViewResource(
        android.R.layout.simple_spinner_dropdown_item);
    splitSpinner.setAdapter(adapter);

    // set the listeners
    billAmountEditText.setOnEditorActionListener(this);
    billAmountEditText.setOnKeyListener(this);
    percentSeekBar.setOnSeekBarChangeListener(this);
    percentSeekBar.setOnKeyListener(this);
    roundingRadioGroup.setOnCheckedChangeListener(this);
    roundingRadioGroup.setOnKeyListener(this);
    splitSpinner.setOnItemSelectedListener(this);

    // get SharedPreferences object
    savedValues = getSharedPreferences("SavedValues", MODE_PRIVATE);
}
```

Figure 6-10   The Java code for the activity (part 2 of 6)

The Java code for the activity   **Page 3**

```java
@Override
public void onPause() {
    // save the instance variables
    Editor editor = savedValues.edit();
    editor.putString("billAmountString", billAmountString);
    editor.putFloat("tipPercent", tipPercent);
    editor.putInt("rounding", rounding);
    editor.putInt("split", split);
    editor.commit();

    super.onPause();
}

@Override
public void onResume() {
    super.onResume();

    // get the instance variables
    billAmountString = savedValues.getString("billAmountString", "");
    tipPercent = savedValues.getFloat("tipPercent", 0.15f);
    rounding = savedValues.getInt("rounding", ROUND_NONE);
    split = savedValues.getInt("split", 1);

    // set the bill amount on its widget
    billAmountEditText.setText(billAmountString);

    // set the tip percent on its widget
    int progress = Math.round(tipPercent * 100);
    percentSeekBar.setProgress(progress);

    // set rounding on radio buttons
    // NOTE: this executes the onCheckedChanged method,
    // which executes the calculateAndDisplay method
    if (rounding == ROUND_NONE) {
        roundNoneRadioButton.setChecked(true);
    }
    else if (rounding == ROUND_TIP) {
        roundTipRadioButton.setChecked(true);
    }
    else if (rounding == ROUND_TIP) {
        roundTotalRadioButton.setChecked(true);
    }

    // set split on spinner
    // NOTE: this executes the onItemSelected method,
    // which executes the calculateAndDisplay method
    int position = split - 1;
    splitSpinner.setSelection(position);
}
```

Figure 6-10   The Java code for the activity (part 3 of 6)

The calculateAndDisplay method works much like it did in chapter 3. However, it contains new code that rounds the tip and total if necessary. In addition, it contains new code that calculates the split amount, shows or hides the Per Person label and amount when necessary, and formats the Per Person amount.

The event handler for the EditText widget is executed when the user clicks the Done button on the soft keyboard that's displayed for the EditText widget. This works as described in chapter 3.

The event handler for the SeekBar widget is executed when the user moves the thumb on the seek bar. Here, the onProgressChanged method displays immediate feedback to the user by updating the tip percent whenever the user changes the progress value on the seek bar. Then, when the user finishes changing the progress value on the seek bar, the onStopTrackingTouch method calculates a new tip percent and calls the calculateAndDisplay method. This method uses the new tip percent to calculate and display new tip and total amounts.

The event handler for the RadioGroup widget is executed when the user selects a new radio button. Here, the onCheckedChanged method begins by using the second parameter to check which radio button is selected. Then, it sets the rounding variable to the appropriate constant value. For example, if the No Rounding radio button is selected, this code sets the rounding variable to the ROUND_NO constant. After it sets the rounding variable, this code calls the calculateAndDisplay method. This method uses the rounding variable to determine what type of rounding the app uses for the calculation.

The event handler for the Spinner widget is executed when the user selects a new item. Here, the onItemSelected method begins by using the second parameter of the method to set the value of the split variable. Since the position of the first item is 0, this code adds a value of 1 to get the value for the split variable. After that, this event handler calls the calculateAndDisplay method. This method uses the split variable to determine how many ways to split the bill.

The event handler for the Key event is executed when the user presses a hardware key and the focus is on the EditText, SeekBar, or RadioGroup widgets. Here, the onKey method begins by using the second parameter of the method to execute code depending on the hardware key. If the user pressed the Enter key on a hard keyboard or the Center key on a D-Pad, this code hides the soft keyboard, calls the calculateAndDisplay method, and consumes the event.

If the user presses the Right or Left keys on a D-Pad and the focus is on the SeekBar widget, this code calls the calculateAndDisplay method. However, in this case, it does not consume the event. As a result, Android can use this event to update the user interface accordingly.

The Java code for the activity

```java
public void calculateAndDisplay() {
    // get the bill amount
    billAmountString = billAmountEditText.getText().toString();
    float billAmount;
    if (billAmountString.equals("")) {
        billAmount = 0;
    }
    else {
        billAmount = Float.parseFloat(billAmountString);
    }

    // get tip percent
    int progress = percentSeekBar.getProgress();
    tipPercent = (float) progress / 100;

    // calculate tip and total
    float tipAmount = 0;
    float totalAmount = 0;
    if (rounding == ROUND_NONE) {
        tipAmount = billAmount * tipPercent;
        totalAmount = billAmount + tipAmount;
    }
    else if (rounding == ROUND_TIP) {
        tipAmount = StrictMath.round(billAmount * tipPercent);
        totalAmount = billAmount + tipAmount;
        tipPercent = tipAmount / billAmount;
    }
    else if (rounding == ROUND_TOTAL) {
        float tipNotRounded = billAmount * tipPercent;
        totalAmount = StrictMath.round(billAmount + tipNotRounded);
        tipAmount = totalAmount - billAmount;
        tipPercent = tipAmount / billAmount;
    }

    // calculate split amount and show/hide split amount widgets
    float splitAmount = 0;
    if (split == 1) {  // no split - hide widgets
        perPersonLabel.setVisibility(View.GONE);
        perPersonTextView.setVisibility(View.GONE);
    }
    else {                  // split - calculate amount and show widgets
        splitAmount = totalAmount / split;
        perPersonLabel.setVisibility(View.VISIBLE);
        perPersonTextView.setVisibility(View.VISIBLE);
    }

    // display the results with formatting
    NumberFormat currency = NumberFormat.getCurrencyInstance();
    tipTextView.setText(currency.format(tipAmount));
    totalTextView.setText(currency.format(totalAmount));
    perPersonTextView.setText(currency.format(splitAmount));

    NumberFormat percent = NumberFormat.getPercentInstance();
    percentTextView.setText(percent.format(tipPercent));
}
```

Figure 6-10   The Java code for the activity (part 4 of 6)

The Java code for the activity

```
//*****************************************************
// Event handler for the EditText
//*****************************************************
@Override
public boolean onEditorAction(TextView v, int actionId, KeyEvent event) {
    if (actionId == EditorInfo.IME_ACTION_DONE ||
        actionId == EditorInfo.IME_ACTION_UNSPECIFIED) {
        calculateAndDisplay();
    }
    return false;
}


//*****************************************************
// Event handler for the SeekBar
//*****************************************************
@Override
public void onStartTrackingTouch(SeekBar seekBar) {
    // TODO Auto-generated method stub
}

@Override
public void onProgressChanged(SeekBar seekBar, int progress,
        boolean fromUser) {
                percentTextView.setText(progress + "%");
}

@Override
public void onStopTrackingTouch(SeekBar seekBar) {
    calculateAndDisplay();
}


//*****************************************************
// Event handler for the RadioGroup
//*****************************************************
@Override
public void onCheckedChanged(RadioGroup group, int checkedId) {
    switch (checkedId) {
        case R.id.roundNoneRadioButton:
            rounding = ROUND_NONE;
            break;
        case R.id.roundTipRadioButton:
            rounding = ROUND_TIP;
            break;
        case R.id.roundTotalRadioButton:
            rounding = ROUND_TOTAL;
            break;
    }
    calculateAndDisplay();
}
```

Figure 6-10   The Java code for the activity (part 5 of 6)

The Java code for the activity

```
//****************************************************
// Event handler for the Spinner
//****************************************************
@Override
public void onItemSelected(AdapterView<?> parent, View v, int position,
        long id) {
    split = position + 1;
    calculateAndDisplay();
}

@Override
public void onNothingSelected(AdapterView<?> parent) {
    // Do nothing
}

//****************************************************
// Event handler for the keyboard and DPad
//****************************************************
@Override
public boolean onKey(View view, int keyCode, KeyEvent event) {
    switch (keyCode) {
        case KeyEvent.KEYCODE_ENTER:
        case KeyEvent.KEYCODE_DPAD_CENTER:

            calculateAndDisplay();

            // hide the soft keyboard
            InputMethodManager imm = (InputMethodManager)
                    getSystemService(Context.INPUT_METHOD_SERVICE);
            imm.hideSoftInputFromWindow(
                    billAmountEditText.getWindowToken(), 0);

            // consume the event
            return true;
        case KeyEvent.KEYCODE_DPAD_RIGHT:
        case KeyEvent.KEYCODE_DPAD_LEFT:
            if (view.getId() == R.id.percentSeekBar) {
                calculateAndDisplay();
            }
            break;
    }
    // don't consume the event
    return false;
}
}
```

Figure 6-10   The Java code for the activity (part 6 of 6)

Perspective

Now that you've finished this chapter, you should understand how high-level and low-level events work. In addition, you should be able to handle events that occur on the widgets described in this chapter. More importantly, you should have all the skills you need to figure out how to handle events that occur on other kinds of widgets. To do that, you can search the Android API or the Internet to find the event listener that you need. Then, you can use the techniques described in this chapter to handle the event by implementing that listener.

Terms

listener	long click
events	focus
high-level events	anonymous class
low-level events	inner anonymous class

Summary

- A *listener* listens for *events* that can occur on a widget. When an event occurs on a widget, the appropriate method of the listener is executed.

- *High-level events* are events that occur on a specific type of widget. An interface that defines a listener for a high-level event is typically nested within the class that defines the widget. These listeners can only be wired to an appropriate widget.

- *Low level events* are events that occur on all types of widgets. An interface that defines a listener for a low-level event is typically nested within the android.view.View class. These listeners can be wired to any type of widget.

- You can create an instance variable that creates an object from a class that implements the listener interface. Since this class doesn't have a name, it's known as an *anonymous class*.

- You can create an instance of a listener interface without assigning it to an instance variable. This is known as an *inner anonymous class*.

- The CheckedChanged event of a check box or radio button occurs when a check box or radio button is checked or unchecked.

- For a spinner, the onItemSelected method is executed when the spinner is first displayed and whenever a new item is selected. The onNothingSelected method is only executed when the selection disappears.

- For a seek bar, the onStartTrackingTouch method is executed when the user begins to change the value of the seek bar. The onProgressChanged method is executed as the user changes the value of the seek bar. The onStopTrackingTouch method is executed when the user finishes changing the value of the seek bar.

- The KeyEvent class contains constants for almost every possible hardware key on a device including the keys on a hard keyboard or a D-Pad.
- The onKey and onTouch methods both return a Boolean value that indicates whether the method has consumed the event. If this method returns a true value, the event is not passed on to the parents of the current widget.

Exercise 6-1   Use anonymous classes for event listeners

In this exercise, you'll modify the Tip Calculator app that's presented in this chapter so it uses anonymous classes for event listeners.

1. Start Eclipse and import the project named ch06_ex1_TipCalculator.
2. Open the .java file for the activity and review its code. Note that the activity class implements five listener interfaces.
3. Test the app to make sure the EditText, SeekBar, RadioGroup, and Spinner widgets all work correctly.
4. Modify the code so it uses an anonymous class (not an inner anonymous class) for the OnEditorActionListener. To do that, modify the declaration for the class so it doesn't implement this interface, create an instance variable for an object that implements this interface, and set that object as the listener.
5. Test the app to make sure the EditText widget still works correctly.
6. Repeat steps 4 and 5 for the listeners for the SeekBar, RadioGroup, and Spinner widgets.

Exercise 6-2   Improve the listener for the Key events

In this exercise, you'll modify the Tip Calculator app that's presented in this chapter so it uses the OnKeyListener interface to handle Key events instead of using the OnEditorActionListener interface to handle Key events.

Test the Enter key and D-Pad keys

1. Start Eclipse and import the project named ch06_ex2_TipCalculator.
2. Run the app on a device or emulator that has a hardware keyboard and a D-Pad.
3. Enter a bill amount and use the Enter key on the keyboard and the Center key on the D-Pad to finish the entry. This should work correctly.
4. Use the D-Pad to move the focus to the seek bar for the tip percent. Then, use the Left and Right keys on the D-Pad to increase or decrease the tip amount. This should work correctly.

Improve the listener for the Key events

5. Enter a new bill amount and use the Down key on the D-Pad to move the focus to the seek bar. Note that this does not update the tip and total amounts or hide the soft keyboard.

6. Modify the code for the onKey method so the Down key on the D-Pad executes the calculateAndDisplay method, hides the soft keyboard, and moves the focus to the next widget. To get this to work correctly, the Down key should not consume the event, but the Center key should continue to consume the event.

7

How to work with themes and styles

In chapter 3, you learned how to manually set the properties of the widgets in your app to control how they look. For example, you set the textSize property for every widget in the app so that each widget uses the same text size. Now, this chapter shows how to use themes and styles to automatically set the properties for widgets.

Themes and styles allow you to separate the design from the content, which is generally considered a best practice. In addition, they can help you keep the appearance of your app consistent, they can help you reduce code duplication, and they can make your app easier to develop and maintain.

An introduction to themes and styles

By default, an Android app uses built-in themes and styles. This topic begins by taking a look at these built-in themes. To start, a *style* is a collection of properties that specify how a widget looks. A *theme* is a style that's applied to an entire activity or app. In other words, the main difference between a style and a theme is where it is applied: a style is applied to a widget, and a theme is applied to an app or activity.

Three themes

Figure 7-1 shows three themes that are commonly used. Each theme provides a different look for the title bar and widgets of the app. The first is commonly used for all APIs prior to Android 3.0 (API 11). The second is commonly used with API 11 and later. And the third is commonly used with Android 4.0 (API 14) and later.

To understand why different themes are used by different versions of Android, consider the history of Android themes. To start, the earliest versions of Android included the theme named Theme.Light. This theme includes formatting for all widgets. In addition, it includes formatting for a *title bar* that displays a title across the top of the activity.

Android 3.0 (API 11) introduced the Holo theme. As a result, the theme named Theme.Holo.Light is available with API 11 and later. However, it isn't available prior to API 11.

Android 4.0 (API 14) introduced the *action bar*, which is essentially a title bar that can also display buttons that allow the user to perform actions. In addition, API 14 added the DarkActionBar theme to the Holo theme. As a result, you can use the theme named Theme.Holo.Light.DarkActionBar with API 14 and later. However, it isn't available prior to API 14. Although the action bar shown in this figure doesn't include buttons, you'll learn how to add buttons to an action bar in chapter 8.

Theme.Light

Theme.Holo.Light

Theme.Holo.Light.DarkActionBar

Description

- A *style* is a collection of properties that specify formatting for a widget.
- A *theme* is a style that's applied to an entire activity or app. Android includes several built-in themes. When a theme is applied, every widget applies each property of the theme that it supports.
- The theme named Theme.Light is commonly used for all APIs prior to Android 3.0 (API 11).
- The theme named Theme.Holo.Light is available for Android 3.0 and later.
- The theme named Theme.Holo.Light.DarkActionBar is available for Android 4.0 (API 14) and later.
- The Holo theme was introduced in Android 3.0. This theme isn't available for older Android devices.
- Prior to Android 4.0, an activity can include a *title bar* that displays the title of the activity.
- With Android 4.0 and later, an activity can include an *action bar* that displays the activity's title and, optionally, buttons that allow the user to perform actions.

Figure 7-1   Three commonly used themes

The theme framework
that's generated by Eclipse

When you create a new project for an Android app, Eclipse automatically generates three styles.xml files that set appropriate themes for the different API levels. These styles.xml files are shown in figure 7-2. To start, the styles.xml file in the res\values directory applies to all API levels 1-10. Then, the styles.xml file in the res\values-v11 applies to API 11 (Android 3.0) and later. In turn, the styles.xml file in the res\values-v14 directory applies to API 14 (Android 4.0) and later.

Within the styles.xml files, a style element can specify a theme or a style. Within a style element, the name attribute specifies the name for the style or theme, and the parent attribute specifies the style or theme that the current style element inherits. In this figure, all of the style elements define themes.

In the styles.xml file in the res\values directory, the first style element defines a theme named AppBaseTheme that inherits the theme named Theme.Light. Then, the second style element defines a theme named AppTheme that inherits the theme named AppBaseTheme.

In the styles.xml file in the res\values-v11 directory, the first style element defines a theme named AppBaseTheme that inherits the theme named Theme.Holo.Light. Since this theme has the same name as a theme in the previous styles.xml file, this theme overrides the theme in the previous styles.xml file. As a result, API 11 and later use a different theme than previous APIs.

In the styles.xml file in the res\values-v14 directory, the first style element uses the same technique to use a different theme for APIs 14 and later. More specifically, APIs 14 and later use the theme named Theme.Holo.Light.DarkActionBar.

In the AndroidManifest.xml file, the application element specifies the theme for the app. In this figure, the theme attribute specifies the theme named AppTheme. This applies an appropriate built-in theme for each API level. However, if you want to customize these themes, you can add item elements to the appropriate style element for each theme to customize that theme. You'll learn how to do that later in this chapter.

The styles.xml file in the...

res\values directory

```
<resources>
    <style name="AppBaseTheme" parent="android:Theme.Light">
        <!-- API 1-10 theme customizations go here -->
    </style>
    <style name="AppTheme" parent="AppBaseTheme">
        <!-- Theme customizations NOT specific to a particular API -->
    </style>
</resources>
```

res\values-v11 directory

```
<resources>
    <style name="AppBaseTheme" parent="android:Theme.Holo.Light">
        <!-- API 11-13 theme customizations can go here. -->
    </style>
</resources>
```

res\values-v14 directory

```
<resources>
    <style name="AppBaseTheme"
            parent="android:Theme.Holo.Light.DarkActionBar">
        <!-- API 14+ theme customizations can go here. -->
    </style>
</resources>
```

The application element of the AndroidManifest.xml file

```
<application
    android:icon="@drawable/ic_launcher"
    android:label="@string/app_name"
    android:theme="@style/AppTheme"
    android:allowBackup="true">
```

Description

- When you create a new project for an Android app, Eclipse automatically generates three styles.xml files that set appropriate themes for the different API levels.
- The styles.xml file in the res\values directory is applied to all API levels unless it is overridden by a styles.xml file in a res\values-vXX directory.
- A style element can specify a theme or a style.
- The name attribute of a style element specifies the name for the style or theme.
- The parent attribute of a style element specifies the style or theme that the current style element inherits.
- The style element named AppTheme inherits the AppBaseTheme style.
- The style element named AppBaseTheme inherits a theme that's appropriate for the API level.
- The AndroidManifest.xml file uses its application element to specify the theme for the application.

Figure 7-2   The theme framework that's generated by Eclipse

How to work with styles

Now that you understand how the built-in themes work, you're ready to learn how to work with styles. To start, you can define a style. Then, you can apply that style to a widget.

How to define a style

Figure 7-3 shows how to define a style. To start, you typically add a style element to the XML file named styles.xml in the res/values/ directory of your project. If necessary, you can create this file.

The first example uses a style element to create a style named TextView. This name is appropriate for a style that's going to be applied to most TextView widgets in the app. Within the style element, this style uses a single item element to specify a *property* of the style. More specifically, this item element sets the textSize property of the style to 20sp.

The second example uses a style element to create a style named EditText. This style inherits the TextView style that was defined in the first example. As a result, the EditText style also specifies a text size of 20sp.

The third example uses a style element to create a style named Label. This style inherits the style named TextView that was defined in the first example. In addition, it uses two item elements to specify 10dp of padding and to make the text bold.

The fourth example uses a style element to create a style named TextView. Label. Since this name begins with TextView followed by a period, it automatically inherits the TextView style defined in the first example. When you use this syntax, you should be aware that you can only use it with user-defined styles. In other words, you can't use it to inherit built-in Android styles.

The fifth example shows how to use the syntax in the fourth example to inherit multiple styles. This style inherits the TextView and Label styles and adds another style named Indent that indents the left margin by 10dp.

When you work with styles, you can set the same properties that are available from the Graphical Layout editor. As a result, if you have already used the Graphical Layout editor to set the properties for your app, you look in the XML file for the layout to find the names of the properties and their settings. Then, if you want, you create styles to set those properties more consistently. In this figure, for instance, the examples show how to use a style to set the textSize, textStyle, padding, and layout_marginLeft properties. However, you can use a style to set any of the properties that are available from the Graphical Layout editor.

When a style inherits another style, its properties override the properties of the inherited style. Let's assume, for example, that I create a style that inherits the Label style. And let's assume that this new style has only a single item element setting the textSize property to 15sp. This would override the textSize property of the Label style, but it wouldn't override the textStyle and padding properties of the Label style.

A style that overrides one property

```
<style name="TextView">
    <item name="android:textSize">20sp</item>
</style>
```

A style that inherits a user-defined style

```
<style name="EditText" parent="@style/TextView" />
```

A style that inherits a user-defined style and overrides two properties

```
<style name="Label" parent="@style/TextView">
    <item name="android:textStyle">bold</item>
    <item name="android:padding">10dp</item>
</style>
```

Another way to code the previous style

```
<style name="TextView.Label">
    <item name="android:textStyle">bold</item>
    <item name="android:padding">10dp</item>
</style>
```

How to inherit multiple user-defined styles

```
<style name="TextView.Label.Indent">
    <item name="android:layout_marginLeft">10dp</item>
</style>
```

Description

- To define a style, you typically add a style element to the XML file named styles.xml in the res\values\ directory of your project. If necessary, you can create this file.

- The name attribute of a style element is required and must specify the name of the style.

- The parent attribute of a style element is optional. However, it can be used to specify a style that the current style inherits. The inherited style can be a built-in Android style or a user-defined style.

- A style element can contain one or more item elements. Each item element specifies a *property* of the style. If a style inherits another style, its properties override the properties of the inherited style.

- A style can specify any properties that are available from the Graphical Layout editor, such as textSize, textStyle, padding, margin, textColor, and background.

- When working with user-defined styles, you can also inherit other styles by coding the name of the inherited style, followed by a period, followed by the name of your new style. You can use this syntax to inherit multiple styles.

Figure 7-3   How to define a style

How to apply a style

Figure 7-4 shows how to apply styles like the ones that were defined in the previous figure. One way to do that is to use the Graphical Layout editor. In the Properties view, you click the button to the right of the Style property in the Properties window. When you do, a dialog box like the one shown in this figure is displayed. Then, you can expand the Style item and select the style you want to apply. In this figure, for example, I have selected the style named TextView.Label.

Another way to apply a style is to edit the XML for the layout. To do that, you just add a style attribute to the widget and use it to specify the name of the style that's in the styles.xml file. In this figure, for instance, the first XML example applies the style named TextView.Label to a TextView widget. Then, the second XML example applies the style named TextView to the TextView widget.

Unlike most attributes, the style attribute does not use the android: namespace prefix. In this figure, for example, the id and text attributes use that prefix, but the style attribute doesn't.

The Reference Chooser dialog for a style

A TextView widget with the TextView.Label style

```
<TextView
    style="@style/TextView.Label"
    android:id="@+id/percentLabel"
    android:layout_width="wrap_content"
    android:layout_height="wrap_content"
    android:layout_alignLeft="@+id/billAmountLabel"
    android:layout_below="@+id/billAmountLabel"
    android:text="@string/tip_percent_label" />
```

A TextView widget with the TextView style

```
<TextView
    style="@style/TextView"
    android:id="@+id/percentTextView"
    android:layout_width="wrap_content"
    android:layout_height="wrap_content"
    android:layout_alignBaseline="@+id/percentLabel"
    android:layout_alignLeft="@+id/billAmountEditText"
    android:padding="5dp"
    android:text="@string/tip_percent" />
```

Description

- From the Graphical Layout editor, you can specify a style for a widget by using the Style property to select a style.

- In XML, you can apply a style by using the style attribute to specify the name of the style in the styles.xml file. This attribute doesn't use the android: namespace prefix.

Figure 7-4   How to apply a style

How to create a style sheet

A *style sheet* is a collection of styles that can be applied throughout an application. To create a style sheet, you store the styles for your application in a styles.xml file like the one shown in figure 7-5.

The first two style elements define the custom themes named AppBaseTheme and AppTheme. As you learned earlier in this chapter, these themes apply the appropriate built-in theme to your app depending on API level.

The next four style elements define the custom styles that are used to apply formatting to all of the widgets of the Tip Calculator app. To start, the styles named TextView, EditText, and Button are designed to be applied to their corresponding widgets. Then, the style named TextView.Label is designed to be applied to the TextView widgets that label other widgets.

A styles.xml file in the res\values directory with four user-defined styles

```xml
<resources xmlns:android="http://schemas.android.com/apk/res/android">

    <!-- Two custom themes -->
    <style name="AppBaseTheme" parent="android:Theme.Light">
        <!-- API 1-10 theme customizations go here -->
    </style>

    <style name="AppTheme" parent="AppBaseTheme">
        <!-- Theme customizations NOT specific to a particular API -->
    </style>


    <!-- Four custom styles -->
    <style name="TextView">
        <item name="android:textSize">20sp</item>
    </style>

    <style name="EditText" parent="@style/TextView" />

    <style name="Button" parent="@style/TextView">
        <item name="android:textStyle">bold</item>
    </style>

    <style name="TextView.Label">
        <item name="android:textStyle">bold</item>
        <item name="android:padding">10dp</item>
    </style>
</resources>
```

Description

- A *style sheet* is a collection of styles that can be applied throughout an application.

Figure 7-5   How to create a style sheet

How to work with themes

Now that you know how to work with styles, you're ready to take another look at themes. In particular, you're ready to learn how to modify the built-in themes by overriding the properties that you want to change. In addition, you're ready to learn how to use some of the built-in themes that aren't used by default.

How to modify a theme

Figure 7-6 shows how to modify a built-in theme. Here, the theme named AppTheme contains three item elements. Each of these item elements uses its name attribute to specify an attribute that corresponds with a style from a built-in theme.

After specifying the style that you want to override, each of these item elements uses its body to specify the custom style that overrides the built-in style. In this figure, the first item element overrides the TextView style in the built-in theme with the custom TextView style defined in this styles.xml file.

This custom TextView style inherits the built-in style named Widget.TextView. As a result, you only need to override the properties that you want to change. In this case, the custom TextView style only overrides the textSize property. Similarly, the custom styles named EditText and Button inherit their corresponding built-in styles.

Since the first three custom styles defined in this figure override the corresponding styles in the built-in theme, they're applied automatically to any TextView, EditText, or Button widgets in the app. In other words, you don't need to manually apply these styles. The fourth custom style, on the other hand, needs to be applied manually.

If you want to modify a custom theme, you often need to do some research and experimentation to figure out which attributes and styles to modify. To find the attribute you want to override, you can view the standard attributes for a theme by viewing the API documentation for the Theme array that's defined in the R.styleable class. Then, you can find the styles that you want to inherit by viewing the styles.xml file that's available from the Android SDKs on your computer. This file is available from the data\res\values directory that corresponds with each Android API. Alternately, you can view the API documentation for the R.styles class. However, since styles aren't well documented, it's often helpful to view the styles.xml file that contains the source code for the styles.

A styles.xml file in the res\values directory that customizes a theme

```xml
<resources xmlns:android="http://schemas.android.com/apk/res/android">

    <style name="AppBaseTheme" parent="android:Theme.Light">
        <!-- API 1-10 theme customizations go here -->
    </style>

    <style name="AppTheme" parent="AppBaseTheme">
        <!-- Theme customizations NOT specific to a particular API -->

        <!-- Set new styles for three widgets in the theme -->
        <item name="android:textViewStyle">@style/TextView</item>
        <item name="android:editTextStyle">@style/EditText</item>
        <item name="android:buttonStyle">@style/Button</item>
    </style>

    <!-- These three styles are applied automatically -->
    <style name="TextView" parent="@android:style/Widget.TextView">
        <item name="android:textSize">20sp</item>
    </style>

    <style name="EditText" parent="@android:style/Widget.EditText">
        <item name="android:textSize">20sp</item>
    </style>

    <style name="Button" parent="@android:style/Widget.Button">
        <item name="android:textSize">20sp</item>
        <item name="android:textStyle">bold</item>
    </style>

    <!-- This style needs to be applied manually -->
    <style name="EditText.Label">
        <item name="android:textStyle">bold</item>
        <item name="android:padding">10dp</item>
    </style>
</resources>
```

The API documentation for the attributes in the R.styleable.Theme array

```
http://developer.android.com/reference/android/R.styleable.html#Theme
```

The styles.xml file for Android 4.2 (API 17)

```
\android-sdks\platforms\android-17\data\res\values\styles.xml
```

Description

- Within a theme, the name attribute of an item element can specify an attribute that corresponds to a style in the built-in theme. Then, the body of the item element can specify a custom style that overrides the built-in style.

- When you create a custom style that overrides a style from a built-in theme, you typically begin your style by inheriting the built-in Android style. Then, you can use the custom style to override only the properties that you want to change.

Figure 7-6   How to modify a theme

How to modify the text appearance for a theme

Android uses a series of built-in styles to control the appearance of the text for a theme. As a result, if you want to modify the text appearance for all widgets on a theme, you may want to modify the text appearance styles.

Figure 7-7 starts by listing some of the built-in text appearance styles. Android primarily uses these styles to control the size and color of the text. Here, the Small, Medium, and Large styles can be appended to the TextAppearance style to change the size of the text. Or, the Inverse style can be applied to the TextAppearance style to use the inverse color scheme.

The styles.xml file in this figure works much like the styles.xml file in the previous figure. However, it uses the TextAppearance styles instead of the styles for the different types of widgets. Here, the code overrides the TextAppearance. Small and TextAppearance.Small.Inverse styles and sets the textSize property to 20sp.

You can use the Graphical Layout editor to determine the default text appearance style that's used for each widget in your app. To do that, display your activity in the Graphical Layout editor, click on the widget, expand the Text View group in the Properties view, and check the value of the Text Appearance property. Then, if necessary, you can override that property to change its text size or color scheme.

Some built-in styles for controlling text appearance

Name	Description
TextAppearance	Displays standard text appearance and color scheme.
TextAppearance.Inverse	Displays standard text appearance with inverse color scheme.
TextAppearance.Small	Displays small font size.
TextAppearance.Small.Inverse	Displays small font size and inverse color scheme.
TextAppearance.Medium	Displays medium font size.
TextAppearance.Large	Displays large font size.

A styles.xml file that modifies the TextAppearance styles

```xml
<resources xmlns:android="http://schemas.android.com/apk/res/android">

    <style name="AppBaseTheme" parent="android:Theme.Light">
        <!-- API 1-10 theme customizations go here -->
    </style>

    <style name="AppTheme" parent="AppBaseTheme">
        <!-- Theme customizations NOT specific to a particular API -->

        <!-- Set new styles for the theme text apperance -->
        <item name="android:textAppearanceSmall">
                    @style/TextAppearanceSmall</item>
        <item name="android:textAppearanceSmallInverse">
                    @style/TextAppearanceSmallInverse</item>
    </style>


    <!-- These styles are applied automatically -->
    <style name="TextAppearanceSmall"
            parent="@android:style/TextAppearance.Small">
        <item name="android:textSize">20sp</item>
    </style>
    <style name="TextAppearanceSmallInverse"
            parent="@android:style/TextAppearance.Small.Inverse">
        <item name="android:textSize">20sp</item>
    </style>
</resources>
```

Description

- Android themes use a series of styles to control the appearance of the text on all widgets in the theme.

Figure 7-7   How to modify the text appearance for a theme

A summary of built-in themes

Figure 7-8 lists some of the built-in themes that are available from Android. By now, you should already be familiar with the three themes that were presented earlier in this chapter. As a result, you shouldn't have much trouble understanding how the rest of these themes work.

If you want to view a complete list of all built-in themes for an activity, display the activity in the Graphical Layout editor. Then, use the drop-down list that's available from the Themes button to view the themes. This list includes the themes shown in this figure as well as many other themes, including any user-defined themes that are available.

Android 4.0 (API 14) introduced a family of DeviceDefault themes. These themes provide a way for manufacturers to inherit and override a built-in Android theme without changing the built-in theme itself. For example, Google's Nexus devices use the DeviceDefault themes to override the Holo themes. As a result, the unmodified Holo themes are available on the device, and the customized DeviceDefault themes are also available on the device.

This provides the app developer with a choice. On one hand, the app developer can choose the DeviceDefault themes. This makes the appearance of the app more consistent with other apps on that device. However, the appearance of the app may vary when viewed on different devices.

On the other hand, the app developer can choose the Holo themes. This makes the appearance of the app more consistent across multiple devices. However, the appearance of the app might not be consistent with other apps on the same device since those apps might use one of the DeviceDefault themes.

Some built-in themes

Name	Displays the activity...
Theme.Light	With a white background.
Theme.Black	With a black background.
Theme.Light.NoTitleBar	With a white background and no title bar.
Theme.Black.NoTitleBar	With a black background and no title bar.
Theme.Dialog	As a dialog box with a black background.
Theme.Holo	Using the Holo theme with a black background.
Theme.Holo.NoActionBar	Using the Holo theme with a black background and no action bar.
Theme.Holo.Dialog	As a dialog box using the Holo theme with a black background.
Theme.Holo.Light	Using the Holo theme with a white background.
Theme.Holo.Light.Dialog	As a dialog box using the Holo theme with a white background.
Theme.Holo.Light.NoActionBar	Using the Holo theme with a white background and no action bar.
Theme.Holo.Light.DarkActionBar	Using the Holo theme with a white background and a dark action bar.
Theme.DeviceDefault	Using a customized version of one of the Android themes.

Description

- To view a list of all built-in themes available to an activity, display the activity in Graphical Layout editor and use the drop-down list that's available from the Themes button to view the themes.
- Android 4.0 (API 14) introduced the family of DeviceDefault themes. The DeviceDefault themes provide a way for manufacturers to provide a default theme for a device without having to modify the other built-in Android themes such as the Holo theme.

Figure 7-8   A summary of built-in themes

How to apply themes

You can use the AndroidManifest.xml file to apply themes to an entire applicaton or to a specific activity. To apply a theme to the entire application, you use the theme attribute of the application element to specify the theme. For instance, figure 7-2 shows how to specify a custom theme named AppTheme for the entire application.

To apply a theme to a specific activity, you use the theme attribute of the activity element to specify the theme. For instance, the first example in figure 7-9 specifies a built-in theme named Theme.Dialog for an activity. This displays the activity as a dialog box using a black background.

The second example in this figure works similarly to the first. However, it specifies a custom theme named DialogTheme. This custom theme uses the older built-in theme named Theme.Dialog for APIs prior to Android 3.0, and it uses a newer built-in theme named Theme.Holo.Light.Dialog for Android 3.0 and later. For Android 3.0 and later, this displays the activity as a dialog box like the one shown in this figure.

When you specify a theme at both the application and activity levels, the theme at the activity level overrides the theme at the application level. In fact, it's common to specify a default theme at the application level that applies to most of the activities in the app. Then, you can specify a theme at the activity level for any activities that need to override the default theme for the app.

When you apply a style to a widget, that style's properties override the properties that are specified at the application or activity level. Similarly, if you set a property directly on a widget, it overrides any properties that are specified by a style.

Theme.Holo.Light.Dialog

An AndroidManifest.xml file that uses a built-in theme

```
<activity
    android:name="com.murach.dialogtest.DialogActivity"
    android:theme="@android:style/Theme.Dialog"
    android:label="@string/about_title" >
```

Another AndroidManifster.xml file that uses a user-defined theme

```
<activity
    android:name="com.murach.dialogtest.DialogActivity"
    android:theme="@style/DialogTheme"
    android:label="@string/about_title" >
```

Two styles.xml files for a user-defined theme

In the res\values directory

```
<style name="DialogTheme" parent="android:Theme.Dialog">
</style>
```

In the res\values-v11 directory

```
<style name="DialogTheme" parent="android:Theme.Holo.Light.Dialog">
</style>
```

Description

- You can use the AndroidManifest.xml file to apply a theme to the entire application or to a specific activity.
- To apply a theme to the entire application, use the theme attribute of the application element to specify the theme as shown in figure 7-2.
- To apply a theme to a specific activity, use the theme attribute of the activity element to specify the theme.
- You can use the theme attribute to specify a built-in theme or a custom theme.
- Since the Holo theme was introduced in Android 3.0 (API 11), you often need to use one theme prior to API 11 and another related theme after API 11. To do that, you can create a custom theme and inherit the appropriate built-in themes for each API level.
- When you specify a theme for the application and activity levels, the theme at the activity level overrides the theme at the application level.

Figure 7-9   How to apply themes

How to work with colors

As you work with themes and styles, you'll find that some properties allow you to specify colors. For example, you can specify a color for the textColor and background properties of most widgets. Most of the time, the colors in the default theme are appropriate for your app. However, if you need to modify the colors in your app, figure 7-10 shows how.

How to define colors

If you want to work with colors, it's generally considered a best practice to begin by adding a colors.xml file to the res/values directory. Then, you can add a color element for each color that you want to define as shown in the first example. Within the color element, you can use the name attribute to specify a name for the color. In this figure, for example, the colors.xml file creates colors named primary, secondary, tertiary, background, dark, and light. Together these colors provide a color scheme that you can use for your app.

Within the body of a color element, you can use a *hexadecimal*, or *hex*, value to specify an RGB value for the color. This works very similarly to using hex values to specify RGB colors with HTML and CSS. If you aren't familiar with how this works, you can search the Internet to learn more about it.

How to apply colors

The second example shows three ways to apply a color to the textColor property of a widget. First, you can use a hex value to specify a color. Second, you can use a name for a custom color. Third, you can use a name for a built-in Android color. Of these techniques, using a custom color can make it easier to apply colors consistently and it can make your app easier to maintain if you decide to change colors later.

The third example shows how to use a style to apply a color. To do that, you use an item element within a style to set the color for a property. When you use this technique, you can use hex values or named styles, though it's generally preferable to use named styles whenever possible.

The fourth example shows how to apply a color by overriding the default colors of a theme. To do that, you use an item element within a theme to override one of the color attributes for the theme. When you use this technique, you must use a named style, not a hex value.

If you want to learn more about the colors that Android uses in its built-in themes, you can view the attrs.xml and colors.xml files that define the default attributes and colors for Android. To find these files, you can look in the data\res\values directories for the Android APIs on your computer.

A colors.xml file in the res/values directory

```xml
<?xml version="1.0" encoding="utf-8"?>
<resources>
    <color name="primary">#141315</color>
    <color name="secondary">#736C6B</color>
    <color name="tertiary">#DDE0CE</color>
    <color name="background">#A6D39D</color>
    <color name="dark">#000000</color>
    <color name="light">#FFFFFF</color>
</resources>
```

How to apply colors to a widget

Using a hexadecimal value
```
android:textColor="#141315"
```

Using a name from the colors.xml file
```
android:textColor="@color/primary"
```

Using the name of a built-in Android color
```
android:textColor="@android:color/darker_gray"
```

How to apply colors to a style

```xml
<style name="TextView" parent="@android:style/Widget.TextView">
    <item name="android:textColor">@color/primary</item>
</style>
```

How to apply colors to a theme

```xml
<style name="AppTheme" parent="AppBaseTheme">
    <!-- Set new colors for the theme -->
    <item name="android:windowBackground">@color/background</item>
    <item name="android:textColorPrimary">@color/primary</item>
    <item name="android:textColorSecondary">@color/secondary</item>
</style>
```

The colors.xml file for Android 4.2 (API 17)

```
\android-sdks\platforms\android-17\data\res\values\colors.xml
```

Description

- To define names for colors, you can add a colors.xml file to the res/values directory.
- To specify a color, you can use *hexadecimal*, or *hex*, values to specify an RGB value.
- To apply a color, you can use a hex value, a name for a user-defined color, or a name for a built-in Android color.
- To learn more about built-in Android colors, view the colors.xml file for one of the Android SDKs that are installed on your system.

Figure 7-10   How to work with colors

Perspective

Now that you've finished this chapter, you should understand how to work with themes and styles. For small apps, you may decide that the default themes and styles are adequate. In that case, you can manually format each widget on the app.

For larger apps, you may want to create a style sheet to apply formatting. Or, you may want to use a custom theme to override some properties of a built-in theme. Both of these techniques separate the design from the content, which is generally considered a best practice. In addition, they help to apply formatting consistently, reduce code duplication, and make it easier to develop and maintain an app.

Although this chapter shows you the basics for working with colors, there's more to learn about them. For example, you can use XML to define color gradients. In addition, you can specify colors that include a level of transparency.

Terms

style	property
theme	style sheet
title bar	hexadecimal
action bar	hex

Summary

- A *style* is a collection of properties that specify formatting for a widget.
- A *theme* is a style that's applied to an entire activity or app. Android includes several built-in themes.
- Prior to Android 4.0, an activity can include a *title bar* that displays the title of the activity. With Android 4.0 and later, an activity can also include an *action bar*.
- A style element can specify a theme or a style and can contain one or more item elements. Each item element specifies a *property* of the style.
- The AndroidManifest.xml file uses its application element to specify the theme for the application.
- A *style sheet* is a collection of styles that can be applied throughout an application.
- Android 4.0 (API 14) introduced the family of DeviceDefault themes.
- To define names for colors, you can add a colors.xml file to the res/values directory.
- To specify a color, you can use *hexadecimal*, or *hex*, values to specify an RGB value.

- To apply a color, you can use a hex value, a name for a user-defined color, or a name for a built-in Android color.
- To learn more about built-in Android colors, view the colors.xml file for one of the Android SDKs that are installed on your system.

Exercise 7-1   Use built-in themes

In this exercise, you'll modify the Tip Calculator app so it uses built-in themes to display a dark background.

1. Start Eclipse and import the project named ch07_ex1_TipCalculator.
2. Open the styles.xml files in the values, values-v11, and values-v14 directories. These files each specify a built-in theme for the API level.
3. Open the AndroidManifest.xml file and review its code. It sets the custom theme named AppTheme as the theme for the application.
4. Open the layout for the app and view it in the Graphical Layout editor. Use the Android Version button to view the layout for API levels 8 and 17. The theme should look different for both of these APIs.
5. Edit the three styles.xml files so the app uses the theme named Theme.Black prior to API 11 and uses Theme.Holo.Black for API 11 and later.
6. Open the layout for the app and view it in the Graphical Layout editor. These themes should display a dark background for all APIs.

Exercise 7-2   Use styles

In this exercise, you'll modify the styles for the Tip Calculator app presented in this chapter.

1. Start Eclipse and import the project named ch07_ex2_TipCalculator.
2. Open the styles.xml file in the values directory. This file defines several custom styles.
3. Open the layout for the app and view its XML. Each widget applies one of the custom styles specified in the styles.xml file.
4. For the first TextView widget, delete the style attribute that applies the style. This should remove the style's formatting from that widget. Then, restore this style attribute.
5. For the last TextView widget, change the style from TextView.Label to TextView.Label.Indent. This should indent the Total label. Then, change the style back to the TextView.Label style.
6. In the styles.xml file, change the textSize attribute of the TextView style to 16sp. This should change the text size for all widgets on the layout.

7. In the styles.xml file, modify the TextView style so it sets the layout_width, layout_height, and padding attributes appropriately for TextView and EditText widgets. Then, switch to the XML for the layout and delete these attributes from that file. This should reduce code duplication.

8. In the styles.xml file, change the padding attribute of the TextView style to 8dp. This should decrease the space between the widgets.

9. In the styles.xml file, modify the Button style so it sets the layout_width and layout_height attributes to 40dp. Then, switch to the XML for the layout and delete these attributes from that file. This should make both buttons a little smaller, and it should reduce code duplication.

10. Add two new TextView widgets to the form for a Per Person label and amount. Set the id, text, and alignment attributes for these widgets appropriately. Then, apply the TextView.Label style to the label and the TextView style to the amount. This is an easy way to apply consistent formatting to these new widgets.

Exercise 7-3   Modify a theme

In this exercise, you'll modify the theme that's used by the Tip Calculator app presented in this chapter.

1. Start Eclipse and import the project named ch07_ex3_TipCalculator.

2. Open the styles.xml file in the values directory. This file modifies the theme to automatically apply custom styles for the TextView, EditText, and Button widgets. This file also includes a custom style named TextView.Label that can be applied manually.

3. Open the layout for the app and view its XML. This XML only applies the TextView.Label style.

4. Add a RadioGroup widget to the layout. Note that the radio buttons within the group use a different text size than the other widgets on the layout.

5. Modify the styles.xml file so it modifies the TextAppearance styles for the theme as shown in figure 7-7 instead of modifying the styles for individual widgets. This should set the textSize property to 20sp for the TextAppearance style and its Small, Medium, and Inverse variations. When you're done, all widgets should use a textSize of 20sp.

8

How to work with menus and preferences

An Android app often includes menus that allow users to perform common tasks and preferences that allow users to customize its settings. Often, a menu includes an item that allows users to display the preferences screen for changing settings. In this chapter, you'll learn how to add menus and preferences to an app.

How to work with menus

A *menu* can contain one or more *items*. These items often provide a way for the user to perform actions and navigate through an app. Since a menu is hidden until the user activates it, menus can improve an app by hiding functionality that isn't typically needed. This frees screen space and reduces clutter on the screen.

An introduction to menus

In Android, the most common type of menu is known as the *options menu*. This type of menu can be displayed by an activity as shown in figure 8-1.

If a device has a physical Menu button, the user can use that button to display the options menu across the bottom of the activity. In this figure, for example, the first activity displays a menu with two items across the bottom of the activity.

If a device doesn't have a physical Menu button, an *action overflow icon* is displayed on the right side of the action bar to display the options menu. In this figure, for example, the second activity displays an action overflow icon that provides a way to display the options menu on the top right side of the activity.

Although it isn't shown in this figure, the options menu works a little differently in versions of Android prior to Android 3.0 (API 11). For these versions of Android, the options menu displays six icons across the bottom of the screen. Then, if this menu contains more than six items, the sixth item (the More item) allows the user to access the overflow items.

An activity with an options menu that has two items

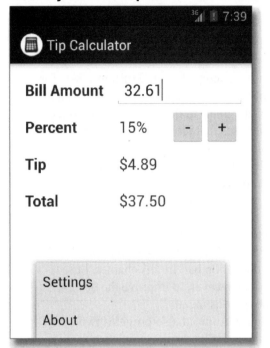

The same options menu displayed from an action overflow icon

Description

- An activity can include an *options menu* that includes one or more *menu items*.
- If the device has a physical Menu button, the user can use that button to display the options menu across the bottom of the screen.
- If a device doesn't have a physical Menu button, an *action overflow icon* is displayed on the right side of the action bar to display the options menu.
- Prior to Android 3.0 (API 11), the options menu displays six items across the bottom of the screen. If this menu contains more than six items, the sixth item (the More item) allows the user to access the overflow items.

Figure 8-1   An introduction to menus (part 1 of 2)

For Android 3.0 (API 11) and higher, the action bar can display some or all of the items from the options menu as *action items*. It's considered a good design guideline to use action items for a small number of commonly used actions. For example, it's common to add action items such as search, refresh, add, edit, delete, and so on.

Action items can be displayed as text or as icons. In part 2 of figure 8-1, for example, the first activity displays both of the items in the options menu as text. Then, the second activity displays both of the items in the options menu as icons. If you specify an icon for an item, Android uses that icon in the action bar, but it uses text for that item in the options menu.

Any menu items that aren't displayed on the action bar are displayed in the options menu. In this figure, for example, the last activity shows the Settings icon and the overflow icon in the action bar, and it shows the About icon in the options menu.

It's generally considered bad design to put items that the user doesn't commonly use in the action bar. For example, it's not good design to put functions such as settings, about, and help in the action bar. In this chapter, I use these items to show how action items work. However, to improve the design, I eventually move these items back into the options menu.

Although they aren't shown in this figure, Android also provides two other types of menus. First, a *floating context menu* can be displayed when a user performs a long click on a widget. This menu appears as a floating list of menu items. Second, a *popup menu* can be displayed when a user clicks on a widget or action item. This type of menu usually appears as a drop-down list below the widget, and it's only available for Android 3.0 (API 11) and higher.

If your app targets Android 3.0 (API 11) or higher, it's considered a good practice to use *contextual action mode* instead of a floating context menu. This mode displays action items that apply to the selected item or items.

A task bar with two action items displayed as text

A task bar with two action items displayed as icons

A task bar with one action icon and an overflow action icon

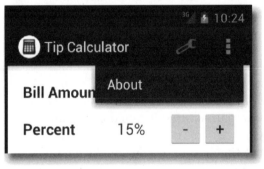

Description

- For Android 3.0 (API 11) and higher, the action bar can include menu items. Typically, these items are for a small number of actions that are important to the app.
- Any items that aren't displayed on the action bar are displayed in the options menu.
- By default, if you specify an icon for an item, Android uses that icon in the action bar, but it uses text for that item in the options menu.
- A *floating context menu* appears as a floating list of menu items when a user performs a long click on the widget. Alternately, for Android 3.0 (API 11) and later, *contextual action mode* can display action items that apply to the selected item or items.
- A *popup menu* usually appears as a list below a widget or action item. Popup menus are only available for Android 3.0 (API 11) and higher.

Figure 8-1   An introduction to menus (part 2 of 2)

How to define a menu

Figure 8-2 shows how to define a menu. To do that, you add an XML file for the menu to the project's res\menu directory. In this figure, for example, the menu is stored in an XML file with a name that clearly indicates that the menu is for the Tip Calculator activity.

The XML file for the menu begins with a root element named menu. Within the menu element, the XML can specify one or more item elements that define the items in the menu. In this figure, for example, the XML defines a menu with the two items shown in the previous figure.

Both item elements define four attributes. To start, the id attribute specifies an ID that can be used to access the item. Then, the title attribute specifies the text for the item that's displayed in the menu. This text is stored in the strings.xml file. Next, the icon attribute specifies an icon for the item that's usually displayed in the action bar. The file for this icon is stored in the res\drawable-xhdpi directory.

The showAsAction attribute specifies whether the item is shown in the action bar. For the first item, this attribute has been set to a value of "ifRoom". As a result, this item is shown in the action bar if there is room. If there isn't enough room, this item is shown in the options menu. For the second item, this attribute has been set to "never". As a result, this item is never shown in the action bar, even if there is room.

When an item is displayed in the action bar, Android's default behavior is to display the icon for the item. Conversely, when an item is displayed on the options menu, Android's default behavior is to display the text for the item, not the icon. Most of the time, that's what you want. As a result, you don't usually need to override Android's default behavior.

If you want to use your own icons, you should start by copying the file for the icon into the appropriate res\drawable directory in your project. In this figure, for example, both icons are for extra-high density screens, so they are stored in the res\drawable-xhdpi directory. Here, both icons are PNG files, and their names begin with a prefix of "ic_" to indicate that they are icons. This prefix isn't required, but it's commonly used.

To find an icon for your item, you can search through the standard icons that are available from the different versions of Android. To do that, you can look through the Android SDKs that are installed on your system, find the Android platform you want to use, and browse through its data\res\drawable folders. Or, you can search the Internet for the appropriate icons.

For this project, I have only supplied icons for the drawable-xhdpi folder. This works because Android automatically scales the icons for other densities. However, for a production app, you typically provide icons that are sized appropriately for all screen densities (ldpi, mdpi, hdpi, and xhdpi).

The file that contains the XML for the menu

```
res\menu\activity_tip_calculator.xml
```

The XML for the menu

```xml
<menu xmlns:android="http://schemas.android.com/apk/res/android">
    <item android:id="@+id/menu_settings"
          android:title="@string/menu_settings"
          android:icon="@drawable/ic_settings"
          android:showAsAction="ifRoom" />
    <item android:id="@+id/menu_about"
          android:title="@string/menu_about"
          android:icon="@drawable/ic_about"
          android:showAsAction="never" />
</menu>
```

Some attributes of a menu item

Name	Description
`title`	Specifies the text for the item.
`icon`	Specifies the icon for the item.
`showAsAction`	Specifies whether the item is shown on the action bar. Typically this attribute is set to a value of "always", "never", or "ifRoom".
`orderInCategory`	Specifies an int value for the sequence of the item where the lowest number is displayed first.

The location of the icon files for the items

```
res\drawable-xhdpi\ic_settings.png
res\drawable-xhdpi\ic_about.png
```

A directory that has standard icons for Android 4.2 (API 17)

```
\android-sdks\platforms\android-17\data\res\drawable-xhdpi
```

Description

- To provide an icon for a menu item, copy the icon file into the appropriate res\drawable directories in your project. Then, use the icon attribute to identify the name of the icon file.

Figure 8-2   How to define a menu

How to display an options menu

The first code example in figure 8-3 shows how to display an options menu. To do that, this code overrides the onCreateOptionsMenu method. This method is a lifecycle method of an activity that Android calls before it needs to display the menu for the first time.

Within the onCreateOptionsMenu method, the first statement gets a MenuInflater object and uses that object to convert, or *inflate*, the XML for the menu items into Java objects. This stores these Java objects in the Menu parameter. Then, the second statement returns a true value to display the menu.

Android only calls the onCreateOptionsMenu method the first time it displays the options menu. As a result, you can't use this method to update the menu each time it's displayed. For most apps, you don't need to do that anyway. However, if you need to update the menu every time it's displayed, you can override the onPrepareOptionsMenu method to do that. To learn more about that method, you can look it up in the API documentation or you can search the Internet.

How to handle option menu events

The second code example in figure 8-3 also shows how to handle the events that occur when a user selects an item from the options menu. To do that, this code overrides the onOptionsItemSelected method. Android calls this method when the user selects an item from the options menu.

Within the onOptionsItemSelected method, this code uses a switch statement to determine which menu item was selected. In this figure, for example, the switch statement gets the ID of the menu item and uses it to determine which menu item was selected.

If the ID of the menu item matches one of the menu items shown earlier in this chapter, this code displays a toast that indicates the name of the item. In addition, it returns a true value. This indicates that the method has consumed the event and stops further processing of the event.

If the ID of the menu item doesn't match one of the menu items shown earlier in this chapter, this code passes the menu item to the onOptionsItemSelected method of the superclass. This allows the superclass to processes the menu item.

The code that displays the menu

```
@Override
public boolean onCreateOptionsMenu(Menu menu) {
    getMenuInflater().inflate(R.menu.activity_tip_calculator, menu);
    return true;
}
```

The code that handles the menu item events

```
@Override
public boolean onOptionsItemSelected(MenuItem item) {
    switch (item.getItemId()) {
        case R.id.menu_settings:
            Toast.makeText(this, "Settings", Toast.LENGTH_SHORT).show();
            return true;
        case R.id.menu_about:
            Toast.makeText(this, "About", Toast.LENGTH_SHORT).show();
            return true;
        default:
            return super.onOptionsItemSelected(item);
    }
}
```

Description

- To display an options menu, you can override the onCreateOptionsMenu method.

- Within the onCreateOptionsMenu method, you typically use a MenuInflater object to convert, or *inflate*, the XML for the menu items into Java objects and store them in the Menu parameter.

- The onCreateOptionsMenu method must return a true value to display the menu.

- To handle the event that's generated when a user selects an item from the options menu, you can override the onOptionsItemSelected method.

- Within the onOptionsItemSelected method, you can use a switch statement to determine which menu item was selected.

- The onOptionsItemSelected method can return a true value to indicate it has consumed the event and to stop further processing. Or, it can return a false value to allow processing to continue.

- Android only calls the onCreateOptionsMenu method the first time it displays the options menu. To update the menu every time it is displayed, you can override the method named onPrepareOptionsMenu.

Figure 8-3   How to display an options menu and handle its events

How to start a new activity

When an app contains more than one activity, menus are often used to navigate between those activities. To do that, you need to create an *intent*, which is an object that provides a description of an operation to be performed. In this case, you need to create an intent for an activity. Then, you can pass this intent to the startActivity method to start the activity as shown in figure 8-4.

The first code example shows two techniques for starting a new activity. The first technique uses two statements. Here, the first statement creates an Intent object by passing two arguments to the constructor of the Intent class. The first argument specifies the application context, and the second argument specifies name of the class for the activity. Then, the second statement passes the Intent object to the startActivity method that's available from the Activity class.

The second technique for starting an activity works the same as the first, but it uses a single statement. Since this yields shorter code that's still easy enough to read, the second technique is commonly used.

The second code example shows how to use menu items to start activities. Here, the first item starts the Settings activity. This activity allows the user to change the settings for an app, and you'll learn about it later in this chapter. Then, the second item starts an About activity. This activity uses a dialog box to display some information about the app. For this to work, both of these activities must be declared in the project's AndroidManifest.xml file, just as the Tip Calculator activity is declared in this file.

Code that starts a new activity

Two statements

```
Intent settings =
    new Intent(getApplicationContext(), SettingsActivity.class)
startActivity(settings);
```

One statement

```
startActivity(new Intent(getApplicationContext(),
    SettingsActivity.class));
```

Code that uses menu items to start new activities

```
@Override
public boolean onOptionsItemSelected(MenuItem item) {
    switch (item.getItemId()) {
        case R.id.menu_settings:
            startActivity(new Intent(getApplicationContext(),
                SettingsActivity.class));
            return true;
        case R.id.menu_about:
            startActivity(new Intent(getApplicationContext(),
                AboutActivity.class));
            return true;
        default:
            return super.onOptionsItemSelected(item);
    }
}
```

Description

- An *intent* provides a description of an operation to be performed. Intents are commonly used with the startActivity method to start activities.

- To create an intent for an activity within the app, pass two arguments to the constructor of the Intent class. The first argument specifies the application context, and the second argument specifies the name of the class for the activity.

- To start a new activity, create an Intent object for the activity and pass that object to the startActivity method.

Figure 8-4   How to start a new activity

How to work with preferences

Preferences, or *settings*, provide a way for users to customize how an app works. Android provides Preference APIs that allow you to build an interface that's consistent with the user experience in other Android apps including built-in apps such as the system settings.

An introduction to preferences

Figure 8-5 shows a Settings activity that provides two preferences. The first preference has a name, a description, and a box that allows you to check or uncheck the preference. The second preference also has a name and a description. If you click on this preference, it displays a dialog box like the one shown in this figure that allows you to select one of three options.

The Settings activity

The dialog for the "Rounding?" item

Description

- An app often allows the user to change *preferences*, or *settings*, for the app.
- Android provides Preference APIs that allow you to build an interface that's consistent with the user experience in other Android apps including built-in apps such as the system settings.

Figure 8-5   An introduction to preferences

How to define preferences

For most activities, you use subclasses of the View class to build the user interface. In this book, for example, the Tip Calculator activity uses the TextView, EditText, and Button classes to build its user interface. When working with preferences, you use subclasses of the Preference class to build the user interface as shown in figure 8-6.

To start, you add an XML file to the res\xml directory. This XML file is typically named preferences.xml, but you can use any name you want. In this figure, the root element for the XML file is a PreferenceScreen element. This element defines a screen that displays one or more preferences.

Within the PreferenceScreen element, the CheckBoxPreference element defines the first preference shown in the previous figure. Here, the key attribute specifies the key that's used to access the preference. Then, the next three attributes specify the name, description, and default value for the preference. The values for the title and summary attributes are strings that are stored in the strings.xml file.

The ListPreference element defines the second preference shown in the previous figure. Here, the key, title, summary, and defaultValue attributes work the same as they do for the CheckBoxPreference element. However, the ListPreference element includes three more attributes. Here, the dialogTitle attribute specifies the title for the dialog box that's used to display the list of options. Then, the entries and entryValues attributes specify the names and values for the list. These arrays are also stored in the strings.xml file.

When you use the Preference API, each preference has a corresponding key-value pair that Android saves in a default shared preferences file, which is available to other activities in the app. When a user changes a setting, Android automatically updates this file. Then, you can read these settings from other activities and use them to modify the behavior of the app.

Although this chapter only shows how to use the CheckBoxPreference and ListPreference classes, other subclasses of the Preference class exist. For example, it's common to use an EditTextPreference element to allow the user to enter text such as a username or password. For more information, about these subclasses, look up the Preference class in the API documentation and follow the links to view its subclasses.

The file that contains the XML for the preferences

```
res\xml\preferences.xml
```

The XML for the preferences

```xml
<?xml version="1.0" encoding="utf-8"?>
<PreferenceScreen
        xmlns:android="http://schemas.android.com/apk/res/android">
    <CheckBoxPreference
        android:key="pref_remember_percent"
        android:title="@string/remember_percent_title"
        android:summary="@string/remember_percent_summary"
        android:defaultValue="true" />
    <ListPreference
        android:key="pref_rounding"
        android:title="@string/rounding_title"
        android:summary="@string/rounding_summary"
        android:dialogTitle="@string/rounding_title"
        android:entries="@array/rounding_keys"
        android:entryValues="@array/rounding_values"
        android:defaultValue="@string/rounding_default" />
</PreferenceScreen>
```

Some attributes that apply to all Preference items

Name	Description
key	Specifies the ID that's used to access the preference.
title	Specifies the title for the preference.
summary	Specifies the summary for the preference.
defaultValue	Specifies the default value for the preference.

Some attributes that apply to a ListPreference

Name	Description
dialogTitle	Specifies the title of the dialog box that sets the value of the preference.
entries	Specifies the names of the entries in the list.
entryValues	Specifies the values of the entries in the list.

Description

- Instead of using View objects to build the user interface, settings are built using various subclasses of the Preference class that you declare in an XML file. This file is typically named preferences.xml, but you can use any name you want.

- For a list of commonly used Preference objects, see the API documentation for the Preference class.

- Each preference has a corresponding key-value pair that the system saves in the default shared preferences file. Whenever the user changes a setting, the system automatically updates that file.

Figure 8-6   How to define preferences

How to display preferences in an activity

Figure 8-7 shows a technique for displaying preferences in an activity that was commonly used in the early days of Android. As of Android 3.0 (API 11), this technique has been deprecated. However, it still works for all versions of Android. As a result, if your app targets Android 3.0 and higher, you should use the newer technique described in the next chapter. However, if your app targets an earlier Android API, you can still use the technique shown in this figure.

The code example shows a class named SettingsActivity that inherits the PreferenceActivity class. This class only includes a single method, the onCreate method. Within this method, the second statement calls the addPreferencesFromResource method of the PreferenceActivity class. This method adds the preferences defined in the XML file shown in the previous figure to the activity defined by this class.

The onCreate method has two annotations: @SuppressWarnings and @Override. Here, the @SuppressWarnings annotation suppresses all deprecation warnings for this method. As a result, Eclipse doesn't display a warning for this project even though the addPreferencesFromResource method has been deprecated. This annotation shows that the programmer understands the warning but has decided to use the deprecated method anyway.

The SettingsActivity class

```
package com.murach.tipcalculator;

import android.os.Bundle;
import android.preference.PreferenceActivity;

public class SettingsActivity extends PreferenceActivity {

    @SuppressWarnings("deprecation")
    @Override
    public void onCreate(Bundle savedInstanceState) {
        super.onCreate(savedInstanceState);

        addPreferencesFromResource(R.xml.preferences);
    }
}
```

Description

- Since the user-interface is built using Preference objects instead of View objects, the activity needs to extend the PreferenceActivity class.

- You can use the addPreferencesFromResource method of your activity to add the preferences defined in the XML file to the activity. This method was deprecated with Android 3.0 (API 11).

- You must declare all activities in the AndroidManifest.xml file.

Figure 8-7   How to display preferences with an activity

How to display preferences in a fragment

Figure 8-8 shows how to display preferences in a *fragment*, which is a class that you can use to define part of a user interface. Then, you can use an activity to display one or more fragments. Also, if necessary, you can reuse a fragment in multiple activities.

Fragments were introduced in Android 3.0 (API 11). They are intended to provide more flexible user interfaces that can work on both smaller screens available from phones and larger screens available from tablets and other devices. Unfortunately, since the PreferenceFragment class is only available from Android 3.0 and later, you can't use this technique with earlier versions of Android such as Android 2.2 (API 8).

The first code example shows a fragment named SettingsFragment that inherits the PreferenceFragment class. This class works much like the class shown in the previous figure. However, the addPreferenceFromResource method that's available from the PreferenceFragment class has not been deprecated. As a result, there's no need to suppress warnings about deprecation.

The second code example shows how to add a fragment to an activity. Here, the fragment named SettingsFragment is added to the activity named SettingsActivity. To do that, this code uses the getFragmentManager method to get a FragmentManager object. From that object, this code calls the beginTransaction, replace, and commit methods. These methods start a transaction, replace the content of the activity with the fragment, and commit the transaction. Here, the replace method uses this Android resource:

```
android.R.id.content
```

to identify the content of the activity, and it creates a new SettingsFragment object to replace that content.

The SettingsFragment class

```
package com.murach.tipcalculator;

import android.os.Bundle;
import android.preference.PreferenceFragment;

public class SettingsFragment extends PreferenceFragment {

    @Override
    public void onCreate(Bundle savedInstanceState) {
        super.onCreate(savedInstanceState);

        // Load the preferences from an XML resource
        addPreferencesFromResource(R.xml.preferences);
    }
}
```

The SettingsActivity class

```
package com.murach.tipcalculator;

import android.app.Activity;
import android.os.Bundle;

public class SettingsActivity extends Activity {

    @Override
    public void onCreate(Bundle savedInstanceState) {
        super.onCreate(savedInstanceState);

        // Display the fragment as the main content
        getFragmentManager().beginTransaction()
                .replace(android.R.id.content, new SettingsFragment())
                .commit();
    }
}
```

Description

- You can use a *fragment* to define part of the user interface for an activity.
- An activity can display one or more fragments.
- A fragment can be reused in multiple activities.
- Fragments are available from Android 3.0 (API 11) and higher.
- Since the user-interface is built using Preference objects instead of View objects, your fragment needs to extend the PreferenceFragment class.
- From a fragment, you can use the addPreferencesFromResource method to add the preferences defined in the XML file to the activity.
- From an activity, you can use a FragmentManager object to replace the content of the activity with the specified fragment.

Figure 8-8   How to display preferences with a fragment

How to get preferences

Figure 8-9 shows how to get preferences that have automatically been written to a file by the Preference API. In the Tip Calculator activity, for example, you need to be able to get preferences so you can use them to modify that app's behavior.

The first example shows how to define the instance variables for the preferences. Here, the first instance variable is for the SharedPreferences object, and the next two are for the Remember Tip Percent and Rounding preferences.

The second example shows how to set the default values in the preferences file the first time the app starts on a device. To do that, this code calls the setDefaultValues method of the PreferenceManager class. This writes the default values that are stored in the XML file that defines the preferences to the default file for shared preferences. Here, the third parameter controls whether to reset the default values that are specified in the XML file. If false, the default values are set only if this method has never been called on this device, which is usually what you want. Since you only need to run this code once, it's often stored in the onCreate method of the activity.

The third example shows how to get the SharedPreferences object. To do that, this code calls the getDefaultSharedPreferences method of the PreferenceManager class. Here, the parameter uses the *this* keyword to specify the current activity as the context. This code is often stored in the onCreate method of an activity.

The fourth example shows how to get the values for the preferences from the SharedPreferences object. To do that, this code calls the appropriate getXxx method from the object and passes a key for the preference that matches the key specified in the XML file that defines the preferences. In addition, this code specifies a default value that's only used if the preference can't be retrieved. To make sure that your preferences are current, you typically store this code in the activity's onResume method.

In the fourth example, the first statement calls the getBoolean method and passes the key for the Remember Tip Percent preference. In most cases, this returns a Boolean value that indicates whether this preference is checked or unchecked. However, if this method isn't able to get a value from the object, the default value of true is returned.

Then, the second statement calls the getString method and passes the key for the Rounding preference. In most cases, this returns a string that indicates the selected option. Since the Tip Calculator activity uses an int value to keep track of rounding, this code converts the string for the rounding option to an int value.

Step 1: Define the instance variables for the preferences

```
private SharedPreferences prefs;
private boolean rememberTipPercent = true;
private int rounding = ROUND_NONE;
```

Step 2: Set the default values in the preferences file (onCreate)

```
PreferenceManager.setDefaultValues(this, R.xml.preferences, false);
```

Step 3: Get the SharedPreferences object (onCreate)

```
prefs = PreferenceManager.getDefaultSharedPreferences(this);
```

Step 4: Get the preferences (onResume)

```
rememberTipPercent = prefs.getBoolean("pref_remember_percent", true);
rounding = Integer.parseInt(prefs.getString("pref_rounding", "0"));
```

Some get methods of the SharedPreferences object

Name	Description
`getBoolean(key, default)`	Gets a Boolean value for the specified key. If the value doesn't exist, it gets the specified default value.
`getString(key, default)`	Gets a String value for the specified key.
`getInt(key, default)`	Gets an int value for the specified key.
`getLong(key, default)`	Gets a long value for the specified key.
`getFloat(key, default)`	Gets a float value for the specified key.

Description

- You can use the default SharedPreferences object to get preferences that have been automatically saved by the Preferences API.

- To set the default values the first time the app starts on a device, you can use the setDefaultValues method of the PreferenceManager class. Here, the third parameter controls whether to reset the default values that are specified in the XML file. If false, the default values are set only if this method has never been called in the past.

- To get the default SharedPreferences object, you can call the getDefaultSharedPreferences method of the PreferenceManager class.

- To get a preference from a SharedPreferences object, you can use the appropriate getXxx method for that preference. When you do, you specify a key that corresponds with the key in the XML file for the preferences. In addition, you specify a default value that's used if no value is retrieved for that preference.

Figure 8-9   How to get preferences

How to use preferences

Once you get preferences, you can use them to change the way your app works as shown in figure 8-10. Typically, you begin by using an if/else statement to check the value for the preference. Then, you execute the code that's appropriate for the value of the preference.

The first code example checks the value of the Remember Tip Percent preference, which has been stored in a variable named rememberTipPercent. If this variable contains a true value, this code remembers the tip percent by reading the last saved tip percent from a SharedPreferences object named prefs. Otherwise, it does not remember the tip percent by setting the tip percent to a default value of 15%.

The second code example checks the value of the Rounding preference, which has been stored in a variable named rounding. If this variable is equal to the constant named ROUND_NONE, the code calculates the tip and total without any rounding. If this variable is equal to the constant named ROUND_TIP, the code rounds the tip. If this variable is equal to the constant named ROUND_TOTAL, the code rounds the total.

The code in the second example may change the tip percent when the tip or total is rounded. However, this code shouldn't change the tip percent that's specified by the user. To provide for that, this code uses a variable named tipPercentToDisplay to store the tip percent that's displayed by the calculation. That way, the tip percent that's specified by the user is stored separately in the tipPercent variable. When there is no rounding, these variables contain the same value, but they may contain different values after the rounding is applied.

Use the "Remember Tip Percent" preference in the onResume method

```
if (rememberTipPercent) {
    tipPercent = prefs.getFloat("tipPercent", 0.15f);
} else {
    tipPercent = 0.15f;
}
```

Use the "Rounding" preference in the calculateAndDisplay method

```
float tipPercentToDisplay = 0;
if (rounding == ROUND_NONE) {
    tipAmount = billAmount * tipPercent;
    totalAmount = billAmount + tipAmount;
    tipPercentToDisplay = tipPercent;
}
else if (rounding == ROUND_TIP) {
    tipAmount = StrictMath.round(billAmount * tipPercent);
    totalAmount = billAmount + tipAmount;
    tipPercentToDisplay = tipAmount / billAmount;
}
else if (rounding == ROUND_TOTAL) {
    float tipNotRounded = billAmount * tipPercent;
    totalAmount = StrictMath.round(billAmount + tipNotRounded);
    tipAmount = totalAmount - billAmount;
    tipPercentToDisplay = tipAmount / billAmount;
}
```

Description

- Once you get the preferences from the default preference file, you can use them to change the way your app works.

Figure 8-10   How to use preferences

More skills for working with preferences

So far in this chapter, you have learned some basic skills for working with preferences. These skills apply to most apps. Now, you're ready to learn some more skills for working with preferences. These skills may apply to apps that have more preferences, or apps that have more complex requirements for how their preferences work.

How to group preferences

If your app has a large number of preferences, you may want to organize them in groups as shown in figure 8-11. Here, the Settings activity has three preferences that have been grouped in two categories. The first category has a title of TIP PERCENT, and it contains the Remember Tip Percent and the Default Tip Percent preference. The second category has a title of ROUNDING, and it contains the Rounding preference.

The code example shows how to create the TIP PERCENT category. To do that, this code nests a CheckBoxPreference element and a ListPreference element within a PreferenceCategory element. These preferences are for the Remember Tip Percent and Default Tip Percent preferences. Here, the PreferenceCategory has a title attribute that specifies the title for the category. As usual, this string is stored in the project's strings.xml file.

A Settings activity that uses categories

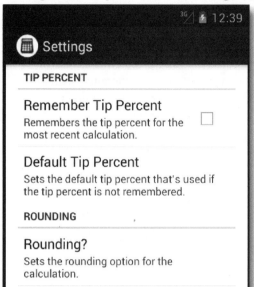

The XML for the preferences

```xml
<?xml version="1.0" encoding="utf-8"?>
<PreferenceScreen
    xmlns:android="http://schemas.android.com/apk/res/android">
    <PreferenceCategory
        android:title="@string/percent_category_title"
        android:key="pref_percent_category">
        <CheckBoxPreference
            android:key="pref_forget_percent"
            android:title="@string/forget_percent_title"
            android:summary="@string/forget_percent_summary"
            android:defaultValue="false" />
        <ListPreference
            android:key="pref_default_percent"
            android:title="@string/default_percent_title"
            android:summary="@string/default_percent_summary"
            android:dependency="pref_forget_percent"
            android:dialogTitle="@string/default_percent_title"
            android:entries="@array/default_percent_keys"
            android:entryValues="@array/default_percent_values"
            android:defaultValue="@string/default_percent_default" />
    </PreferenceCategory>
    ...
</PreferenceScreen>
```

Description

- You can group preferences by nesting one or more Preference elements within a PreferenceCategory element.

Figure 8-11   How to group preferences

How to enable and disable preferences

For some apps, you may want to disable a preference when a corresponding CheckBoxPreference element is unchecked. To do that, you can use a preference's dependency attribute to link it to the corresponding preference as shown in figure 8-12. Here, the Settings activity has two preferences where the second preference is disabled when the first preference is unchecked. Conversely, the second preference is enabled when the first preference is checked.

A Settings activity that uses dependencies

The XML for the preferences

```xml
<?xml version="1.0" encoding="utf-8"?>
<PreferenceScreen
    xmlns:android="http://schemas.android.com/apk/res/android">
    <PreferenceCategory
        android:title="@string/percent_category_title"
        android:key="pref_percent_category">
        <CheckBoxPreference
            android:key="pref_forget_percent"
            android:title="@string/forget_percent_title"
            android:summary="@string/forget_percent_summary"
            android:defaultValue="false" />
        <ListPreference
            android:key="pref_default_percent"
            android:title="@string/default_percent_title"
            android:summary="@string/default_percent_summary"
            android:dependency="pref_forget_percent"
            android:dialogTitle="@string/default_percent_title"
            android:entries="@array/default_percent_keys"
            android:entryValues="@array/default_percent_values"
            android:defaultValue="@string/default_percent_default" />
    </PreferenceCategory>
    ...
</PreferenceScreen>
```

The dependency attribute

Name	Description
dependency	Specifies the key of the preference that the current preference depends upon.

Description

- You can enable or disable a preference by using its dependency attribute to link the preference to a corresponding CheckBoxPreference element.

Figure 8-12   How to enable or disable a preference

How to use Java to work with preferences

So far in this chapter, you've used the default behavior provided by the Preference API. Now, figure 8-13 presents a Java class that shows how to work with preferences. This is often useful if you need to override the default behavior of the Preference API. For example, the last figure showed how to use the default behavior to disable the Default Tip Percent preference when the Forget Tip Percent preference is unchecked. Now, this figure shows how to disable the Default Tip Percent preference when the Remember Tip Percent preference is checked.

The SettingsFragment class in this figure handles the event that occurs when the user changes any of the preferences defined by the XML file. To do that, this class implements the OnSharedPreferenceChangeListener. Then, it registers this listener in the onResume method, and it unregisters this listener in the onPause method. As a result, Android executes the onSharedPreferenceChanged method whenever the user changes a preference.

The onCreate method gets the default SharedPreferences object. Then, the onResume method uses this object to get the Boolean value for the Remember Tip Percent preference. Finally, the onResume method passes this Boolean value to the setDefaultPercentPreference method.

The setDefaultPercentPreference method enables or disables the preference. To start, the first statement uses the findPreference method that's available from the current class to get a Preference object for the Default Tip Percent preference. Then, an if/else statement checks whether the rememberPercent parameter is true. If so, this statement uses the setEnabled method to disable the Default Tip Percent preference. Otherwise, it uses the setEnabled method to enable this preference.

The onSharedPreferenceChanged method begins by checking if the Remember Tip Percent preference was changed. If so, this code gets the value of this preference and passes it to the setDefaultPercentPreference method, which enables or disables the Default Tip Percent preference.

A class that works with preferences

```java
package com.murach.tipcalculator;

import android.content.SharedPreferences;
import android.content.SharedPreferences.OnSharedPreferenceChangeListener;
import android.os.Bundle;
import android.preference.Preference;
import android.preference.PreferenceFragment;
import android.preference.PreferenceManager;

public class SettingsFragment extends PreferenceFragment
implements OnSharedPreferenceChangeListener {

    private SharedPreferences prefs;
    private boolean rememberPercent;

    @Override
    public void onCreate(Bundle savedInstanceState) {
        super.onCreate(savedInstanceState);
        addPreferencesFromResource(R.xml.preferences);
        prefs = PreferenceManager.getDefaultSharedPreferences(getActivity());
    }

    @Override
    public void onResume() {
        super.onResume();
        rememberPercent = prefs.getBoolean("pref_remember_percent", true);
        this.setDefaultPercentPreference(rememberPercent);
        prefs.registerOnSharedPreferenceChangeListener(this);
    }

    private void setDefaultPercentPreference(boolean rememberPercent)
    {
        Preference defaultPercent = findPreference("pref_default_percent");
        if (rememberPercent) defaultPercent.setEnabled(false);
        else                 defaultPercent.setEnabled(true);
    }

    @Override
    public void onPause() {
        prefs.unregisterOnSharedPreferenceChangeListener(this);
        super.onPause();
    }

    @Override
    public void onSharedPreferenceChanged(SharedPreferences prefs,
            String key) {
        if (key.equals("pref_remember_percent")) {
            rememberPercent = prefs.getBoolean(key, true);
        }
        this.setDefaultPercentPreference(rememberPercent);
    }
}
```

Figure 8-13   How to use Java to work with preferences

Perspective

The skills presented in this chapter should be enough to get you started with menus and preferences. However, Android provides many more features for working with menus and preferences. If necessary, you can use the Menu and Preference APIs to create more complex menus and preferences. With menus, for example, you can group menu items, and you can use Java code to work with menus dynamically. With preferences, you can create a preference that leads to another screen of preferences, and you can build custom preferences by extending one of the Preference classes.

Terms

menu	contextual action mode
items	inflate
options menu	intent
action overflow icon	preferences
action items	settings
floating context menu	fragment
popup menu	

Summary

- An activity can include an *options menu* that includes one or more *menu items*.

- If the device doesn't have a physical Menu button, an *action overflow icon* is displayed on the right side of the action bar to display the options menu.

- For Android 3.0 (API 11) and higher, the action bar can include menu items.

- Any items that aren't displayed on the action bar are displayed in the options menu.

- By default, if you specify an icon for an item, Android uses that icon in the action bar, but it uses text for that item in the options menu.

- A *floating context menu* appears as a floating list of menu items when a user performs a long click on the widget. Alternately, for Android 3.0 (API 11) and later, *contextual action mode* can display action items that apply to the selected item or items.

- A *popup menu* usually appears as a list below a widget or action item. Popup menus are only available for Android 3.0 (API 11) and higher.

- Within the onCreateOptionsMenu method, you typically use a MenuInflater object to convert, or *inflate*, the XML for the menu items into Java objects and store them in the Menu parameter.

- To handle the event that's generated when a user selects an item from the options menu, you can override the onOptionsItemSelected method and use a switch statement to determine which menu item was selected.

- An *intent* provides a description of an operation to be performed. Intents are commonly used with the startActivity method to start activities.

- Android provides Preference APIs that allow you to build an interface that's consistent with the user experience in other Android apps.

- Each preference has a corresponding key-value pair that the system saves in the default shared preferences file. Whenever the user changes a setting, the system automatically updates that file.

- You can use a *fragment* to define part of the user interface for an activity.

- You can use the default SharedPreferences object to get preferences that have been automatically saved by the Preferences API.

- You can group preferences by nesting one or more Preference elements within a PreferenceCategory element.

- You can enable or disable a preference by using its dependency attribute to link the preference to a corresponding CheckBoxPreference element.

Exercise 8-1   Experiment with menus and settings

In this exercise, you'll experiment with the menus and settings that are available from the Tip Calculator app.

1. Start Eclipse and import the project named ch08_ex1_TipCalculator.

2. Run the app. If the device or emulator has a physical Menu button, use it to display the menu. Otherwise, use the action overflow icon to display the menu. This menu should have two items: Settings and About.

3. Select the Settings item to display the Settings activity. Then, click the Back button to return to the main activity.

4. Select the About item to display the About activity. This activity should appear as a dialog box. Then, click outside of the dialog box to return to the main activity.

5. Open the XML file in the menu directory and modify it so the Settings item is displayed in the action bar if there is enough room. This should display an icon for the Settings item in the action bar, but the About item should remain in the options menu.

6. Run the app. Then, use the main activity to set the tip percent to 17%.

7. Select the Settings action item to display the Settings activity. Then, remove the check from the Remember Tip Percent box. Next, click the Back button to return to the main activity. The tip percent should be reset to its default value of 15%.

8. Use the main activity to set the tip percent to 17%.

9. Change the orientation of the activity. Since the app isn't remembering the tip percent, this should reset the tip percent to 15%.

10. Select the Settings action item to display the Settings activity.

11. Use the Settings activity to select the "Remember Tip Percent" check box.

12. Use the Settings activity to change the default rounding for the app so that the total is always rounded.

13. Use the main activity to increase or decrease the tip. To do that, you may need to click on the Increase (+) and Decrease (-) buttons several times. This should automatically round the total and adjust the tip percent accordingly.

14. Open the XML file for the menu and delete the icon attribute from the Settings item. This should display text for the Settings item in the action bar.

15. View the files in the res\drawable-xhdpi directory of the project. This directory should include a second Settings icon.

16. Open the XML file for the menu and modify it so the Settings item uses the new Settings icon. This should display the new Settings icon in the action bar.

Exercise 8-2   Work with menus

In this exercise, you'll modify the menus used by the Tip Calculator app presented in this chapter.

1. Start Eclipse and import the project named ch08_ex2_TipCalculator.

2. Open the XML file in the menu directory. Then, add a Help item to the end of this menu.

3. Open the .java file for the Tip Calculator activity. Then, modify the code so it displays a toast that indicates that the Help feature has *not* been implemented.

4. Add a Refresh item to the beginning of the menu. This item should display the icon named ic_refresh that's in the project's res\drawable-xhdpi directory, but only if there's enough room on the action bar.

5. Modify the .java file for the Tip Calculator activity so the Refresh item calls the calculateAndDisplay method.

6. Modify the onClick method so it does *not* call the calculateAndDisplay method. Instead, add some code that updates and displays the tip percent, but not the tip amount or total. That way, you need to use the Refresh item to update the tip amount and total after you click on the buttons.

7. Create an XML file for a menu for the Preferences activity. This menu should have two items: Tip Calculator and About.

8. Modify the Preferences activity so it displays the menu.

9. Modify the Preferences activity so it handles the two items. The Tip Calculator item should display the Tip Calculator activity, and the About item should display the About activity.

Exercise 8-3   Work with preferences

In this exercise, you'll add another setting to the Tip Calculator app presented in this chapter.

1. Start Eclipse and import the project named ch08_ex3_TipCalculator.

2. Open the preferences.xml file in the xml directory. Add a ListPreference for a setting named Default Tip Percent. This setting should allow the user to specify a default tip percent of 10%, 15%, or 20%.

3. Open the .java file for the Tip Calculator activity. Modify this code so it gets the default tip percent and uses it if the user has decided not to remember the tip percent.

4. Switch to the preferences.xml file. Modify this code so it puts the Remember Tip Percent and Default Tip Percent preferences in a category named Tip Percent. Also, create a category named Rounding for the Rounding preference.

5. Run the app. It should work correctly. However, the default tip percent is only used if the app does *not* remember the tip percent. As a result, there's no need to set the default tip percent unless you uncheck the Remember Tip Percent box.

6. Open the .java file for the Settings fragment. Modify this code so it disables the Default Tip Percent preference when the Remember Tip Percent preference is unchecked.

9

How to work with fragments

Android introduced fragments in version 3.0 (API 11). You can use fragments to create user interfaces that work well on both small and large screens. For example, fragments allow you to create a user interface that works well on a small screen such as a phone while also taking advantage of the extra space that's available from a large screen such as a tablet.

An introduction to fragments

A *fragment* is a class that you can use to define part of a user interface. Then, you can use an activity to display one or more fragments. In this context, fragments are sometimes referred to as *panes*.

Single-pane and multi-pane layouts

On a small screen, an activity typically only displays a single fragment. This is known as a single-pane layout. In the first example of figure 9-1, for instance, the Tip Calculator activity displays the Tip Calculator fragment, and the Settings activity displays a Settings fragment. At this point, the app looks and acts as if it was not using fragments.

However, on a large screen, an activity can display multiple fragments. This is known as a multi-pane layout. In the second example, for instance, a large screen is in portrait and landscape orientation. As a result, the Tip Calculator activity displays both the Tip Calculator and Settings fragments.

In landscape orientation, the Tip Calculator activity displays the Tip Calculator fragment on the left side of the screen. Then, it displays the Settings fragment on the right side of the screen.

In portrait orientation, the Tip Calculator activity displays the Tip Calculator fragment at the top of the screen. Then, it displays the Settings fragment below the Tip Calculator fragment. This makes both fragments immediately available to the user.

On both of the large screens in this figure, there's plenty of room to display the soft keyboard. As a result, Android often displays this keyboard automatically. If there is enough room, Android displays an enhanced version of the soft keyboard that includes a separate keypad for some special characters. This makes it easier for the user to enter the bill amount.

Two activities displaying two fragments

One activity displaying two fragments in both landscape and portrait orientation

Description

- You can use a *fragment* to define part of the user interface for an activity. A fragment is sometimes referred to as a *pane*.
- On a small screen, an activity typically only displays a single fragment. This is known as a *single-pane layout*.
- On a large screen, an activity can display multiple fragments. This is known as a *multi-pane layout*.
- A fragment can be reused in multiple activities.

Figure 9-1   Single-pane and multi-pane layouts

How to use support libraries

Fragments were introduced in Android 3.0 (API 11). As a result, fragments are not available by default from earlier versions of Android such as Android 2.2 (API 8).

Fortunately, when you create a new project using Eclipse, it usually makes fragments available to all versions of Android by adding a *support library* to your project. A support library is a JAR file that contains packages and classes that make new APIs available to older versions of Android.

If Eclipse doesn't automatically add the support library to your project, you can manually add the support library to your project as shown in figure 9-2. To do that, you can download the JAR file shown in this figure and copy it into your project's libs directory. Then, your app can use fragments and still run on all versions of Android.

Unfortunately, the latest version of the support library for API 4 does not include support for the PreferenceFragment class. As a result, if you use this class, your app must target API 11 and higher. One workaround is to use the Fragment or FragmentList class instead of the PreferenceFragment class. However, if you do that, you must write all of the code that displays and saves the preferences yourself.

A JAR file for the support library for API 4 and higher

```
libs\android-support-v4.jar
```

How to install the Android support library

1. Start the Android SDK Manager as described in appendix A (PC) or B (Mac).

2. Expand the Extras category and view the Android Support Library package. If the package isn't already installed, install it. This downloads all support files into the directory shown below.

The SDK directory that contains the Android support library

```
\android-sdks\extras\android\support\
```

How to make the support library available to a project

3. Locate the JAR file for the support library and copy it into your project's libs directory.

Description

- Fragments are available from Android 3.0 (API 11) and higher.

- Android provides *support libraries* that you can add to your project. These libraries make new APIs such as fragments available on older versions of Android.

- The support library for API 4 makes the Fragment class compatible with all modern versions of Android.

- The support library for API 4 does not include support for the PreferenceFragment class. As a result, if you use this class, your app must target Android API 11 and higher.

Figure 9-2   How to use support libraries

The lifecycle methods of a fragment

The Fragment class defines methods that are called by the Android operating system at different points in the lifecycle of an activity. This is shown by the diagram in figure 9-3. Most of these methods are similar to the lifecycle methods of an activity. For example, a fragment includes the onCreate, onPause, and onResume methods that are available from an activity.

The lifecycle of the activity that contains the fragment affects the lifecycle of the fragment. In other words, when Android calls a lifecycle method for an activity, it also calls a similar lifecycle method for each fragment within that activity. For example, when Android calls the onPause method of an activity, it also calls the onPause method for each fragment within that activity. However, a fragment has some extra lifecycle methods that aren't available from an activity. Of these methods, the most commonly used method is the onCreateView method.

Android calls the onCreateView method when it's time to display the layout for the fragment. This occurs after the onCreate method but before the onStart method. Conversely, Android calls the onDestroyView method when it's time to remove the fragment's layout from the activity. This occurs after the onStop method but before the onDestroy method.

The onAttach and onDetach methods are another pair of lifecycle methods that are occasionally used to work with a fragment. Android calls the onAttach method when the fragment has been associated with an activity, and it calls the onDetach method when the fragment has been disassociated from the activity.

The onActivityCreated method does not have a corresponding lifecycle method. Android calls this method after it has returned from its call to the onCreate method of the activity.

When an activity is in the resumed state, it can add and remove fragments. However, when the activity leaves the resumed state, any fragments within the activity must respond to the lifecycle methods of the activity such as the onPause, onStop, and onDestroy methods.

The lifecycle methods of a fragment

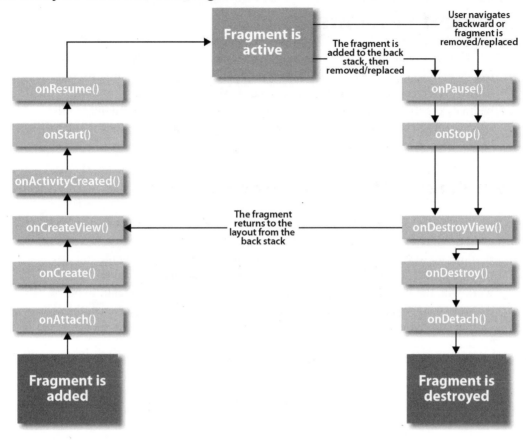

Description

- A fragment has lifecycle methods that are similar, but not identical to the lifecycle methods of an activity.

Figure 9-3   The lifecycle methods of a fragment

How to use single-pane layouts for small screens

Now that you understand how fragments work, you're ready to learn how to use them in a single-pane layout. This type of layout is useful for small screen devices such as phones. Since many devices have small screens, you often begin by creating a single-pane layout. Then, once you have that layout working correctly, you can add one or more multi-pane layouts for devices that have larger screens as shown later in this chapter.

How to create the layout for a fragment

Figure 9-4 shows how to create the layout for a fragment. To start, the XML code for a fragment works the same as the XML code for an activity. As a result, if you want to convert an existing layout for an activity to a fragment, you can rename the XML file. In Eclipse, you can use the Graphical Layout editor to work with the layout of a fragment just as you can use the Graphical Layout editor to work with the layout for an activity. In this figure, for instance, the Graphical Layout editor shows the layout for the Tip Calculator fragment.

The layout for a fragment

Description

- The XML for a fragment works the same as the XML for an activity.
- You can use the Graphical Layout editor to work with the layout for a fragment.

Figure 9-4   How to create the layout for a fragment

How to create the class for a fragment

Figure 9-5 shows how to create the class for a fragment. In general, this class works much like the class for an activity. However, there are a few differences. To start, the class for a fragment extends the Fragment class, not the Activity class.

Within the fragment class, the onCreate method typically initializes components of the fragment that aren't related to the layout for the user interface. In this figure, for example, the first two statements work with the preferences for the app. Both of these statements were described in the previous chapter, so you should understand how they work. Then, the third statement uses the setHasOptionsMenu method to indicate that this fragment has an options menu with items that should be added to the options menu for the activity. This is necessary because an activity that contains multiple fragments may need to combine the menu items from multiple fragments in its options menu.

The onCreateView method for a fragment typically contains the code that's used to create and initialize the layout for the fragment. Here, the first parameter of this method is a LayoutInflater object that you can use to inflate a layout. And the second paramter is a ViewGroup object for the parent layout. The layout for the fragment should be attached to this layout.

Within the method, the first statement calls the inflate method from this object to return a View object for the layout. To do that, this statement passes three arguments to the inflate method. The first argument specifies the ID for the layout of the fragment, the second argument specifies the container parameter of the method, and the third argument specifies a false value to indicate that the layout is not the root layout.

After inflating the layout, the code in the onCreateView method can get references to the widgets on the layout. To do that, you can call the findViewById method from the View object that was created by the first statement. In this figure, for example, the second statement gets a reference to the EditText widget for the bill amount.

The onCreateView method can also set listeners for widgets. This code works like the code in the onCreate method of an activity. As a result, you should already understand how it works.

If you need to convert an existing activity to a fragment, you can usually start by cutting code from the activity class and pasting it into the fragment class. For example, to convert the Tip Calculator activity described in the previous chapter to the Tip Calculator fragment described in this chapter, you can begin by creating the class for the Tip Calculator fragment. Then, you can cut and paste the instance variables and methods from the activity into the fragment. Next, you can modify the onCreate and onCreateView methods so they work as shown in this figure.

Method of the Fragment class

Method	Description
setHasOptionsMenu(bool)	Set to true to indicate that this fragment has an options menu and would like to participate in populating the activity's option menu by having Android call this fragment's onCreateOptionsMenu method.

The declaration for the TipCalculatorFragment class

```
public class TipCalculatorFragment extends Fragment
implements OnEditorActionListener, OnClickListener
```

The onCreate method

```
@Override
public void onCreate(Bundle savedInstanceState) {
    super.onCreate(savedInstanceState);

    // set the default values for the preferences
    PreferenceManager.setDefaultValues(getActivity(),
        R.xml.preferences, false);

    // get default SharedPreferences object
    prefs = PreferenceManager.getDefaultSharedPreferences(getActivity());

    // turn on the options menu
    setHasOptionsMenu(true);
}
```

The onCreateView method

```
@Override
public View onCreateView(LayoutInflater inflater, ViewGroup container,
        Bundle savedInstanceState) {

    // inflate the layout for this fragment
    View view = inflater.inflate(R.layout.fragment_tip_calculator,
        container, false);

    // get references to the widgets
    billAmountEditText = (EditText)
            view.findViewById(R.id.billAmountEditText);

    // set the listeners
    billAmountEditText.setOnEditorActionListener(this);

    // return the View for the layout
    return view;
}
```

Description

- The Java code for a fragment works much like the Java code for an activity. However, there are several differences, especially in the onCreate and onCreateView methods.

Figure 9-5   How to create the class for a fragment

How to display a fragment in an activity

Figure 9-6 shows how to display a fragment in an activity. To start, you can create an XML file that contains the layout for the activity. In this figure, for example, that file is named activity_main.xml. This XML begins by defining a linear layout that matches the width and height of the parent. Within the LinearLayout element, this code contains a single fragment element. The name attribute of this element specifies the name of the package and the class for the Tip Calculator fragment shown in the previous figure.

After creating the XML file for the layout, you can use the activity class to display the layout as shown in the second example. In this figure, for instance, the activity class displays the layout named activity_main that was created in the first example. Since the event handling has been moved into the fragment class, this activity class just needs to display the layout.

The activity_main.xml file

```
<LinearLayout xmlns:android="http://schemas.android.com/apk/res/android"
    android:orientation="horizontal"
    android:layout_width="match_parent"
    android:layout_height="match_parent">

    <fragment android:name="com.murach.tipcalculator.TipCalculatorFragment"
            android:id="@+id/main_fragment"
            android:layout_weight="1"
            android:layout_width="0dp"
            android:layout_height="match_parent" />

</LinearLayout>
```

The TipCalculatorActivity class

```
package com.murach.tipcalculator;

import android.app.Activity;
import android.os.Bundle;

public class TipCalculatorActivity extends Activity {

    @Override
    public void onCreate(Bundle savedInstanceState) {
        super.onCreate(savedInstanceState);
        setContentView(R.layout.activity_main);
    }
}
```

Description

- To add a fragment to a layout, add a fragment element and use its name attribute to specify the fully qualified name for the class that defines the fragment.

Figure 9-6   How to display a fragment in an activity

How to create a preference fragment

Figure 9-7 shows how to create a fragment that displays preferences. Since this user interface is built using Preference objects instead of View objects, your fragment needs to extend the PreferenceFragment class instead of the Fragment class.

The first code example shows a fragment named SettingsFragment that inherits the PreferenceFragment class. This code is the same as the code for the SettingsFragment class shown in the previous chapter. As a result, you should already understand how it works.

How to display a preference fragment in an activity

Once you've created a preference fragment like the one shown in the first example, it's easy to display it in an activity. To do that, you can define a layout for the activity that uses a fragment element to display the fragment as shown in the second example. Then, you can define an activity class that displays that layout as shown in the third example.

The SettingsFragment class

```
package com.murach.tipcalculator;

import android.os.Bundle;
import android.preference.PreferenceFragment;

public class SettingsFragment extends PreferenceFragment {

    @Override
    public void onCreate(Bundle savedInstanceState) {
        super.onCreate(savedInstanceState);

        // Load the preferences from an XML resource
        addPreferencesFromResource(R.xml.preferences);
    }
}
```

The activity_settings.xml file

```
<LinearLayout xmlns:android="http://schemas.android.com/apk/res/android"
    android:orientation="horizontal"
    android:layout_width="match_parent"
    android:layout_height="match_parent">

    <fragment android:name="com.murach.tipcalculator.SettingsFragment"
              android:id="@+id/settings_fragment"
              android:layout_weight="1"
              android:layout_width="0dp"
              android:layout_height="match_parent" />

</LinearLayout>
```

The SettingsActivity class

```
package com.murach.tipcalculator;

import android.app.Activity;
import android.os.Bundle;

public class SettingsActivity extends Activity {

    @Override
    public void onCreate(Bundle savedInstanceState) {
        super.onCreate(savedInstanceState);

        // set the view for the activity using XML
        setContentView(R.layout.activity_settings);
    }
}
```

Description

- Since the user interface is built using Preference objects instead of View objects, your fragment needs to extend the PreferenceFragment class instead of the Fragment class. Then, it can use the addPreferencesFromResource method to add the preferences defined in the XML file to the fragment.

Figure 9-7   How to work with preference fragments

How to use multi-pane layouts for large screens

Now that you understand how to use fragments in a single-pane layout, you're ready to learn how to use them in a multi-pane layout. This type of layout is commonly used for devices that have large screens such as tablets.

How to add multiple fragments to a layout

Figure 9-8 shows how to add multiple fragments to a layout. To do that, you can add two or more fragment elements to a layout file.

The first code example is designed to display two fragments when the screen is in landscape mode. To do that, this code uses a linear layout with its orientation set to horizontal. Then, it uses two fragment elements to display the Tip Calculator and Settings fragments. Both of these fragments have their layout_weight attributes set to a value of 1. As a result, they split the width of the screen equally.

The second code example is designed to display two fragments when the screen is in portrait mode. To do that, this code uses a linear layout with its orientation set to vertical. Then, it uses two fragment elements to display the Tip Calculator and Settings fragments. Both of these fragments have their layout_weight attributes set to a value of 1. As a result, they split the height of the screen equally.

Two layout files for the main activity

res\layout\activity_main_twopane_land.xml

```xml
<LinearLayout xmlns:android="http://schemas.android.com/apk/res/android"
    android:orientation="horizontal"
    android:layout_width="match_parent"
    android:layout_height="match_parent">

    <fragment android:name="com.murach.tipcalculator.TipCalculatorFragment"
            android:id="@+id/main_fragment"
            android:layout_weight="1"
            android:layout_width="0dp"
            android:layout_height="match_parent" />

    <fragment android:name="com.murach.tipcalculator.SettingsFragment"
            android:id="@+id/settings_fragment"
            android:layout_weight="1"
            android:layout_width="0dp"
            android:layout_height="match_parent" />

</LinearLayout>
```

res\layout\activity_main_twopane_port.xml

```xml
<LinearLayout xmlns:android="http://schemas.android.com/apk/res/android"
    android:orientation="vertical"
    android:layout_width="match_parent"
    android:layout_height="match_parent">

    <fragment android:name="com.murach.tipcalculator.TipCalculatorFragment"
            android:id="@+id/main_fragment"
            android:layout_weight="1"
            android:layout_height="0dp"
            android:layout_width="match_parent" />

    <fragment android:name="com.murach.tipcalculator.SettingsFragment"
            android:id="@+id/settings_fragment"
            android:layout_weight="1"
            android:layout_width="match_parent"
            android:layout_height="0dp" />

</LinearLayout>
```

Description

- To add more than one fragment to a layout, add two or more fragment elements to a layout file.

Figure 9-8   How to add multiple fragments to a layout

How to detect large screens

When the Tip Calculator app starts, it launches the Tip Calculator activity, which displays the activity_main layout that's in the layout directory. However, for large screens, you probably want to display one of the two-pane layouts described in the previous figure. To do that, you can detect large screens by creating a values directory that includes a large or xlarge qualifier as shown in figure 9-9.

In this figure, the first example is stored in this directory:

```
res\values-large-land
```

This directory uses two qualifiers. First, it uses the large qualifier to indicate that this directory contains a layout for devices with large screens. Second, it uses the land qualifier to indicate that this directory contains a layout for devices that are in landscape orientation.

Within this directory, the layout.xml file contains code that specifies which layout to use. To do that, it uses an item element to create an *alias*, which is a file that points to another resource file. You can often use an alias to reduce code duplication.

To specify an alias for a layout, code an item element. Then, set the name attribute to the name of the alias and set the type attribute of the item to a value of "layout". Next, use the body of the element to point to the correct layout that's in the project's layout directory. In the first example, the alias named activity_main points to the layout named activity_main_twopane_land. As a result, if the screen is large and in landscape orientation, the Tip Calculator app displays the layout named activity_main_twopane_land, instead of the activity_main layout, which is what you want.

The second example works the same as the first example. However, the values directory uses the port qualifier. Then, the alias points to a layout that's appropriate for portrait orientation. As a result, this code is only used when the screen is in portrait orientation.

When you use the large qualifier, you should be aware that it only works with Android 3.0 and higher. Since most tablets use Android 3.0 and higher, that's usually fine.

Two qualifiers

Qualifier	Screen size	Typical device
large	At least 640dp x 480dp	5" and 7" tablets
xlarge	At least 960dp x 720dp	10" tablet

The layout files for devices with large screens

res\values-large-land\layout.xml

```
<?xml version="1.0" encoding="utf-8"?>
<resources>
    <item name="activity_main" type="layout">
        @layout/activity_main_twopane_land
    </item>
</resources>
```

res\values-large-port\layout.xml

```
<?xml version="1.0" encoding="utf-8"?>
<resources>
    <item name="activity_main" type="layout">
        @layout/activity_main_twopane_port
    </item>
</resources>
```

Description

- For devices with small screens, Android uses a layout in the layout directory.
- To detect large screens, you can create a values directory that uses the large or xlarge qualifiers.
- To detect landscape and portrait orientations, you can create a values directory that uses the land or port qualifiers.
- Within a values directory, you can use an *alias* to point to a layout file that's stored in a layout directory. This provides a way to avoid duplicating code in multiple layout files.
- To specify an alias for a layout, code an item element. Then, set its name attribute to the name of the layout that you want to replace and set its type attribute to a value of "layout". Next, use the body of the element to point to the correct XML file in the project's layout directory.
- The large and xlarge qualifiers work with Android versions 3.0 and later. As a result, they aren't available from earlier versions of Android.

Figure 9-9   How to detect large screens

How to detect screen width

One problem with the large qualifier is that it typically selects a wide range of screen sizes. For example, it typically selects a 5" tablet as well as a 7" tablet. As a result, the developer still faces the difficult task of designing user interfaces that work with a wide range of screen sizes. To give developers more control, Android added the smallest-width qualifier in version 3.2 (API 13).

Since earlier versions of Android don't recognize the smallest-width qualifier, it's typical for a project to use the qualifiers described in the previous figure as well as the smallest-width qualifier described in figure 9-10. That way, Android uses the smallest-width qualifier for API 13 and higher, and it uses the large qualifier for API 11 and 12.

The smallest-width qualifier works much like the large qualifier described in the previous figure. In this figure, the first example is stored in this directory:

```
res\values-sw600dp-land
```

This directory uses a smallest-width qualifier of "sw600dp" to indicate that this layout should be used for screens with a minimum width of 600dp. This is the typical smallest width of a 7" tablet. In addition, this directory uses the land qualifier to specify that this layout should be used for screens in landscape orientation.

If necessary, you can define several smallest-width qualifiers like this:

```
res\values-sw480dp
```

and this:

```
res\values-sw600dp
```

and this:

```
res\values-sw720dp
```

Then, Android selects the layout based on the screen's smallest width. For example, a screen that's 700dp x 500dp would use the the layout specified by the values-sw480dp directory. Similarly, a screen that's 1024dp x 600dp would use the layout specified by the values-sw600dp directory.

Smallest-width qualifiers examples

Qualifier	Typical screen size	Typical device
sw480dp	640dp x 480dp	5" tablet
sw600dp	1024dp x 600dp	7" tablet
sw720dp	960dp x 720dp	10" tablet

The layout files for devices with a minimum screen size

res\values-sw600dp-land\layout.xml
```
<?xml version="1.0" encoding="utf-8"?>
<resources>
    <item name="activity_main" type="layout">
        @layout/activity_main_twopane_land
    </item>
</resources>
```

res\values-sw600dp-port\layout.xml
```
<?xml version="1.0" encoding="utf-8"?>
<resources>
    <item name="activity_main" type="layout">
        @layout/activity_main_twopane_port
    </item>
</resources>
```

Description

- The smallest-width qualifier allows you to select screens whose smallest width is at least as wide as specified width.

- To detect devices with screens that have a minimum width, you can create a values directory that uses the smallest-width qualifier. For example, a qualifier of sw600dp specifies a device that has a minimum smallest width of 600dp.

- The smallest-width qualifier was introduced in Android 3.2 (API 13). As a result, it isn't available with earlier versions of Android.

- Since earlier versions of Android don't recognize the smallest-width qualifier, you typically use the large qualifier as well.

Figure 9-10   How to detect screen width

How to control the soft keyboard

In figure 9-11, the EditText element contains a requestFocus element. As a result, when the app starts on a device with a large screen, Android moves the focus into this EditText widget. Then, if there is enough room, Android displays the soft keyboard for that EditText widget. If that isn't what you want, you can delete the requestFocus element.

Another common problem is that Android doesn't always display the correct action button on the soft keyboard by default. For example, it may display a Next button instead of a Done button. To fix this problem, you can use the imeOptions attribute of an EditText element to specify the correct action button for the soft keyboard. In this figure, for example, the imeOptions attribute specifies a Done button for the EditText widget.

Some values for the imeOptions attribute

Setting	The EditText widget displays a...
`actionDone`	Done button
`actionNext`	Next button
`actionPrevious`	Previous button
`actionGo`	Go button
`actionSeach`	Search button

Attributes of an EditText widget that can control the soft keyboard

```
<EditText
    android:id="@+id/billAmountEditText"
    android:layout_width="wrap_content"
    android:layout_height="wrap_content"
    android:layout_alignBaseline="@+id/billAmountLabel"
    android:layout_marginLeft="5dp"
    android:layout_toRightOf="@+id/billAmountLabel"
    android:ems="8"
    android:inputType="numberDecimal"
    android:imeOptions="actionDone"
    android:text="@string/bill_amount"
    android:textSize="20sp" >

    <requestFocus />
</EditText>
```

Description

- If you don't want Android to display the soft keyboard when the app starts, you can delete the requestFocus element from the body of the EditText element.
- If Android does not display the correct action button on the soft keyboard, you can use the imeOptions attribute of the EditText element to specify the correct action button on the soft keyboard.

Figure 9-11   How to control the soft keyboard

Other skills for working with fragments

So far, this chapter has presented the skills that you typically need for working with fragments. However, in some situations, you may need the skills presented in the next two figures to work with fragments.

How to get a reference to a fragment

Figure 9-12 shows how to get a reference to a fragment object. Once you have a reference to a fragment object, you can call any of its public methods. Or, if you aren't able to get a reference, you can take the appropriate action such as displaying the appropriate menu.

The onCreateOptionsMenu of the Tip Calculator fragment displays the appropriate options menu depending on whether the Settings fragment has been created. To do that, the first statement calls the getFragmentManager method to return a FragmentManager object. This method can be called from within an activity or a fragment. In this case, it's called from within a fragment.

After getting the FragmentManager object, this code calls the findFragmentById method from that object to find the Settings fragment. Then, it casts the Fragment object that's returned to the SettingsFragment type.

After the first statement, this code checks whether the manager was able to get the fragment. If so, this code displays a menu that includes a Settings item. Otherwise, it displays a menu that does not include a Settings item, which is usually what you want if the Settings fragment is already displayed.

The onSharedPreferenceChanged method of the Settings fragment calls a method from the Tip Calculator fragment. To do that, the first statement attempts to get the Tip Calculator fragment. Since this code works like the code in the first example, you shouldn't have much trouble understanding it.

After the first statement, this code checks whether the manager was able to get the fragment. If so, it calls the onResume method of this fragment to refresh the user interface so it reflects the latest changes to the settings. Without this statement, the user interface of the Tip Calculator fragment won't refresh when you use the Settings fragment to change settings.

Method of the Activity and Fragment classes

Method	Description
`getFragmentManager()`	Returns a FragmentManager object.

Method of the FragmentManager class

Method	Description
`findFragmentById(id)`	Returns a Fragment object for the fragment with the specified ID. If a Fragment object doesn't exist for that ID, this method returns a null value.

The onCreateOptionsMenu method of the Tip Calculator fragment

```
@Override
public void onCreateOptionsMenu(Menu menu, MenuInflater inflater) {

    // attempt to get the fragment
    SettingsFragment settingsFragment = (SettingsFragment)
            getFragmentManager()
            .findFragmentById(R.id.settings_fragment);

    // if the fragment is null, display the appropriate menu
    if (settingsFragment == null) {
        inflater.inflate(R.menu.fragment_tip_calculator, menu);
    } else {
        inflater.inflate(R.menu.fragment_tip_calculator_twopane, menu);
    }
}
```

The onSharedPreferenceChanged method of the Settings fragment

```
@Override
public void onSharedPreferenceChanged(SharedPreferences prefs,
        String key) {

    // attempt to get the fragment
    TipCalculatorFragment tipFragment =
        (TipCalculatorFragment) getFragmentManager()
            .findFragmentById(R.id.main_fragment);

    // if the fragment is not null, call a method from it
    if (tipFragment != null) {
        tipFragment.onResume();
    }
}
```

Description

- You can use the FragmentManager object to get a reference to a Fragment object.

Figure 9-12   How to get a reference to a fragment

How to replace one fragment with another

Figure 9-13 shows how to use Java code to replace the content of a container with a fragment. Here, the fragment named SettingsFragment replaces any other fragments that might be in the content container for the activity.

To do that, this code uses the getFragmentManager method to get a FragmentManager object. From that object, this code calls the beginTransaction, replace, and commit methods. These methods start a transaction, replace the content of the activity with the fragment, and commit the transaction. Here, the replace method uses this Android resource:

```
android.R.id.content
```

to identify the container for the content of the activity, and it creates a new SettingsFragment object to replace that content.

The code in this figure accomplishes the same task as the code shown in the second and third examples of figure 9-6. So, which approach is better? This is largely a matter of personal choice. In this book, I use the approach taken in figure 9-6 because I think XML is easier to understand when you're first getting started. However, this figure shows you often have a choice between using XML or Java to accomplish the same task. As your applications grow more complex, you may want or need to use Java to manage your fragments.

Method of the FragmentManager class

Method	Description
beginTransaction()	Begins a transaction and returns a FragmentTransaction object.

Methods of the FragmentTransaction class

Method	Description
replace(id, fragment)	Replaces the fragment in the container with the specified ID with an instance of the specified Fragment object.
commit()	Finishes the transaction.

The SettingsActivity class

```
package com.murach.tipcalculator;

import android.app.Activity;
import android.os.Bundle;

public class SettingsActivity extends Activity {

    @Override
    public void onCreate(Bundle savedInstanceState) {
        super.onCreate(savedInstanceState);

        // display the fragment as the main content
        getFragmentManager().beginTransaction()
                .replace(android.R.id.content, new SettingsFragment())
                .commit();
    }
}
```

Description

- You can use a FragmentManager object to replace one fragment with another fragment.

Figure 9-13   How to replace one fragment with another

Perspective

The skills presented in this chapter should be enough to get you started with fragments. However, Android apps can use fragments in many ways. As you progress through this book, you'll encounter other examples of apps that use fragments. For example, chapter 14 shows how to use fragments in a way that's common for many apps. As a result, studying that chapter should help to broaden your understanding of fragments.

Terms

fragment	multi-pane layout
panes	support library
single-pane layout	alias

Summary

- You can use a *fragment* to define part of the user interface for an activity. A fragment is sometimes referred to as a *pane*.
- On a small screen, an activity typically only displays a single fragment. This is known as a *single-pane layout*.
- On a large screen, an activity can display multiple fragments. This is known as a *multi-pane layout*.
- Fragments are available from Android 3.0 (API 11) and higher.
- Android provides *support libraries* that you can add to your project. These libraries make new APIs such as fragments available on older versions of Android.
- The XML and Java code for a fragment works much like the XML and Java code for an activity.
- To add a fragment to a layout, add a fragment element and use its name attribute to specify the fully qualified name for the class that defines the fragment.
- If the user interface for a fragment uses Preference objects instead of View objects, the fragment needs to extend the PreferenceFragment class instead of the Fragment class.
- To add more than one fragment to a layout, add two or more fragment elements to a layout file.
- To detect large screens, you can create a values directory that uses the large or xlarge qualifiers.
- To detect landscape and portrait orientations, you can create a values directory that uses the land or port qualifiers.

- Within a values directory, a layout can use an *alias* to point to a layout file that's stored in a layout directory. This provides a way to avoid duplicating code in multiple layout files.

- The smallest-width qualifier allows you to select screens whose smallest width is at least as wide as specified width.

- From a fragment or activity, you can call the getFragmentManager method to get a FragmentManager object that you can use to find and work with the fragments in your app.

Exercise 9-1   Test the app

In this exercise, you'll experiment with the fragments that are available from the Tip Calculator app.

1. Start Eclipse and import the project named ch09_ex1_TipCalculator.

2. Expand the libs directory. It should include the android-support-v4.jar that contains the support library.

3. Open the AndroidManifest.xml file. The minimum SDK for this project should be API 11 (Android 3.0). That's because the PreferenceFragment class isn't included in the support library.

4. Expand the src directory. This directory should contain the source code for three activities and two fragments.

5. Expand the res directory. This directory should contain the values directories that use a combination of these qualifiers: large, land, port, and smallest-width.

6. Expand the res\layout directory. This directory should contain five activity layouts. Of these layouts, three are for the "main" activity, the Tip Calculator activity. These layouts include both two-pane layouts described in this chapter.

7. Expand the res\values-sw600dp-port directory. Then, open the layout.xml file. This file should use an alias to display the two-pane portrait layout for the Tip Calculator activity.

8. Run the app on a phone or an emulator for a phone. This should use the Tip Calculator activity to display the Tip Calculator fragment, and it should use the Settings activity to display the Settings fragment.

9. Run the app on a tablet or an emulator for a tablet. If necessary, you can create an emulator for a tablet as described in chapter 4. This should use the Tip Calculator activity to display both the Tip Calculator and Settings fragments.

10. Switch the orientation. This should still display both fragments, but it should use a different layout.

11. Test the app to make sure it works correctly when displayed in a tablet. It should. If it doesn't, fix any problems you encounter.

Exercise 9-2   Create a new fragment

In this exercise, you'll create a third fragment (the About fragment) and display it.

1. Start Eclipse and import the project named ch09_ex2_TipCalculator.

2. In the res\layout directory add a layout named fragment_about. Then, copy the XML from the activity_about layout into the layout for the fragment.

3. Add a class to the src directory named AboutFragment. Then, write the code for this activity so it displays the fragment_about layout. This class should be a regular fragment, not a preference fragment.

4. Open the activity_about layout in the res\layout directory and modify it so it uses a fragment element to display the AboutFragment class. This fragment element's height and width should wrap the content.

5. Run the app. It should work just as it did before.

6. Open the activity_main_twopane_port layout in the res\layout directory and modify it so it displays the About fragment after the other two fragments.

7. Run the app on a tablet that has a large screen. In portrait orientation, the app should display all three fragments.

8. Open the activity_main_twopane_land layout and modify it so it displays the About fragment below the Settings fragment. To do that, you can nest both of these fragments in a linear layout that has vertical orientation.

9. Run the app on a tablet. The app should display all three fragments in both orientations. In addition, both orientations should allow you to access an Options menu that contains an About item.

10. Open the Java code for the TipCalculatorFragment. In the onCreateOptionsMenu method, modify the code so it doesn't inflate a menu if you're using one of the "two-pane" layouts.

Exercise 9-3   Use the fragment manager

In this exercise, you'll use a different technique to display a fragment in an activity.

1. Start Eclipse and import the project named ch09_ex3_TipCalculator.

2. Open the .java file for the Tip Calculator activity. Modify this code so it uses the FragmentManager object to display the Tip Calculator fragment.

3. Run the app to make sure it still works correctly.

4. Open the .java file for the Settings activity. Modify this code so it uses the FragmentManager object to display the Settings fragment.

5. Run the app to make sure it still works correctly.

Section 3

The News Reader app

In sections 1 and 2, you learned the essential skills for developing a simple Android app, the Tip Calculator app. Now, this section shows how to develop another app, the News Reader app.

Along the way, this section presents many essential Android skills that can be applied to other apps. To start, chapter 10 shows how to build a simple version of a News Reader app that uses threads, files, a simple adapter, and intents. Then, chapter 11 shows how to modify this app so it uses services and notifications. Finally, chapter 12 shows how to modify this app so it uses broadcast receivers.

10

How to work with threads, files, adapters, and intents

In this chapter, you'll learn how to create a News Reader app that downloads news items from the web and displays them so they can be read by the user. To be able to create such an app, you need to learn several new skills including how to work with threads, files, adapters, and intents.

An introduction to the News Reader app

The skills presented in this chapter are necessary to create the News Reader app. To put these skills in context, this chapter begins by showing the user interface for this app. Then, it shows the XML file that contains the data for this app.

The user interface

Figure 10-1 shows the user interface for the News Reader app. When the app starts, the top of the Items activity displays the title of the news feed. Below that, the Items activity displays a list of items in the news feed. Here, each item has a publication date and a title.

If the user clicks on an item in the Items activity, the app displays the Item activity, which displays more information about the clicked item. Here, each item has a title, a publication date, a description, and a link to the original article that says "Read more on thc wcb".

If the user clicks on this link, the app starts the default web browser for the device. Then, it displays the link in that browser. Although this exits the News Reader app, the user can easily use the Back button to navigate back to the News Reader app.

The Items activity

The Item activity

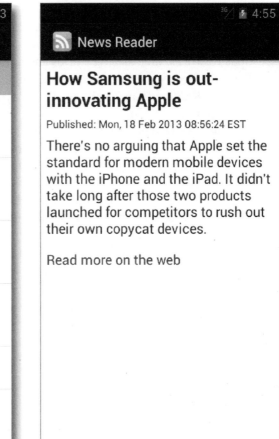

Description

- When the app starts, the Items activity displays the title of the news feed followed by a list of items in the news feed where each item has a publication date and a title.

- If the user clicks on a news item, the Item activity displays more information about that item, including a link that can display the original article in a web browser.

Figure 10-1   The user interface for the News Reader app

The XML for an RSS feed

Figure 10-2 shows the URL for the *RSS (Rich Site Summary) feed* that's used by the News Reader app. A web site can use an RSS feed to publish frequently updated works, such as blog entries and news headlines. Then, an app can read that RSS feed. Since RSS uses a standardized XML file format, an app that can read one RSS feed can usually read other types of RSS feeds with little or no modification.

The RSS feed that's returned by the URL in this figure is actually much more complex than the XML shown in this figure. However, the XML shown in this figure shows the parts of the RSS feed that are used by the News Reader app.

To start, there are title and pubDate elements that come before any item elements. These elements store the title and publication data for the feed.

After the elements for the feed, there are a series of item elements. Each item element contains title, link, description, and pubDate elements. These elements store the title, link, description, and publication date for each item in the feed.

The URL for the RSS feed

```
http://rss.cnn.com/rss/cnn_tech.rss
```

Simplified XML for the RSS feed

```xml
<rss xmlns:media="http://search.yahoo.com/mrss/"
    xmlns:feedburner="http://rssnamespace.org/feedburner/ext/1.0"
    version="2.0">
<channel>
  <title>CNN.com - Technology</title>
  <pubDate>Mon, 18 Feb 2013 14:11:56 EST</pubDate>
  <item>
    <title>How Samsung is out-innovating Apple</title>
    <link>http://rss.cnn.com/~r/rss/cnn_tech/~3/N9m_DSAe5rY/index.html
      </link>
    <description>There's no arguing that Apple set the standard for
      modern mobile devices with the iPhone and the iPad. It didn't take
      long after those two products launched for competitors to rush out
      their own copycat devices.
    </description>
    <pubDate>Mon, 18 Feb 2013 08:56:24 EST</pubDate>
  </item>
  <item>
    <title>Nude scammers blackmail men online</title>
    <link>http://rss.cnn.com/~r/rss/cnn_tech/~3/gbgkR_pqCvg/index.html
      </link>
    <description>Who would have thought that getting naked and naughty
      with a stranger online could have negative consequences
      </description>
    <pubDate>Mon, 18 Feb 2013 14:11:05 EST</pubDate>
  </item>
  <item>
    <title>Iceland wants to ban Internet porn</title>
    <link>http://rss.cnn.com/~r/rss/cnn_tech/~3/oVgnzhw61PU/index.html
      </link>
    <description>Iceland is working on banning Internet pornography,
      calling explicit online images a threat to children.
      </description>
    <pubDate>Fri, 15 Feb 2013 12:39:43 EST</pubDate>
  </item>
  ...
  ...
</channel>
</rss>
```

Description

- An *RSS* (*Rich Site Summary*) *feed* can be used to publish frequently updated works, such as blog entries and news headlines.

- Since RSS uses a standardized XML file format, the feed can be published once and viewed by many different apps.

Figure 10-2   The XML for an RSS feed

How to work with threads

The News Reader app reads data from the Internet, it writes data to a file, and it reads data from that file. In an Android app, you should perform tasks like these in their own threads.

How threads work

As figure 10-3 explains, a *thread* is a single flow of execution through an app. By default, an Android app uses a single thread, called the *UI thread*. This thread displays the user interface for the app and listens for events that occur when the user interacts with the user interface.

The user interface for an app should always be responsive to the user. As a result, you don't want to do any processing on the UI thread that takes more than a second or so. For example, downloading data from the Internet may take several seconds or even minutes depending on the amount of data and speed of the connection. As a result, you should never perform this type of processing on the UI thread. Instead, you should create a second thread.

Similarly, if you want to perform a file I/O operation such as reading or writing a file, you should do that processing in a separate thread. That's because I/O operations are thousands of times slower than CPU operations. So any program that reads data from a disk spends much of its time waiting for that information to be retrieved.

The first diagram shows how an app might work when executed as a single thread. First, the UI thread displays the user interface and listens for events. Then, an event occurs that causes the program to perform an I/O operation. Since this I/O operation runs in the UI thread, the UI thread can't listen for events and respond to the user while the I/O thread is running. In other words, if the user clicks on a button while the I/O operation is running, the UI thread can't respond to that event. When the I/O operation completes, the user interface becomes responsive again.

The second diagram shows how this app benefits from using a separate thread to perform the I/O operation. This allows the two tasks to overlap. As a result, the UI thread can continue listening for events while the I/O operation runs. The result is a user interface that's always responsive.

How using threads improves user interface responsiveness

One thread

UI thread | Listen | Perform I/O | Listen |

Time ⟶

Two threads

UI thread | Listen | Listen | Listen |

Start thread ↓   Update UI ↑

I/O thread | Perform I/O |

Time ⟶

Description

- A *thread* is a single sequential flow of control within a program. A thread often completes a specific task.

- By default, an Android app uses a single thread, called the *UI thread*, to display the user interface. Any task that can slow or stop the responsiveness of the UI thread should be run in a separate thread.

Figure 10-3   How threads work

How to execute asynchronous tasks

An *asynchronous task* is a task that runs in a separate thread in the background and does not need to be synchronized with other threads. Since Android apps often need to perform asynchronous tasks, the Android framework includes a class named AsyncTask that makes it easy to perform these types of tasks. This class handles the code that creates and manages the thread for you. As a result, you just need to create a class that inherits the AsyncTask class and override the appropriate methods.

Since an asynchronous class is usually closely related to an activity class, it's usually coded as a nested inner class of the activity class. In part 1 of figure 10-4, for example, the DownloadFeed class defines an asynchronous task that's closely related to the ItemsActivity class. As a result, the DownloadFeed class is nested within the ItemsActivity class.

The ItemsActivity class begins by defining a constant for a string that contains the URL for the RSS feed described earlier in this chapter. Then, this class defines an onCreate method. This method starts by displaying the layout for the news items. Then, it creates an instance of the DownloadFeed object and calls its execute method to start its thread. In addition, this code uses the execute method to pass the URL string to the DownloadFeed object.

The DownloadFeed class extends the AsyncTask class. This class uses *generics*, which is a feature of Java that allows a class to operate on various types of objects. In particular, the declaration of the AsyncTask class allows it to work with three generic types: (1) parameters, (2) progress, and (3) result. As the programmer, you can decide what types of objects you want to use with this class. In this figure, the DownloadFeed class uses an AsyncTask class like this:

```
AsyncTask<String, Void, String>
```

to indicate that it accepts one or more string parameters, doesn't handle progress values, and returns a string.

Within the DownloadFeed class, the doInBackground method contains the code that runs in the background thread. This method accepts the array of String objects as specified by the first generic type. Since this activity only passes one string value, the first statement in this method gets that string from the params variable. At this point, you are ready to download the data from the feed specified by the URL. You'll learn how to do that later in this chapter. Finally, this method returns a message in a String object as specified by the third generic type.

After the doInBackground method finishes, the onPostExecute method runs. This method accepts a String object as specified by the third generic type. This method contains code that updates the user interface. For now, this method uses a toast to display the result string that's returned by the doInBackground method. Later in this chapter, you'll see this method can be used to update the user interface.

An activity with a nested AsyncTask class

```
package com.murach.newsreader;

import android.os.AsyncTask;
import android.os.Bundle;
import android.app.Activity;
import android.content.Context;
import android.widget.Toast;

public class ItemsActivity extends Activity {

    private static String URL_STRING =
            "http://rss.cnn.com/rss/cnn_tech.rss";

    @Override
    protected void onCreate(Bundle savedInstanceState) {
        super.onCreate(savedInstanceState);
        setContentView(R.layout.activity_items);

        new DownloadFeed().execute(URL_STRING);
    }

    class DownloadFeed extends AsyncTask<String, Void, String> {
        @Override
        protected String doInBackground(String... params) {
            // get the parameter
            String urlString = params[0];

            // download the feed and write it to a file

            // return a message
            return "Feed downloaded";
        }

        @Override
        protected void onPostExecute(String result) {
            Context context = ItemsActivity.this;
            Toast.makeText(context, result, Toast.LENGTH_LONG).show();
        }
    }
}
```

Description

- An *asynchronous task* is a task that runs in a separate thread in the background and does not need to be synchronized with other threads.
- It's common to use a nested inner class to create an asynchronous thread. Within the inner class, you can access the context for the activity by coding the name of the outer class, a period, and the *this* keyword.

Figure 10-4   How to execute asynchronous tasks (part 1 of 2)

When you nest one class within another, the inner class can access the object for the outer class by coding the name of the outer class, a period, and the *this* keyword. In this figure, for example, the onPostExecute method of the inner class uses this code:

```
ItemsActivity.this
```

to get the ItemsActivity object. Then, it stores the ItemsActivity object in a Context object, which is later used by the makeText method of the Toast object.

Part 2 of this figure shows more details about how the AsyncTask class works. To start, it shows several examples for how you could declare a class that extends the AsyncTask class.

The first example declares a DownloadFeed class that extends the AsyncTask class. This class accepts one or more string parameters, uses int values to update the progress of the thread, and returns a string. In this case, the doInBackground method would accept an array of String objects and return a String object. Similarly, the onPostExecute method would accept the String object returned by the doInBackground method. For primitive types such as the int type, you must use the corresponding wrapper class such as the Integer class.

The second example accepts a URL object as a parameter, doesn't update the progress of the thread, and returns a string. This shows that you can use any object type as a generic type. In addition it shows that you can use the Void keyword if your class doesn't use one of the generic types.

The third example shows that you can use the Void keyword for all three generic types. This can make sense when the class for the asynchronous task works with instance variables. In that case, it often doesn't need to accept parameters or return data. In this case, the doInBackground method would accept an array of Void objects and return a Void object. To do that, this method can return a null value.

The fourth example declares a ReadFeed class that extends the AsyncTask class. This class accepts one or more string parameters, uses int values to update the progress of the thread, and returns an RSSFeed object. The code for the RSSFeed class is shown later in this chapter.

This figure also shows the four methods of the AsyncTask class that you can override. Of these four methods, the code within the doInBackground method executes in a background thread. As a result, you should put any network or I/O code in this thread.

The other three methods execute in the UI thread. As a result, you can use them to update the user interface. Typically, you use the onPreExecute method to display a ProgressBar widget that displays the progress of the background thread. Then, you can use the onProgressUpdate method to update the ProgressBar widget every time the background thread uses the publishProgress method to update its progress. Finally, you can use the onPostExecute method to update the user interface when the background thread finishes.

The generic types for the AsyncTask class

```
AsyncTask<Params, Progress, Result>
```

Possible class declarations

```
class DownloadFeed extends AsyncTask<String, Integer, String> { ... }
class DownloadFeed extends AsyncTask<URL, Void, String> { ... }
class DownloadFeed extends AsyncTask<Void, Void, Void> { ... }
class ReadFeed extends AsyncTask<String, Integer, RSSFeed> { ... }
```

The AsyncTask class

Method	Is executed...
`onPreExecute()`	On the UI thread before the task is executed.
`doInBackground(Params...)`	On the background thread immediately after the onPreExecute method finishes. An array of parameters is passed to this method. This method returns a result that's passed to the onPostExecute method. If this method uses the publishProgress method, it passes progress values to the onProgressUpdate method.
`onProgressUpdate(Progress...)`	On the UI thread after a call to the publishProgress method is made in the doInBackground method.
`onPostExecute(Result)`	On the UI thread after the doInBackground method finishes.

Description

- The AsyncTask class uses *generics* to allow a class to operate on various types of objects. The AsyncTask class provides for three generic types: (1) parameters, (2) progress, and (3) result.

- The AsyncTask class provides an easy way to perform a background task and publish results on the UI thread without having to manually manipulate threads.

Figure 10-4   How to execute asynchronous tasks (part 2 of 2)

How to execute timed tasks

Android apps sometimes need to start a separate thread to execute a timed task, which is a task that executes after a specified delay or at a specified interval. For the News Reader app, for example, you may want to check for updates every hour. For a Stopwatch app, you may want to update the user interface every second. To do that, you can use the Timer and TimerTask classes as shown in figure 10-5. These Java classes provide a way to start a thread that executes tasks at a specified interval.

The code example in this figure shows a startTimer method that you can use to start a timed task. This method begins by creating a variable named startMillis that stores the number of milliseconds for the current time. To get these milliseconds, this code calls the currentTimeMillis method from the System class.

After getting the starting time, this code creates a new TimerTask object named task. To be able to create this object, you must override the abstract run method that's declared in the TimerTask class. Then, you can store the code that executes the task within this method. In this figure, for example, the run method contains two statements. The first statement gets the number of milliseconds that have elapsed since the timer was started. Then, the second statement passes the number of elapsed milliseconds to a method named updateView. This method updates the user interface and is shown in the next figure.

After creating the TimerTask object, this code creates a Timer object and uses it to execute the TimerTask object. To start, the first statement creates a Timer object that starts a *daemon thread*, which is a thread that ends when the app ends. This is the most common type of thread for Android apps. However, you can also start a regular thread that would continue running even if the app ended. Then, the second statement calls the schedule method of the Timer object to execute the task immediately (after 0 milliseconds) and to continue executing it every second (1000 milliseconds). As a result, the task continues to execute every second until the timer is cancelled or until the app ends.

Sometimes, you may want to execute a task only once. To do that, you don't specify a repeat interval for the schedule method. For example, to execute a task once after a delay of 2 hours, you could use this code:

```
task.schedule(task, 1000 * 60 * 60 * 2);
```

This code calculates the number of milliseconds for 2 hours by multiplying the number of milliseconds in 1 second (1000) by the number of seconds in a minute (60) by the number of minutes in an hour (60) by the number of hours (2).

The classes used to work with timed tasks

```
java.util.Timer
java.util.TimerTask
```

A method that starts a timed task

```java
private void startTimer() {
    final long startMillis = System.currentTimeMillis();
    TimerTask task = new TimerTask() {

        @Override
        public void run() {
            long elapsedMillis =
                System.currentTimeMillis() - startMillis;
            updateView(elapsedMillis);
        }
    };
    Timer timer = new Timer(true);
    timer.schedule(task, 0, 1000);  // execute every second
}
```

The TimerTask class

Constructor/Method	Description
TimerTask()	Creates a new TimerTask object.
run()	An abstract method of the TimerTask class. You can override this method and execute the code for the task within it.

The Timer class

Constructor/Method	Description
Timer(isDaemon)	Creates a Timer object and specifies whether its thread is a *daemon thread* that ends when the app ends.
schedule(task, delay)	Executes the specified TimerTask object once, after the specified delay in milliseconds.
schedule(task, delay, interval)	Executes the specified TimerTask object repeatedly, after the specified delay in milliseconds at the specified interval in milliseconds.
cancel()	Cancels this timer, discarding any currently scheduled tasks.

Description

- You can use the TimerTask and Timer classes to create a thread that executes tasks after a specified delay or at a specified interval.

Figure 10-5   How to execute timed tasks

How to update the UI thread

If you create a timed task, that task runs in its own thread in the background. As the background thread runs, you need a way to update the UI thread. To do that, you can use the post method that's available from a View object as shown in figure 10-6. This method accepts any object that implements the Runnable interface.

The updateView method in this figure begins by declaring that it accepts a parameter for a long value. This parameter contains the number of elapsed milliseconds. In addition, this parameter must be declared as final so it can be used within the UI thread.

Within this method, the first statement calls the post method on a TextView widget. Within the parentheses for this method, this code uses an anonymous class to create an instance of a Runnable object.

This anonymous class begins by calculating the number of elapsed seconds by dividing the number of elapsed milliseconds by 1000. Then, the anonymous class implements the run method of the Runnable interface. This method contains a single statement that uses the setText method of a TextView widget to update the text that's displayed on the widget. This text includes the number of seconds that have elapsed since the timed task was started.

A method that updates the UI thread

```
private void updateView(final long elapsedMillis) {
    // UI changes need to be run on the UI thread
    messageTextView.post(new Runnable() {

        int elapsedSeconds = (int) elapsedMillis/1000;

        @Override
        public void run() {
            messageTextView.setText("Seconds: " + elapsedSeconds);
        }
    });
}
```

The View class

Method	Description
post(runnable)	Executes the specified Runnable object.

The Runnable interface

Method	Description
run()	Contains the statements that are run within the thread.

Description

- You can use the post method of a View object to execute any Runnable object on the UI thread.

Figure 10-6   How to update the UI thread

How to work with files

Now that you know how to create and execute a thread, you're ready to learn how to write the code that's executed by the thread. This code often reads and writes files. For example, the News Reader app needs to read an RSS feed from the Internet and write that data to an XML file. Then, it needs to read that XML file and parse its data into a series of Java objects.

How to download a file from the Internet

Figure 10-7 shows how to download data from the Internet and store it in a file on an Android device. To do that, you get an input stream from the Internet, you get an output stream to a file on your device, and you use standard Java I/O methods to read data from the input stream and write that data to the output stream.

The code example begins by declaring a constant that stores the name, but not the path, of the XML file that stores the data for the news feed. Then, this code uses a try/catch statement to handle any I/O exceptions that may be thrown by the I/O code. If an IOException is thrown, the catch clause handles the exception by displaying it in the LogCat view.

Within the try clause, the first statement creates a URL object from a string that points to the URL for the RSS feed described earlier in this chapter. Then, the second statement calls the openStream method on the URL object to return an InputStream object that you can use to read the input stream.

To be able to read data from the Internet, you must include the INTERNET permission in the manifest file for your app. To do that, you can add a permission like the one shown in this figure at the same indentation level as the application element.

After getting the InputStream object, this code calls the openFileOutput method to get a FileOutputStream object. Calling the openFileOutput method is an application-level operation. However, you can call this method from an activity and Android automatically calls the method from the application level. This is true even if you call this method from an inner class of an activity.

The openFileOutput method accepts two parameters. The first parameter specifies the filename, but not the path, of the file. You can't specify a path because Android automatically uses the default path for the app. This path uses the package name of the app to give each app its own directory for storing files. If the filename doesn't exist in this directory, Android automatically creates the file for you and opens it. Otherwise, it opens the existing file.

The second parameter specifies the mode that Android uses when it opens the file. For this parameter, you can use any of the MODE_XXX constants that are available from the Context class. However, the MODE_PRIVATE constant is used in most cases. This constant only allows the current app to work with the file, which is usually what you want. In addition, if a file with the specified name already exists, this mode overwrites the existing file, which is also usually what you want.

The classes used to download a file from the Internet

```
java.net.URL
android.content.Context
```

How to download a file

```
final String FILENAME = "news_feed.xml";
try{
    // get the input stream
    URL url = new URL("http://rss.cnn.com/rss/cnn_tech.rss");
    InputStream in = url.openStream();

    // get the output stream
    FileOutputStream out =
            openFileOutput(FILENAME, Context.MODE_PRIVATE);

    // read input and write output
    byte[] buffer = new byte[1024];
    int bytesRead = in.read(buffer);
    while (bytesRead != -1)
    {
        out.write(buffer, 0, bytesRead);
        bytesRead = in.read(buffer);
    }
    out.close();
    in.close();
}
catch (IOException e) {
    Log.e("News reader", e.toString());
}
```

The URL class

Method	Description
`openStream()`	Returns an InputStream object for the specified URL.

The INTERNET permission in the AndroidManifest.xml file

```
<uses-permission android:name="android.permission.INTERNET" />
```

The Context class

Method	Description
`openFileOutput(filename, mode)`	Returns a FileOutputStream object for the specified file. If the file doesn't already exist, this method creates it. For the second parameter, use MODE_PRIVATE to overwrite any existing files.

Description

- You can use Java and Android APIs to download a file by reading input from the Internet and writing output to the file system.

Figure 10-7   How to download a file from the Internet

The MODE_PRIVATE constant stores a value of 0. As a result, the code in figure 10-7 can also be written like this:

```
FileOutputStream out = openFileOutput(FILENAME, 0);
```

Although this code is shorter, I prefer using the MODE_PRIVATE constant because it makes the code easier to read.

How to parse an XML file

Figure 10-8 shows how to read the XML file that was written in the previous figure and how to use *SAX (Simple API for XML)* to parse this XML into a series of Java objects. To do that, the code in this figure uses Java classes from the javax.xml.parsers and org.xml.sax packages.

The code example in this figure begins by declaring a constant for the name of the XML file that stores the data for the news feed. Then, the second statement declares an RSSFeed object named feed. The RSSFeed class is a custom class that's used to store the data for the news feed. The code for this class is shown later in this chapter. For now, all you need to know is that an RSSFeed object can store the data for a news feed.

After declaring the RSSFeed object, this code uses a try/catch statement to handle any exceptions that may be thrown by the code. If an exception is thrown, the catch block displays it in the LogCat view.

Within the try block, the first three statements get an XMLReader object that can be used to read an XML file. This is boilerplate code that you can cut and paste to get an XMLReader object. Then, this code creates an object from the RSSFeedHandler class. This class is another custom Java class that's shown later in this chapter. For now, all you need to know is that it contains the code that's used to parse the XML file. After creating the RSSFeedHandler object, the next statement sets this object as the content handler in the XMLReader object.

After setting up the content handler, this code uses the openFileInput method to get an input stream for the file. This method works much like the openFileOutput method described in the previous figure. However, it only requires one parameter, the filename.

Now that the XMLReader object and the input stream have been created, this code can parse the data. To do that, it creates an InputSource object from the FileInputStream object. Then, it calls the parse method from the XMLReader object and passes the InputSource object to it. This parses the data in the XML file and stores it in an RSSFeed object. Finally, this code calls the getFeed method of the RSSHandler class to return the RSSFeed object.

The classes used to work with SAX

```
javax.xml.parsers.SAXParser
javax.xml.parsers.SAXParserFactory

org.xml.sax.InputSource
org.xml.sax.XMLReader
```

How to parse an XML file

```java
final String FILENAME = "news_feed.xml";
RSSFeed feed;
try {
    // get the XML reader
    SAXParserFactory factory = SAXParserFactory.newInstance();
    SAXParser parser = factory.newSAXParser();
    XMLReader xmlreader = parser.getXMLReader();

    // set content handler
    RSSFeedHandler theRssHandler = new RSSFeedHandler();
    xmlreader.setContentHandler(theRssHandler);

    // get the input stream
    FileInputStream in = openFileInput(FILENAME);

    // parse the data
    InputSource is = new InputSource(in);
    xmlreader.parse(is);

    // get the content handler and return it
    feed = theRssHandler.getFeed();
}
catch (Exception e) {
    Log.e("News reader", e.toString());
}
```

The Context class

Method	Description
openFileInput(filename)	Returns a FileInputStream object for the specified file.

Description

- You can use *SAX* (*Simple API for XML*) to parse XML files.

Figure 10-8   How to parse an XML file

The RSSFeedHandler class

Figure 10-9 shows the RSSFeedHandler class. This class begins by extending the DefaultHandler class, which contains methods that you can override to handle the events that occur when the XML file is parsed. In particular, the DefaultHandler class defines the startDocument, endDocument, startElement, endElement, and characters methods that are overridden in the RSSFeedHandler class.

Within the RSSFeedHandler class, the first two statements define instance variables for the RSSFeed and RSSItem objects that are used to store the data for the feed and for the items within the feed. The code for the RSSFeed and RSSItem classes is presented in the next two figures. For now, all you need to know is that both classes have a default constructor as well as get and set methods that allow you to store data.

The next six statements define the Boolean instance variables that this class uses to determine when the various elements of the XML file are being parsed.

The getFeed method returns the RSSFeed object. This method can be called after the parser finishes parsing the XML file.

The startDocument method is executed when the parser starts reading the XML document. The code in this method creates the RSSFeed and RSSItem objects. As a result, you can use these objects in the other methods of this class.

The endDocument method is executed when the parser finishes reading this document. In this class, this method doesn't include any statements. However, if you needed to perform some processing when the parser finishes reading the document, you could add it here.

The startElement method is executed after the parser reads a start element such as <item>. The third parameter of this method contains the qualified name of the element. Within this method, an if/else statement checks whether the element is needed by the News Reader app. If so, this code takes the appropriate action. For the element named item, this code creates a new RSSItem object that can be used to store the data for the item. Then, it exits this method. For the element named title, this code sets isTitle variable to true. Then, it exits the method. And so on.

The RSSFeedHandler class   **Page 1**

```
package com.murach.newsreader;

import org.xml.sax.helpers.DefaultHandler;
import org.xml.sax.*;

public class RSSFeedHandler extends DefaultHandler {
    private RSSFeed feed;
    private RSSItem item;

    private boolean feedTitleHasBeenRead = false;
    private boolean feedPubDateHasBeenRead = false;

    private boolean isTitle = false;
    private boolean isDescription = false;
    private boolean isLink = false;
    private boolean isPubDate = false;

    public RSSFeed getFeed() {
        return feed;
    }

    public void startDocument() throws SAXException {
        feed = new RSSFeed();
        item = new RSSItem();
    }

    public void endDocument() throws SAXException { }

    public void startElement(String namespaceURI, String localName,
            String qName, Attributes atts) throws SAXException {

        if (qName.equals("item")) {
            item = new RSSItem();
            return;
        }
        else if (qName.equals("title")) {
            isTitle = true;
            return;
        }
        else if (qName.equals("description")) {
            isDescription = true;
            return;
        }
        else if (qName.equals("link")) {
            isLink = true;
            return;
        }
        else if (qName.equals("pubDate")) {
            isPubDate = true;
            return;
        }
    }
```

Figure 10-9   The RSSFeedHandler class (part 1 of 2)

The endElement method is executed after the parser reads an end element such as </item>. Like the startElement method, the third parameter contains the qualified name of the element. Within this method, an if statement checks whether the element is named item. If so, it adds the current RSSItem object to the RSSFeed object.

The characters method is executed when the parser reads the characters within an element. This method contains three parameters. Within this method, the first statement converts these three parameters into a string. Then, a nested if/else statement stores this string in the appropriate RSSFeed or RSSItem object.

This code is a little tricky because both the feed and each item have elements named title and pubDate. As a result, the code for the title element begins by checking whether the feed title has already been read. If not, it sets the title in the RSSFeed object. Then, it sets the appropriate Boolean variable to indicate that the feed title has been read. Otherwise, this code sets the title in the RSSItem object. Finally, this code sets the isTitle variable to a false value to indicate that the title element is no longer being parsed.

The code for the pubDate element works similarly to the code for the title element. However, the code for the link and description elements are simpler since they only exist in the item element.

The RSSFeedHandler class

```
public void endElement(String namespaceURI, String localName,
        String qName) throws SAXException
{
    if (qName.equals("item")) {
        feed.addItem(item);
        return;
    }
}

public void characters(char ch[], int start, int length)
{
    String s = new String(ch, start, length);
    if (isTitle) {
        if (feedTitleHasBeenRead == false) {
            feed.setTitle(s);
            feedTitleHasBeenRead = true;
        }
        else {
            item.setTitle(s);
        }
        isTitle = false;
    }
    else if (isLink) {
        item.setLink(s);
        isLink = false;
    }
    else if (isDescription) {
        item.setDescription(s);
        isDescription = false;
    }
    else if (isPubDate) {
        if (feedPubDateHasBeenRead == false) {
            feed.setPubDate(s);
            feedPubDateHasBeenRead = true;
        }
        else {
            item.setPubDate(s);
        }
        isPubDate = false;
    }
}
}
```

Figure 10-9   The RSSFeedHandler class (part 2 of 2)

The RSSFeed class

Figure 10-10 shows the code for the RSSFeed class. This class stores the data for the RSS feed. This data includes the feed's title and publication date as well as an ArrayList object that stores RSSItem objects. Most of the code for this class is standard Java code that lets you set and get data. As a result, you shouldn't have much trouble understanding it.

However, the code within the getPubDateMillis method uses the SimpleDateFormat object to parse the string for the date into a Date object. Here, the parse method accepts a string in this format:

```
Mon, 18 Feb 2013 14:11:56 EST
```

and converts that string into a Date object. Then, this code calls the getTime method of the Date object to return the milliseconds for the time.

The RSSItem class

Figure 10-11 shows the code for the RSSItem class. This class stores the data for each item in the RSS feed. This data includes the item's title, description, link, and publication date. Most of the code for this class is standard Java code that lets you set and get this data. As a result, you shouldn't have much trouble understanding it.

However, the code within the getPubDateFormatted method uses two SimpleDateFormat objects to change the format for the date. To start, the parse method of the first SimpleDateFormat object converts the string for the date into a Date object. Then, the format method of the second SimpleDateFormat object formats the Date object so it uses this format:

```
Monday, 5:56 AM (Feb 18)
```

As a result, the News Reader app can use this user-friendly date format instead of the unwieldy date format that's stored in the XML file.

The RSSFeed class

```
package com.murach.newsreader;

import java.text.ParseException;
import java.text.SimpleDateFormat;
import java.util.ArrayList;
import java.util.Date;

public class RSSFeed {
    private String title = null;
    private String pubDate = null;
    private ArrayList<RSSItem> items;

    private SimpleDateFormat dateInFormat =
            new SimpleDateFormat("EEE, dd MMM yyyy HH:mm:ss Z");

    public RSSFeed() {
        items = new ArrayList<RSSItem>();
    }

    public void setTitle(String title) {
        this.title = title;
    }

    public String getTitle() {
        return title;
    }

    public void setPubDate(String pubDate) {
        this.pubDate = pubDate;
    }

    public long getPubDateMillis() {
        try {
            Date date = dateInFormat.parse(pubDate.trim());
            return date.getTime();
        }
        catch (ParseException e) {
            throw new RuntimeException(e);
        }
    }

    public int addItem(RSSItem item) {
        items.add(item);
        return items.size();
    }

    public RSSItem getItem(int index) {
        return items.get(index);
    }

    public ArrayList<RSSItem> getAllItems() {
        return items;
    }
}
```

Figure 10-10   The RSSFeed class

The RSSItem class

```java
package com.murach.newsreader;

import java.text.ParseException;
import java.text.SimpleDateFormat;
import java.util.Date;

public class RSSItem {

    private String title = null;
    private String description = null;
    private String link = null;
    private String pubDate = null;

    private SimpleDateFormat dateOutFormat =
        new SimpleDateFormat("EEEE h:mm a (MMM d)");

    private SimpleDateFormat dateInFormat =
        new SimpleDateFormat("EEE, dd MMM yyyy HH:mm:ss Z");

    public void setTitle(String title) {
        this.title = title;
    }

    public String getTitle() {
        return title;
    }

    public void setDescription(String description) {
        this.description = description;
    }

    public String getDescription() {
        return description;
    }

    public void setLink(String link) {
        this.link = link;
    }

    public String getLink() {
        return link;
    }

    public void setPubDate(String pubDate) {
        this.pubDate = pubDate;
    }

    public String getPubDate() {
        return pubDate;
    }
```

Figure 10-11   The RSSItem class (part 1 of 2)

The RSSItem class

```
public String getPubDateFormatted() {
    try {
        Date date = dateInFormat.parse(pubDate.trim());
        String pubDateFormatted = dateOutFormat.format(date);
        return pubDateFormatted;
    }
    catch (ParseException e) {
        throw new RuntimeException(e);
    }
}
}
```

Figure 10-11   The RSSItem class (part 2 of 2)

How to work with adapters

After the News Reader app reads the RSS feed and parses it into a series of Java objects, it needs to display the data that's stored in those objects. To do that, you can store the data in an adapter. Then, you can use a ListView widget to display the data that's stored in the adapter.

How to create the layout for a list view

Figure 10-12 shows the activity_items layout as it's displayed in the Graphical Layout editor. This layout begins by using a TextView widget to display the title of the news feed. Then, it uses a ListView widget to display the items of the news feed.

This figure also shows the listview_item layout as it's displayed in the Graphical Layout editor. This layout uses two TextView widgets to display the publication date and title for each item. The first TextView widget displays the item's publication date in the default font size. The second TextView widget displays the item's title with a larger font size of 24sp. In addition this layout sets the margins for these widgets to control the alignment and spacing of these widgets.

The activity_items layout

The listview_item layout

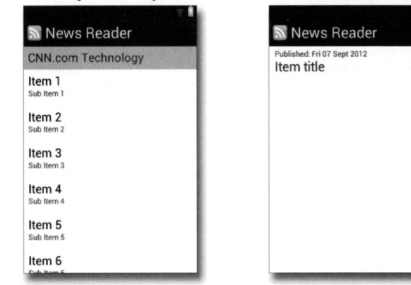

The ListView widget in the activity_items layout

```
<ListView
    android:id="@+id/itemsListView"
    android:layout_width="match_parent"
    android:layout_height="match_parent" />
```

The listview_item layout

```
<?xml version="1.0" encoding="utf-8"?>
<LinearLayout xmlns:android="http://schemas.android.com/apk/res/android"
    android:layout_width="match_parent"
    android:layout_height="match_parent"
    android:orientation="vertical" >

    <TextView
        android:id="@+id/pubDateTextView"
        android:layout_width="wrap_content"
        android:layout_height="wrap_content"
        android:layout_marginLeft="10dp"
        android:layout_marginTop="5dp"
        android:text="@string/item_pub_date" />

    <TextView
        android:id="@+id/titleTextView"
        android:layout_width="wrap_content"
        android:layout_height="wrap_content"
        android:layout_marginLeft="10dp"
        android:layout_marginRight="10dp"
        android:text="@string/item_title"
        android:textSize="24sp" />

</LinearLayout>
```

Figure 10-12   How to create the layout for a list view

How to use an adapter to display data in a list view

In chapter 5, you learned how to use an adapter to display an array of strings in a Spinner widget. Now, figure 10-13 shows how to use another type of adapter to display data in a ListView widget. More specifically, it shows how to use the SimpleAdapter class to display data in the ListView widget described in the previous figure.

The code example begins by calling the getAllItems method of the RSSFeed object to get an ArrayList object that contains RSSItem objects. Then, this code creates an ArrayList object that contains HashMap objects. Here, each HashMap object stores the data for each item in the list. In this figure, for example, each HashMap object stores the publication date and title for each item.

After storing the data for the adapter in the ArrayList variable named data, this code gets the ID for the listview_item layout and stores it in the variable named resource. This resource defines the layout for each item in the ListView widget.

Then, this code creates an array of strings that contain the names of the keys that get the data from the HashMap objects. This is where the data comes from. These keys correspond with the keys that are used to store data in the HashMap objects.

Next, it creates an array of int values that contain the IDs of the widgets in the listview_item layout. This is where the data goes to. As a result, these IDs should exist in the listview_item layout.

After setting up all of the necessary variables, this code creates the SimpleAdapter object. Here, the first parameter uses the *this* keyword to specify the context for the adapter. Then, the next four parameters use the data, resource, from, and to variables that were set up earlier.

Finally, the last statement calls the setAdapter method to set the SimpleAdapter object in the ListView widget. This causes the ListView widget to display the data stored in the SimpleAdapter object.

As you review this figure, you may notice that the adapter accepts a List object that contains Map objects. However, the code in this figure uses an ArrayList object that contains HashMap objects. This works because an ArrayList object is a type of List object, and a HashMap object is a type of Map object.

Code that creates and sets the adapter

```
// get the items for the feed
ArrayList<RSSItem> items = feed.getAllItems();

// create a List of Map<String, ?> objects
ArrayList<HashMap<String, String>> data =
        new ArrayList<HashMap<String, String>>();
for (RSSItem item : items) {
    HashMap<String, String> map = new HashMap<String, String>();
    map.put("date", item.getPubDateFormatted());
    map.put("title", item.getTitle());
    data.add(map);
}

// create the resource, from, and to variables
int resource = R.layout.listview_item;
String[] from = {"date", "title"};
int[] to = {R.id.pubDateTextView, R.id.titleTextView};

// create and set the adapter
SimpleAdapter adapter =
    new SimpleAdapter(this, data, resource, from, to);
itemsListView.setAdapter(adapter);
```

The constructor for the SimpleAdapter class

Parameter	Description
context	The context of the View associated with this SimpleAdapter object.
data	A List object that contains Map objects. The Map objects contain the data for the items in the list. The Map objects should include all keys specified in the parameter named from.
resource	The ID of a layout for each item in the list. The layout file should include the views defined in the parameter named to.
from	An array of column names that are in the Map objects.
to	An array of IDs for the widgets that should display the columns in the parameter named from.

Description

- You can use the SimpleAdapter class to display data in a ListView widget.

Figure 10-13   How to use an adapter to display data in a list view

How to handle events for an adapter

Figure 10-14 shows how to handle the events that occur when the user clicks on one of the items in a list. To do that, you implement the OnItemClickListener interface that's nested in the AdapterView class. Then, you wire this event handler to the appropriate widget. In this figure, for example, the onItemClick method is executed when the user clicks on one of the items in the ListView widget.

Within the onItemClick method, the third parameter provides the position of the selected item. In most cases, this is the only parameter that you need. However, if necessary, the first parameter provides the AdapterView object for the list of items, the second parameter provides a View object for the selected item, and the fourth parameter provides the ID of the selected item.

Within the onItemClick method, the first statement uses the position parameter to get the RSSItem object at the specified position. At this point, you can write additional code that processes the data stored in the RSSItem object. In this next figure, for example, you'll learn how to pass this data to another activity so it can be displayed.

Step 1: Import the interface for the listener

```
import android.widget.AdapterView.OnItemClickListener;
```

Step 2a: Implement the interface for the listener

```
public class ItemsActivity extends Activity
implements OnItemClickListener {
```

Step 2b: Implement the interface for the listener

```
@Override
public void onItemClick(AdapterView<?> parent, View v,
        int position, long id) {

    // get item at position
    RSSItem item = feed.getItem(position);
}
```

Step 3: Set the listeners

```
itemsListView.setOnItemClickListener(this);
```

Description

- The onItemClick method is executed when the user clicks on one of the items in the ListView widget.

- Within the onItemClick method, the third parameter provides the position of the selected item.

Figure 10-14   How to handle events for an adapter

How to work with intents

In the News Reader app, the Items activity displays a list of items. If you click on one of these items, you need to display the data for that item in the Item activity. To do that, you can use an intent to pass data from the Items activity to the Item activity.

Within the Item activity, you can click on a link to display it in the default browser on your device. To do that, you can use an intent to tell your device that you want to view the link in a web browser.

How to pass data between activities

As you learned in chapter 8, an *intent* is an object that provides a description of an operation to be performed. When an intent specifies a specific component within an app such as an activity, it can be referred to as an *explicit intent*. If you want, you can use an explicit intent to pass data from one component to another.

In figure 10-15, for example, the code in the ItemsActivity class creates an explicit intent for the ItemActivity class. Then, it uses the putExtra method of the Intent object to store some data in the intent. Here, it stores a string named title and an int value named position. Next, this code uses the startActivity method to start the activity specified by the intent.

The code in the ItemActivity class begins by using the getIntent method to get the Intent object that was passed to it. Then, it uses the getStringExtra method to get the string named title, and it uses the getIntExtra method to get the int value named position.

Although the code example only shows how to work with a string and an int value, the putExtra method of the Intent class works for most primitive types and arrays of primitive types. In addition, the Intent class includes an appropriate getXxxExtra method for most primitive types. As a result, you can use an Intent object to pass most types of data between activities.

Code in the ItemsActivity class

```
// create the intent
Intent intent = new Intent(this, ItemActivity.class);

// put data in the intent
intent.putExtra("title", item.getTitle());
intent.putExtra("position", position);

// start the intent
this.startActivity(intent);
```

Code in the ItemActivity class

```
// get the intent
Intent intent = getIntent();

// get data from the intent
String pubDate = intent.getStringExtra("title");
int position = intent.getIntExtra("position", 0);
```

The Intent class

Constructor/Method	Description
`Intent(context, class)`	Creates an intent for the specific class in the project.
`putExtra(name, value)`	Stores the specified value or array of values with the specified name.
`getStringExtra(name)`	Gets the string value with the specified name.
`getIntExtra(name, default)`	Gets the int value with the specified name. If no int value exists with the specified name, this method returns the specified default value.

Description

- An *explicit intent* specifies a component such as an activity.
- You can use an explicit intent to pass data from one activity to another.
- The Intent class provides getXxxExtra methods for most primitive types and arrays of primitive types.

Figure 10-15 How to pass data from one activity to another

How to view a URL in a web browser

Figure 10-16 shows how to view a URL in a web browser. To do that, you can use an *implicit intent*, which is an intent that specifies the action you want to perform. Then, Android determines the best app to perform that action.

The first example in this figure begins by creating a string for a URL that links to a news article on the Internet. Then, it uses the parse method of the Uri class to create a Uri object for this link.

After creating the Uri object, this example creates an Intent object. This code passes the ACTION_VIEW constant as the first argument of the constructor, and it passes the Uri object as the second argument. As a result, the Intent object specifies that you want to view the specified URI in a web browser. Then, after you use the startActivity method to start the Intent object, Android determines which web browser to start and uses that browser to view the specified URI.

How to dial or call a phone number

This figure also shows how to use an implicit intent to dial or call a phone number. This is similar to viewing a page in a web browser.

However, there are three primary differences. First, the string that specifies the URI must begin with "tel:", not "http:". Second, the first parameter that's passed to the constructor for the Intent must be the ACTION_DIAL or ACTION_CALL constant. Third, if you use the ACTION_CALL constant, you must add the CALL_PHONE permission to the AndroidManifest.xml file for your project. As with all permissions, this permission must be coded at the same indentation level as the application element.

How to view a URL in a web browser

```
// create a Uri object for the link
String link = "http://rss.cnn.com/~r/rss/cnn_tech/~3/N9m_DSAe5rY/";
Uri uri = Uri.parse(link);

// create the intent and start it
Intent viewIntent = new Intent(Intent.ACTION_VIEW, uri);
startActivity(viewIntent);
```

How to call a phone number

```
// get the Uri for the phone number
String number = "tel:800-111-1111";
Uri callUri = Uri.parse(number);

// create the intent and start it
Intent callIntent = new Intent(Intent.ACTION_DIAL, callUri);
startActivity(callIntent);
```

The Intent class

Constructor	Description
`Intent(action, uri)`	Create an intent with the specified action constant and the specified data Uri.

Constant	Target	Action
`ACTION_VIEW`	Activity	View specified Uri in a web browser.
`ACTION_DIAL`	Activity	Places the specified phone number in the dialer, but lets the user decide whether to make the call.
`ACTION_CALL`	Activity	Call the specified phone number. This method requires the CALL_PHONE permission.

The CALL_PHONE permission in the AndroidManifest.xml file

```
<uses-permission android:name="android.permission.CALL_PHONE" />
```

Description

- An *implicit intent* specifies the action you want to perform. Then, Android determines the best app to perform that action.
- You can use an implicit intent to view a URL in a web browser or to call a phone number.

Figure 10-16   How to view a URL in a web browser

The News Reader app

Now that you've learned the skills necessary to create the News Reader app, you're ready to see how all the pieces fit together.

The activity_items layout

Figure 10-17 shows the XML for the activity_items layout. This layout uses a linear layout with vertical orientation. It begins by displaying a TextView widget for the title of the feed. Then, it uses a ListView widget to display the items of the feed. In turn, the ListView widget uses the listview_layout to display each item. Since this was described earlier in this chapter, you shouldn't have much trouble understanding how it works.

The activity_items layout

```xml
<?xml version="1.0" encoding="utf-8"?>
<LinearLayout xmlns:android="http://schemas.android.com/apk/res/android"
    android:layout_width="match_parent"
    android:layout_height="match_parent"
    android:orientation="vertical" >

    <TextView
        android:id="@+id/titleTextView"
        android:layout_width="match_parent"
        android:layout_height="wrap_content"
        android:background="#FFAC83"
        android:padding="7dp"
        android:text="@string/items_title"
        android:textSize="22sp" />

    <ListView
        android:id="@+id/itemsListView"
        android:layout_width="match_parent"
        android:layout_height="match_parent" />

</LinearLayout>
```

The listview_item layout

```xml
<?xml version="1.0" encoding="utf-8"?>
<LinearLayout xmlns:android="http://schemas.android.com/apk/res/android"
    android:layout_width="match_parent"
    android:layout_height="match_parent"
    android:orientation="vertical" >

    <TextView
        android:id="@+id/pubDateTextView"
        android:layout_width="wrap_content"
        android:layout_height="wrap_content"
        android:layout_marginLeft="10dp"
        android:layout_marginTop="5dp"
        android:text="@string/item_pub_date" />

    <TextView
        android:id="@+id/titleTextView"
        android:layout_width="wrap_content"
        android:layout_height="wrap_content"
        android:layout_marginLeft="10dp"
        android:layout_marginRight="10dp"
        android:text="@string/item_title"
        android:textSize="24sp" />

</LinearLayout>
```

Figure 10-17   The activity_items layout

The ItemsActivity class

Figure 10-18 shows the Java code for the ItemsActivity class. This class uses most of the skills presented in this chapter.

The onCreate method begins by creating a FileIO object from the FileIO class that's shown in the next figure. Then, it gets references to the TextView and ListView widgets for the Items activity. Next, it sets the listener for the ListView widget. Finally, it creates a new object from the inner class named DownloadFeed, and it calls the execute method from that object to start it.

The inner class named DownloadFeed uses a background thread to download the XML file for the RSS feed. After the background thread finishes executing, this class displays a message in the LogCat view. Then, it creates a new object from the inner class named ReadFeed, and it calls the execute method from that object to start it.

The inner class named ReadFeed uses a background thread to read the XML file for the RSS feed and parse it into an RSSFeed object that contains multiple RSSItem objects. After the background thread finishes executing, this class displays a message in the LogCat view. Then, it calls the updateDisplay method of its outer class to update the user interface.

Both of these inner classes specify the Void type for all three generic types of the AsyncTask class. As a result, the doInBackground and onPostExecute methods both use the Void type.

The updateDisplay method uses an adapter to display the items in the feed in the ListView widget. When it finishes, it displays a message in the LogCat view.

The onItemClick method handles the event that occurs when the user clicks on one of the items in the ListView widget. To handle this event, this code starts the Item activity and uses an explicit intent to pass all of the data for the item to that activity.

The ItemsActivity class

Page 1

```java
package com.murach.newsreader;

import java.util.ArrayList;
import java.util.HashMap;

import android.os.AsyncTask;
import android.os.Bundle;
import android.app.Activity;
import android.content.Intent;
import android.util.Log;
import android.view.View;
import android.widget.AdapterView;
import android.widget.ListView;
import android.widget.SimpleAdapter;
import android.widget.TextView;
import android.widget.AdapterView.OnItemClickListener;

public class ItemsActivity extends Activity
implements OnItemClickListener {

    private RSSFeed feed;
    private FileIO io;

    private TextView titleTextView;
    private ListView itemsListView;

    @Override
    protected void onCreate(Bundle savedInstanceState) {
        super.onCreate(savedInstanceState);
        setContentView(R.layout.activity_items);

        io = new FileIO(getApplicationContext());

        titleTextView = (TextView) findViewById(R.id.titleTextView);
        itemsListView = (ListView) findViewById(R.id.itemsListView);

        itemsListView.setOnItemClickListener(this);

        new DownloadFeed().execute();
    }

    class DownloadFeed extends AsyncTask<Void, Void, Void> {
        @Override
        protected Void doInBackground(Void... params) {
            io.downloadFile();
            return null;
        }

        @Override
        protected void onPostExecute(Void result) {
            Log.d("News reader", "Feed downloaded");
            new ReadFeed().execute();
        }
    }
```

Figure 10-18   The ItemsActivity class (part 1 of 3)

The ItemsActivity class

```
class ReadFeed extends AsyncTask<Void, Void, Void> {
    @Override
    protected Void doInBackground(Void... params) {
        feed = io.readFile();
        return null;
    }

    @Override
    protected void onPostExecute(Void result) {
        Log.d("News reader", "Feed read");

        // update the display for the activity
        ItemsActivity.this.updateDisplay();
    }
}

public void updateDisplay()
{
    if (feed == null) {
        titleTextView.setText("Unable to get RSS feed");
        return;
    }

    // set the title for the feed
    titleTextView.setText(feed.getTitle());

    // get the items for the feed
    ArrayList<RSSItem> items = feed.getAllItems();

    // create a List of Map<String, ?> objects
    ArrayList<HashMap<String, String>> data =
            new ArrayList<HashMap<String, String>>();
    for (RSSItem item : items) {
        HashMap<String, String> map = new HashMap<String, String>();
        map.put("date", item.getPubDateFormatted());
        map.put("title", item.getTitle());
        data.add(map);
    }

    // create the resource, from, and to variables
    int resource = R.layout.listview_item;
    String[] from = {"date", "title"};
    int[] to = {R.id.pubDateTextView, R.id.titleTextView};

    // create and set the adapter
    SimpleAdapter adapter =
        new SimpleAdapter(this, data, resource, from, to);
    itemsListView.setAdapter(adapter);

    Log.d("News reader", "Feed displayed");
}
```

Figure 10-18   The ItemsActivity class (part 2 of 3)

The ItemsActivity class   **Page 3**

```
@Override
public void onItemClick(AdapterView<?> parent, View v,
        int position, long id) {

    // get the item at the specified position
    RSSItem item = feed.getItem(position);

    // create an intent
    Intent intent = new Intent(this, ItemActivity.class);

    intent.putExtra("pubdate", item.getPubDate());
    intent.putExtra("title", item.getTitle());
    intent.putExtra("description", item.getDescription());
    intent.putExtra("link", item.getLink());

    this.startActivity(intent);
    }
}
```

Figure 10-18   The ItemsActivity class (part 3 of 3)

The FileIO class

Figure 10-19 shows the Java code for the FileIO class. This class begins by defining a constant named URL_STRING for a string that specifies the URL for the RSS feed. Then, it defines a constant named FILENAME that specifies the name of the file that stores the current RSS feed. Next, it defines a Context object that can store the context for the app.

The constructor for the FileIO class accepts a Context object. Within the constructor, a single statement sets the variable named context for this class to the Context object parameter. That way, the context variable can be used by the downloadFile and readFile methods of this class.

The downloadFile method downloads the XML for the RSS feed from the Internet and stores it in the specified file. Conversely, the readFile method reads the specified file and returns an RSSFeed object. To do that, these methods use the techniques described earlier in this chapter for downloading, writing, and reading a file. However, both of these methods use the FILENAME constant defined earlier in this class. As a result, you can be sure that both of these methods are working with the same file, which is what you want.

The activity_item layout

Figure 10-20 shows the XML for the activity_item layout. This layout uses a scroll view to make sure that the user can scroll up and down if the item is too tall to fit on the screen. Then, it uses a linear layout with vertical orientation to display four TextView widgets for the title, publication date, description, and link of the item. This code sets the color of the fourth TextView widget to blue to indicate to the user that it is a link that can be clicked.

The FileIO class

```java
package com.murach.newsreader;

import java.io.FileInputStream;
import java.io.FileOutputStream;
import java.io.IOException;
import java.io.InputStream;
import java.net.URL;

import javax.xml.parsers.SAXParser;
import javax.xml.parsers.SAXParserFactory;
import org.xml.sax.InputSource;
import org.xml.sax.XMLReader;

import android.content.Context;
import android.util.Log;

public class FileIO {

    private final String URL_STRING = "http://rss.cnn.com/rss/cnn_tech.rss";
    private final String FILENAME = "news_feed.xml";
    private Context context = null;

    public FileIO (Context context) {
        this.context = context;
    }

    public void downloadFile() {
        try{
            // get the URL
            URL url = new URL(URL_STRING);

            // get the input stream
            InputStream in = url.openStream();

            // get the output stream
            FileOutputStream out =
                context.openFileOutput(FILENAME, Context.MODE_PRIVATE);

            // read input and write output
            byte[] buffer = new byte[1024];
            int bytesRead = in.read(buffer);
            while (bytesRead != -1)
            {
                out.write(buffer, 0, bytesRead);
                bytesRead = in.read(buffer);
            }
            out.close();
            in.close();
        }
        catch (IOException e) {
            Log.e("News reader", e.toString());
        }
    }
```

Figure 10-19   The FileIO class (part 1 of 2)

The FileIO class   **Page 2**

```java
        public RSSFeed readFile() {
            try {
                // get the XML reader
                SAXParserFactory factory = SAXParserFactory.newInstance();
                SAXParser parser = factory.newSAXParser();
                XMLReader xmlreader = parser.getXMLReader();

                // set content handler
                RSSFeedHandler theRssHandler = new RSSFeedHandler();
                xmlreader.setContentHandler(theRssHandler);

                // read the file from internal storage
                FileInputStream in = context.openFileInput(FILENAME);

                // parse the data
                InputSource is = new InputSource(in);
                xmlreader.parse(is);

                // set the feed in the activity
                RSSFeed feed = theRssHandler.getFeed();
                return feed;
            }
            catch (Exception e) {
                Log.e("News reader", e.toString());
                return null;
            }
        }
    }
```

Figure 10-19   The FileIO class (part 2 of 2)

The activity_item layout

```xml
<?xml version="1.0" encoding="utf-8"?>
<ScrollView xmlns:android="http://schemas.android.com/apk/res/android"
    android:layout_width="fill_parent"
    android:layout_height="fill_parent"
    android:orientation="vertical" >

    <LinearLayout
        android:layout_width="match_parent"
        android:layout_height="wrap_content"
        android:orientation="vertical" >

        <TextView
            android:id="@+id/titleTextView"
            android:layout_width="match_parent"
            android:layout_height="wrap_content"
            android:paddingLeft="7dp"
            android:paddingRight="7dp"
            android:paddingTop="5dp"
            android:text="@string/item_title"
            android:textSize="24sp"
            android:textStyle="bold" />

        <TextView
            android:id="@+id/pubDateTextView"
            android:layout_width="match_parent"
            android:layout_height="wrap_content"
            android:paddingLeft="7dp"
            android:paddingTop="5dp"
            android:text="@string/item_pub_date" />

        <TextView
            android:id="@+id/descriptionTextView"
            android:layout_width="match_parent"
            android:layout_height="wrap_content"
            android:paddingLeft="7dp"
            android:paddingRight="7dp"
            android:paddingTop="5dp"
            android:text="@string/item_description"
            android:textSize="18sp" />

        <TextView
            android:id="@+id/linkTextView"
            android:layout_width="match_parent"
            android:layout_height="wrap_content"
            android:layout_marginTop="10dp"
            android:paddingLeft="7dp"
            android:paddingTop="5dp"
            android:text="@string/item_link"
            android:textColor="@color/blue"
            android:textSize="18sp" />

    </LinearLayout>
</ScrollView>
```

Figure 10-20   The activity_item layout

The ItemActivity class

Figure 10-21 shows the Java code for the ItemActivity class. When this activity is started, its onCreate method gets all necessary data from the Intent object and displays it on the four TextView widgets defined by the layout for this activity.

If the user clicks on the TextView widget for the link, the onClick method is executed. This method starts by creating a Uri object for the link that's stored in the Intent object. Then, it uses an implicit intent to view the URI in a web browser.

The ItemActivity class

```
package com.murach.newsreader;

import android.net.Uri;
import android.os.Bundle;
import android.view.View;
import android.view.View.OnClickListener;
import android.widget.TextView;
import android.app.Activity;
import android.content.Intent;

public class ItemActivity extends Activity implements OnClickListener {

    @Override
    protected void onCreate(Bundle savedInstanceState) {
        super.onCreate(savedInstanceState);
        setContentView(R.layout.activity_item);

        // get references to widgets
        TextView titleTextView = (TextView)
            findViewById(R.id.titleTextView);
        TextView pubDateTextView = (TextView)
            findViewById(R.id.pubDateTextView);
        TextView descriptionTextView = (TextView)
            findViewById(R.id.descriptionTextView);
        TextView linkTextView = (TextView)
            findViewById(R.id.linkTextView);

        // get the intent and its data
        Intent intent = getIntent();
        String pubDate = intent.getStringExtra("pubdate");
        String title = intent.getStringExtra("title");
        String description =
            intent.getStringExtra("description").replace('\n', ' ');

        // display data on the widgets
        pubDateTextView.setText(pubDate);
        titleTextView.setText(title);
        descriptionTextView.setText(description);

        // set the listener
        linkTextView.setOnClickListener(this);
    }

    @Override
    public void onClick(View v) {
        // get the intent and create the Uri for the link
        Intent intent = getIntent();
        String link = intent.getStringExtra("link");
        Uri viewUri = Uri.parse(link);

        // create the intent and start it
        Intent viewIntent = new Intent(Intent.ACTION_VIEW, viewUri);
        startActivity(viewIntent);
    }
}
```

Figure 10-21   The ItemActivity class

Perspective

Now that you've finished this chapter, you should be able to create an app like the News Reader app that uses threads, files, adapters, and intents. However, there are many possible ways to improve this app. In the next chapter, you'll learn how to use services and notifications to improve this app.

Terms

RSS (Rich Site Summary) feed

thread

UI thread

asynchronous task

generics

daemon thread

SAX (Simple API for XML)

intent

explicit intent

implicit intent

Summary

- An *RSS* (*Rich Site Summary*) *feed* can be used to publish frequently updated works, such as blog entries and news headlines. Since RSS uses a standardized XML file format, the feed can be published once and viewed by many different apps.

- A *thread* is a single sequential flow of control within a program that often completes a specific task.

- By default, an Android app uses a single thread, called the *UI thread*, to display the user interface. Any task that can slow or stop the responsiveness of the UI thread should be run in a separate thread.

- An *asynchronous task* is a task that runs in a separate thread in the background and does not need to be synchronized with other threads. It's common to use a nested inner class to create an asynchronous thread. Within the inner class, you can access the context for the activity by coding the name of the outer class, a period, and the *this* keyword.

- The AsyncTask class uses *generics* to allow a class to operate on various types of objects, and provides for three generic types: (1) parameters, (2) progress, and (3) result. It provides an easy way to perform a background task and publish results on the UI thread without having to manually manipulate threads.

- You can use the TimerTask and Timer classes to create a thread that executes tasks after a specified delay or at a specified interval.

- You can use Java and Android APIs to download a file by reading input from the Internet and writing output to the file system.

- You can use *SAX* (*Simple API for XML*) to parse XML files.

- You can use the SimpleAdapter class to display data in a ListView widget. The onItemClick method is executed when the user clicks on one of the items in the ListView widget.

- An *explicit intent* specifies a component such as an activity, and can be used to pass data from one activity to another. An *implicit intent* specifies the action you want to perform, and can be used to view a URL in a web browser or to call a phone number.

Exercise 10-1   Review the News Reader app

In this exercise, you'll take the News Reader app for a test drive to see how it works.

1. Start Eclipse and import the project named ch10_ex1_NewsReader.

2. Run the app. The Items activity should display a list of news items, and the LogCat view should display messages indicating that the RSS feed has been downloaded, read, and displayed. Scroll down through this list of items to its end.

3. Click on a news item. This should display the news item in the Items activity, but it shouldn't display any new messages in the LogCat view.

4. Click on the "Read more" link. This should start a web browser and display the web page for the news item in that browser.

5. Click on the Back button twice to return to the Items activity. This should not display any new messages in the LogCat view.

6. Change the orientation of the emulator or device. This should display messages in the LogCat view that indicate that the RSS feed has been downloaded, read, and displayed. On the good side, this provides a way for you to refresh the news feed so it has the latest news items. On the other hand, this isn't the best way to check for updates. In the next chapter, you'll learn a better way to handle this issue.

Exercise 10-2   Work with asynchronous tasks

In this exercise, you'll modify the DownloadFeed and ReedFeed classes used by the News Reader app presented in this chapter so they accept parameters and return results.

Modify the DownloadFeed class

1. Start Eclipse and import the project named ch10_ex2_NewsReader.

2. Open the .java file for the Items activity. Then, modify the code for the DownloadFeed class so it uses this declaration:
   ```
   class DownloadFeed extends AsyncTask<String, Void, Void>
   ```

3. Save the file. When you do, Eclipse should display an error that shows that the declaration for the doInBackground method doesn't specify the correct types.

4. Modify the declaration for the doInBackground method so it accepts multiple String objects as a parameter.

5. Modify the code for the doInBackground method so it gets the first String object that's passed to it and uses that string to create the URL object. In other words, this method should not use the URL_STRING constant.

6. Modify the code that creates and executes the DownloadFeed class so it passes the String object for the URL to the doInBackground method.

7. Run the app to make sure this code works correctly.

Modify the ReadFeed class

8. Modify the code for the ReadFeed class so it uses this declaration:

    ```
    class ReadFeed extends AsyncTask<String, Void, RSSFeed>
    ```

9. Save the file. When you do, Eclipse should show that there are errors with both the doInBackground and onPostExecute methods.

10. Modify the declaration for the doInBackground method so it returns an RSSFeed object.

11. Modify the declaration for the onPostExecute method so it accepts an RSSFeed object as a parameter.

12. Modify the code for the doInBackground method so the end of the try clause returns an RSSFeed object instead of setting it in the activity. Also, modify the end of the catch clause so it returns a null value.

13. Modify the code for the onPostExecute method so it sets the feed instance variable in the activity to the RSSFeed parameter. Make sure to set the feed before this method calls the updateDisplay method of the activity.

14. Run the app to make sure this code works correctly.

Exercise 10-3   Modify the News Reader app

In this exercise, you'll make some enhancements to the News Reader app so it reads a different RSS feed and formats it differently.

1. Start Eclipse and import the project named ch10_ex3_NewsReader.

2. Change the URL for the RSS feed to:

    ```
    http://rss.cnn.com/rss/cnn_world.rss
    ```

3. In the activity_items layout, add a TextView widget for the publication date just below the TextView widget for the title.

4. In the ItemsActivity class, add code that uses the TextView widget you just added to display the publication date for the feed.

5. Modify the feed's publication date so it's displayed in this format:

    ```
    Thursday 9:12 AM (Feb 28)
    ```

    To do that, add a getPubDateFormatted method to the RSSFeed class that works like one that's in the RSSItem class.

6. In the listview_item layout, add a TextView widget to display the description for the feed.

7. In the ItemsActivity class, modify the code for the adapter so it also stores the description and displays it in the list of items. When you're done, the Items activity should display the publication date, title, and description for each item.

8. Modify the code that's executed when you click on an item so that clicking on an item displays the item in a web browser. When you're done, the News Reader app should never display the Item activity.

11

How to work with services and notifications

Some types of Android apps need to execute a task in the background, even when the app isn't running. To do that, you can use an Android component known as a service. When a service completes a task, it may want to notify the user. To do that, you can use an Android feature known as a notification.

How to work with the Application object

Before you learn how to work with services, you should learn how to create a custom Application object for your app. This provides two benefits. First, it allows you to store data that applies to the entire app in a central location that's always available to all components of the app including all of the app's activities and services. Second, it allows you to execute code when the application starts, as opposed to executing code each time an activity starts.

How to define the Application object

Figure 11-1 shows how to define a custom Application object. Here, the NewsReaderApp class extends the Application class. Then, it defines an instance variable that can store the milliseconds that correspond with the publication date for the current RSS feed for the app. Next, it defines get and set methods for this variable.

After defining the methods that store the application's data, this class overrides the onCreate method of the Application class. Android calls this method when the app starts, but not when an activity starts. As a result, this method is a good place to put any code that should only be run once, when the app starts. For now, this method just prints a message to the LogCat view. However, later in this chapter, you'll learn how to modify this method so it starts a service.

When storing data in the Application object, keep in mind that the Application object is created when the app starts and remains available as long as any component of the app is running. As a result, you should avoid storing large amounts of data in this object whenever possible. For the News Reader app, for example, you could store an RSSFeed object in the Application object. That way, the current feed would always be available to all components in the app. Since the RSSFeed object only contains text and isn't too large, this might be OK in most cases. However, since the News Reader app only needs to store the milliseconds for the feed, it's probably a better design to only store a long value for the milliseconds as shown in this figure.

A starting point for the NewsReaderApp class

```java
package com.murach.newsreader;

import android.app.Application;
import android.util.Log;

public class NewsReaderApp extends Application {

    private long feedMillis = -1;

    public void setFeedMillis(long feedMillis) {
        this.feedMillis = feedMillis;
    }

    public long getFeedMillis() {
        return feedMillis;
    }

    @Override
    public void onCreate() {
        super.onCreate();
        Log.d("News reader", "App started");
    }
}
```

Description

- To store data and methods that apply to the entire application, you can extend the Application class and add instance variables and methods. The Application object is created when the app starts and remains available until the app ends.

- To run code only once when the application starts, you can override the onCreate method of your custom Application class.

Figure 11-1   How to define the Application object

How to register the Application object

When you extend the Application class, you must register that class. Then, when the application starts, Android creates the Application object from your custom Application class. If you don't register the custom Application class, Android creates the Application object from the Application class, not from your custom class.

To register your custom Application class, open the AndroidManifest.xml file. Then, edit the application element so its name attribute specifies the name of your custom Application class. In figure 11-2, for example, the name attribute of the application element specifies the NewsReaderApp class.

How to use the Application object

Once you have registered your custom application class, you can use the Application object. To do that, you can start by using the getApplication method to get a reference to the Application object. In this figure, the first example declares a variable of the NewsReaderApp type. Then, the second example uses the getApplication method to get a reference to the Application object, and it casts this Application object to the NewsReaderApp type.

Once you have a reference to the Application object, you can use it just as you would use any other object. In this figure, the third example calls the getFeedMillis method from the Application object to get the milliseconds for the publication date of the current RSS feed object for the app.

The application element of the AndroidManifest.xml

```
<application
    android:name=".NewsReaderApp"
    android:allowBackup="true"
    android:icon="@drawable/ic_launcher"
    android:label="@string/app_name"
    android:theme="@style/AppTheme" >
```

How to use the Application object

Step 1: Declare a variable for the Application object
```
NewsReaderApp app;
```

Step 2: Get a reference to the Application object
```
app = (NewsReaderApp) getApplication();
```

Step 3: Use the object
```
long feedMillis = app.getFeedMillis();
```

A method of the Context class

Method	Description
getApplication()	Returns the Application object for your app. You can cast this object to the custom Application class for your app.

Description

- When you extend the Application class, you must register that class. Then, when the application starts, Android creates the Application object from your custom Application class.

- To register your custom Application class, open the AndroidManifest.xml file and edit the application element so its name attribute specifies the name of your custom Application class.

- To get a reference to the Application object, you can use the getApplication method of the Context class.

- Once you have a reference to the Application object, you can use it just as you would use any other object.

Figure 11-2   How to register and use the Application object

How to work with services

A *service* is an Android component that does not provide a user interface. Instead, it performs a task that runs in the background. Since a service continues to run even if the user switches to another application, services should be used for tasks that should run independently of activities. For example, the News Reader app could use a service to download updates from the network, even when another app is running. Similarly, a Music Player app could use a service to play music even while another app is running.

The lifecycle of a service

Figure 11-3 shows two possible lifecyles of a service. The service on the left is an *unbound service*, which is a service that does not return a result to the caller. This type of service is created and started when a component such as an activity calls the startService method to start it. Then, the service runs until it's stopped.

An unbound service is typically stopped in one of three ways. First, another component such as an activity can call the stopService method to stop it. Second, the service can call the stopSelf method to stop itself. Third, Android can stop the service when the user turns off the device or if the device runs low on memory.

The service on the right is a *bound service*, which is a service that can interact with one or more components. This type of service is started when a component calls the bindService method to bind to it. Then, other components can bind to the service at the same time. The service runs as long as at least one component is bound to it. However, when the last component unbinds, Android stops the service.

Although a service runs in the background, a service does not run in a background thread. On the contrary, it runs in the same thread as the component that started it. As a result, if your service is going to perform any long running tasks such as networking or file I/O, you should create a new thread within the service to perform those tasks. Otherwise, your user interface may become unresponsive.

In this chapter, you'll learn how to use a bound service, which is the most common type of service. However, if you need a service to interact with an activity, you can search the Internet to learn more about how to do this. Also, a service can be both bound and unbound. To create such a service, you can implement all necessary lifecycle methods. For example, you can implement the onStartCommand and onBind methods. Then, you can call the startService method to run the service indefinitely, and you can call the bindService command to bind components to the service.

The lifecycle of a service

Figure 11-3 The lifecycle of a service

Description

- A *service* performs tasks in the background and does not provide a user interface. Services should be used for tasks that run independently of activities. For example, a service might download updates from the network, play music, and so on.

- A service continues to run even if the user switches to another app.

- An *unbound service* does not interact with other components such as activities. This type of service runs until it's stopped by another component or by itself.

- A *bound service* can interact with components such as activities. This type of service runs only as long as another component is bound to it. Multiple components can bind to the service at once, but when all of them unbind, the service is destroyed.

How to create a service

Figure 11-4 shows the starting code for an unbound service. This service implements all four methods that you typically need for an unbound service.

To start, this code imports the three classes that are needed to implement a service: the Service, Intent, and IBinder classes. Then, the code declares that the NewsReaderService class extends the Service class. Within this class, the onCreate method is called when the service is first created. As a result, it's a good place to put code that's only executed once.

The onStartCommand method is executed each time a component uses the startService method to start this service. As a result, it's a good place to put code that may need to be executed multiple times.

The onStartCommand method returns an int value that tells Android what to do if the device runs low on memory and Android needs to kill the thread for the service. The Service class provides three constants that you can return from this method.

The START_STICKY constant tells Android to leave the service in the started state. Then, when more memory becomes available, restart the service by calling the onStartCommand method with a null intent for the intent parameter. In other words, the "start" for the service is "sticky", so the service should be restarted as soon as possible. This constant is appropriate for services that should always be running in the background. For example, if a service is started once and executes a task at specified intervals, you probably want to restart it.

The START_NOT_STICKY constant tells Android to stop the service and to not restart it. In other words, the "start" is not "sticky", so the service should not be restarted when memory becomes available. This constant is appropriate for a service that doesn't need to be completed and is called by the app at specified intervals. For example, an app might start a service every 15 minutes to perform a task. If the service gets destroyed while doing that task, it's probably best to let it stay stopped since the app will start it again anyway when the app is restarted.

The START_REDELIVER_INTENT constant tells Android to stop the service. However, when more memory becomes available, Android should restart the service by calling the onStartCommand method with the last delivered intent as the intent parameter. In other words, Android calls the onStartCommand method and "redelivers" the last intent. This constant is appropriate for a service that performs a task that needs to be completed. Android continues to redeliver the intent until the service calls its stopSelf method, which indicates that the service has completed.

The onBind method is executed when another component attempts to bind to the service. This method returns an IBinder object that the component can use to interact with the service. For an unbound service, such as the service shown in this chapter, you can return a null value. However, the onBind method is an abstract method that must be implemented. As a result, you can't completely omit this method, even for an unbound service.

The NewsReaderService class with its lifecycle methods implemented

```
package com.murach.newsreader;

import android.app.Service;
import android.content.Intent;
import android.os.IBinder;

public class NewsReaderService extends Service {

    @Override
    public void onCreate() {
        // Code that's executed once when the service is created.
    }

    @Override
    public int onStartCommand(Intent intent, int flags, int startId) {
        // Code that's executed each time another component
        // starts the service by calling the startService method.
        return START_STICKY;
    }

    @Override
    public IBinder onBind(Intent intent) {
        // Code that's executed each time another component
        // binds to the service by calling the bindService method.
        // This method is required.
        // For an unbound service, you can return a null value.
        return null;
    }

    @Override
    public void onDestroy() {
        // Code that's executed once when the service
        // is no longer in use and is being destroyed.
    }
}
```

Some constants of the Service class

Constant	If Android destroys the thread for the service due to lack of memory, Android should...
START_STICKY	Leave the service in the started state. When memory becomes available, restart the service by calling the onStartCommand method with a null value for the intent parameter.
START_NOT_STICKY	Stop the service and do not restart it. As a result, the service will only restart if the app resumes and executes a method that restarts the service.
START_REDELIVER_INTENT	Stop the service. When memory becomes available, restart the service by calling the onStartCommand method with the last delivered intent for the intent parameter.

Figure 11-4   How to create a service

The onDestroy method is executed when the service is no longer in use and is being destroyed. As a result, it's a good place to clean up any resources that need to be cleaned up.

How to register a service

Before you can use a service, you must register it. The first example in figure 11-5 shows how to do that. To start, you open the manifest file and add a service element at the same indentation level as the activity elements. Then, use the name attribute of the service element to specify the name of the class for your service.

By default, a service is public and can be started by other applications. However, if you want to make a service private, you can add an export attribute to the service element and set it to false as shown in the second example.

How to start and stop a service

Before you can start a service, you must create an Intent object for the service. In this figure, for instance, the third example creates an Intent object for the service that's defined by the NewsReaderService class.

You can start a service by calling the startService method and passing the intent for the service to it. In this figure, the fourth example starts the service. Since the startService method belongs to the Context class, you can call it from any component, including an activity or the Application object.

When you use the startService method to start a service, the service runs indefinitely. In some cases, that's what you want. In other cases, you may want to stop the service.

For instance, if you want to allow the user to stop a service, you can provide a menu item that stops the service. Then, when the user selects that menu item, you can use the stopService method to stop the service as shown in the fifth example.

Or, if you have a service that performs a task, you may want to stop the service after it finishes its task. To do that, you can call the stopSelf method from within the class for the service as shown in the sixth example.

A public service in the AndroidManifest.xml file

```
<service
    android:name=".NewsReaderService">
</service>
```

A private service in the AndroidManifest.xml file

```
<service
    android:name=".NewsReaderService"
    android:exported="false">
</service>
```

Code that uses a service

Create the Intent object for the service

```
Intent serviceIntent = new Intent(this, NewsReaderService.class);
```

Start the service

```
startService(serviceIntent);
```

Stop the service from an activity or other component

```
stopService(serviceIntent);
```

Stop the service from within the service

```
stopSelf();
```

Some methods of the Context class

Method	Description
`startService(intent)`	Start the service that's specified by the intent.
`stopService(intent)`	Stop the service that's specified by the intent.

A method of the Service class

Method	Description
`stopSelf()`	Stops the current service.

Description

- Before you can use a service, you must register it.
- To register your service, add a service element to the AndroidManifest.xml file at the same indentation level as the activity elements and use the name attribute of the service element to specify the name of the class for your service.
- To make a service private, so it can only be accessed by the current app, you can add an exported attribute and set it to a value of false.
- If a component starts a service by calling the startService method, the service runs until another component stops it by calling the stopService method, until it stops itself by calling the stopSelf method, or until the device is turned off.

Figure 11-5   How to register, start, and stop a service

How to use threads with services

As mentioned earlier in this chapter, a service does not run in its own thread by default. As a result, if you want a service to perform a long running task such as accessing a network or file I/O, you should create a new thread within the service. If, for example, you want a thread to execute a single task, you can use the AsyncTask class to create a background thread for that class. Or, if you want a thread to execute a task at a specified interval, you can use the TimerTask and Timer classes to create and schedule a thread as shown in figure 11-6.

The NewsReaderService class shown in this figure begins by extending the Service class. Within the class, the first statement declares an instance variable for the Timer class. This instance variable is used to schedule when the timer task is executed.

Android calls the first three methods of this class at various times in the life-cycle of this service. This code sends messages to the LogCat view to show that they have been executed. However, since no components in the app attempt to bind to this service, the onBind method should never be called. Of these metods, the onCreate method calls the startTimer method, and the onDestroy method calls the stopTimer method.

The startTimer method begins by creating a TimerTask object and overriding its run method. This method creates the thread for the task. As a result, all of the long running code for the task should be placed within this method. For now, this example just sends a message to the LogCat view that indicates that the timer task has been executed. After creating the TimerTask object, the startTimer method creates a Timer object and uses it to schedule the TimerTask object to run after a delay of 10 seconds and to continue to run every 10 seconds after that.

The stopTimer method begins by checking to make sure the Timer object is not null. If so, it uses the Timer object to cancel the thread for the timer task.

A Service class that runs a timed task in its own thread

```
package com.murach.newsreader;

import java.util.Timer;
import java.util.TimerTask;

import android.app.Service;
import android.content.Intent;
import android.os.IBinder;
import android.util.Log;

public class NewsReaderService extends Service {

    private Timer timer;

    @Override
    public void onCreate() {
        Log.d("News reader", "Service created");
        startTimer();
    }

    @Override
    public IBinder onBind(Intent intent) {
        Log.d("News reader", "No binding for this service");
        return null;
    }

    @Override
    public void onDestroy() {
        Log.d("News reader", "Service destroyed");
        stopTimer();
    }

    private void startTimer() {
        // create task
        TimerTask task = new TimerTask() {
            @Override
            public void run() {
                Log.d("News reader", "Timer task executed");
            }
        };

        // create and start timer
        timer = new Timer(true);
        int delay = 1000 * 10;      // 10 seconds
        int interval = 1000 * 10;   // 10 seconds
        timer.schedule(task, delay, interval);
    }

    private void stopTimer() {
        if (timer != null) {
            timer.cancel();
        }
    }
}
```

Figure 11-6 How to use threads with services

How to test a service

In the next chapter, you'll learn how to start a service when the device boots. For now, you can test a service by starting it when the app starts. To do that, you can call the startService method from the onCreate method of the Application object as shown in figure 11-7. That way, Android starts the service once, when the app starts. For now, this a logical place to start the News Reader service. In the next chapter, you'll learn how to start a service when the device boots. That way, Android starts the service once, when the device boots.

Since a service doesn't have a user interface, you can test it by using the Log class to print messages to the LogCat view. Then, after you start the service, you can check the LogCat view to make sure that the service is running and performing its task correctly.

In this figure, the first two messages show that the app has started and the service has been created. Then, the next few messages show that the timer task has been executed. Finally, if the last message is shown, it indicates that the service has been destroyed. This message is typically displayed when the service is stopped because the stopService method has been called by another component or the stopSelf method has been called by the service after it has completed its task.

The onCreate method of the Application object

```
@Override
public void onCreate() {
    super.onCreate();
    Log.d("News reader", "App started");

    // start service
    Intent service = new Intent(this, NewsReaderService.class);
    startService(service);
}
```

The messages that are displayed in the LogCat view

```
App started
Service created
Timer task executed
Timer task executed
Timer task executed
...
Service destroyed
```

Description

- To start a service when the app starts, you can call the startService method from the onCreate method of the Application object. That way, the service is started once, when the app starts.

- To test a service, you can use the Log class to print messages to the LogCat view.

Figure 11-7   How to test a service

How to view all services

Figure 11-8 shows how to use the Settings app to view all of the apps and services that are running on a device or emulator. To start, you can navigate to the Home screen. The easiest way to do that is to press the Home button. Then, if the Settings app is available from the Home screen, you can select it. Otherwise, you can view all of the apps on your device and start the Settings app. From there, click on the Apps item and the Running tab. To see the Running tab, you may need to scroll to the right. In addition, on some systems, you may need to click the Show Running Services item to display the services.

In this figure, the Running tab shows that three apps and two services are running on the device. The Settings app is the app that's currently being used to display the apps and services. However, it doesn't use a service. The News Reader app is also running, and it uses a service. Similarly, the Android Keyboard app is running, and it uses a service as well.

For most devices, it would be unusual for only three apps to be running. To get an idea of how many apps and services typically run on a device, you should take a moment to browse through the apps on a phone or tablet that has been running for a while. Most likely, you'll see a long list of apps and services.

The apps and services that are running on the device

Procedure

1. Navigate to the Home screen.
2. Start the Settings app. This app may be available on Home screen, or you may need to display all apps.
3. Click on the Apps item.
4. Display the Running tab. To do that, you may need to scroll to the right.
5. View the apps and services. On some systems, you may need to click the Show Running Services item.

Description

- You can use the Settings app to view all of the apps and services that are running on a device or emulator.

Figure 11-8   How to view services

How to work with notifications

A *notification* provides a way for a service to display a message even when another app is running. For example, the service for the News Reader app could display a notification that indicates that an updated feed is available.

How notifications work

Figure 11-9 shows how notifications work. To start, when Android first displays a notification, the notification appears as an icon in the *notification area* at the top of the screen. In this figure, the first two screens display a notification in the notification area. Here, the first screen is the News Reader app and the second screen is the Home screen. Both screens display a notification for the News Reader app at the top of the screen.

When one or more notifications have been displayed, the user can pull down on the notification area to open the *notification drawer*. This allows the user to view all notifications. In this figure, the third screen shows the notification drawer when it contains only one notification, a notification that indicates that an updated feed is available from the News Reader app. However, it's possible for the notification area and drawer to display multiple notifications.

Often, a user can perform an action by clicking on the notification in the notification drawer. Typically, this starts the app that displayed the notification. In this figure, if the user selects the notification, it starts the News Reader app and displays the updated feed.

The News Reader app after it has started a notification

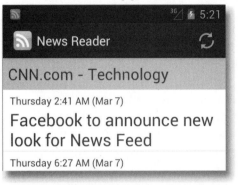

A notification displayed in the notification area

The notification drawer

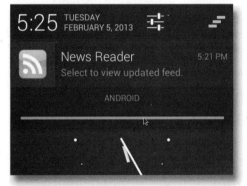

Description

- A *notification* provides a way for a service to display a message even when another app is running.
- When it's first displayed, a notification appears as an icon in the *notification area* at the top of the screen.
- To view a notification, the user can pull down on the notification area to open the *notification drawer*. Then, the user can often execute an action by clicking on the notification. Typically, this starts the app that displayed the notification.

Figure 11-9   How notifications work

How to create a pending intent

To create the action for a notification, you can create a *pending intent*, which is a special type of intent that can be passed to another app and executed by that app at a later time. That way, you can pass this intent to Android, so it can use the intent in its notification system.

Before you learn how to create a pending intent for a notification, you need to understand some terminology. To start, the *back stack* is a stack of all recently used activities. These activities are sorted in the order in which they were used. When the user clicks the Back button, the top activity is removed from the stack, and the next activity is displayed.

All activities in the back stack belong to a *task*, which is a cohesive unit that can contain multiple activities. For example, the News Reader task contains any Items or Item activities that are in the back stack. When the user clicks the Home button or the Task button to the right of that button, the current task moves to the background. However, all of the activities for that task remain on the back stack for that task. This back stack can be referred to as the *task history*. That way, if you navigate back to the task, you can continue the task where you left off.

The code examples in figure 11-10 show how to create a pending intent for a notificaiton. The first step is to create a regular intent for an activity as described in the previous chapter. However, when starting an activity from a notification, it's common to use the addFlags method to control how the activity is displayed in the back stack. In most cases, you want to set the "activity clear top" flag as shown in step 1. That way, if the activity is already in the task, the other activities in the task that are on top of the specified activity are cleared from the stack. This brings the specified activity to the top of the stack, which is usually what you want.

If you don't set this flag, Android creates a new instance of the activity and adds it to the top of the stack for the task. If the task already contains this activity, this creates multiple instances of the same activity. Sometimes this is what you want. For example, you may want to place the same activity in the stack twice if it displays different data.

The second step is to create the pending intent. To do that, you typically create a flag that tells Android what to do if an intent with the same name already exists but hasn't yet been executed. In most cases, you want to use an "update current" flag to keep the current intent but update its data with any data in the new intent. (This is the "extra" data that you can set with the putExtra method.) Then, you pass that flag and the intent you created in step 1 to the getActivity method of the PendingIntent class. This returns a PendingIntent object that you can use in a notification. Here, the second parameter specifies a code that isn't currently used by Android. As a result, you can specify any int value you want for this parameter.

The classes used to work with pending intents

```
android.app.PendingIntent
android.content.Intent
```

How to create a pending intent

Step 1: Create the intent

```
Intent notificationIntent = new Intent(this, ItemsActivity.class)
    .addFlags(Intent.FLAG_ACTIVITY_CLEAR_TOP);
```

Step 2: Create the pending intent

```
int flag = PendingIntent.FLAG_UPDATE_CURRENT;
PendingIntent pendingIntent =
        PendingIntent.getActivity(this, 0, notificationIntent, flag);
```

Some constants and methods of the Intent class

Constant	Description
`FLAG_ACTIVITY_CLEAR_TOP`	If the activity is already running, it closes all activities on top of the specified activity, which brings the specified activity to the top of the task history.
`FLAG_ACTIVITY_NEW_TASK`	Creates a new task for the activity and places it on top of the task history.
Method	**Description**
`addFlags(flags)`	Adds flags that control the behavior of the intent.

Some constants and methods of the PendingIntent class

Constant	Description
`FLAG_UPDATE_CURRENT`	If the pending intent already exists, keep it but replace its extra data with the extra data from the new intent.
Method	**Description**
`getActivity(context, code, intent, flags)`	Returns a PendingIntent object for the specified activity intent with the specified flags.

Description

- A *pending intent* is an intent that can be passed to other apps so that they can execute the intent at a later time.
- The *back stack* is a stack of all recently used activities. These activities are sorted in the order in which they were used. When the user clicks the Back button, the top activity is removed from the stack, and the next activity is displayed.
- A *task* is a cohesive unit that can contain multiple activities. Every activity belongs to a task. A task can move to the background when the user starts a new task or navigates away from the current task.
- The *task history* is the back stack for a particular task.

Figure 11-10   How to create a pending intent

How to create a notification

Once you've created a pending intent for a notifcation, you can create the notification as shown in figure 11-11. When you do, you set the pending intent in the notification.

Android 3.0 made many improvements to notifications. To take advantage of these features while also staying compatible with older versions of Android, you can use the inner Builder class of the NotificationCompat class that's available from the support library that's added to most Android projects by default. This is the same support library that you learned how to use in chapter 9 to use fragments with older versions of Android. As a result, if this library isn't available to your project, you can add it to your project as described in chapter 9.

The first step in creating a notification is to declare all of the variables that you need. In this figure, the first step declares four variables. These variables specify (1) the icon, (2) the ticker text that's displayed when the notification first appears, (3) the title of the notification when it's displayed in the drawer, and (4) the text for the notification when it's displayed in the drawer. Here, the icon is the same launcher icon that's used for the app. Similarly, the title of the notification is the same as the name of the app.

The second step in creating a notification is to create the Notification object. To do that, you create an object from the Builder class that's nested in the NotificationCompat class. Then, you use the various set methods of the Builder object to set the various components of the notification. For example, you can use the setContentIntent method to set the pending intent for the notification. Similarly, you can use the setSmallIcon method to set the icon for the notification. In addition, if you pass a true value to the setAutoCancel method, Android automatically removes the intent if the user selects it in the drawer, which is usually what you want. Finally, you use the build method to build the Notification object. Since all of these methods are designed to be chained, you can chain them together as shown in this figure.

After creating a Notification object like the one shown in this figure, you typically display it. To do that, you use a system service as shown in the next figure.

The classes used to work with notifications

```
android.app.Notification
android.support.v4.app.NotificationCompat
```

How to create a notification

Step 1: Create the variables for the notification

```
int icon = R.drawable.ic_launcher;
CharSequence tickerText = "Updated news feed is available";
CharSequence contentTitle = getText(R.string.app_name);
CharSequence contentText = "Select to view updated feed.";
```

Step 2: Create the Notification object

```
Notification notification =
        new NotificationCompat.Builder(this)
    .setSmallIcon(icon)
    .setTicker(tickerText)
    .setContentTitle(contentTitle)
    .setContentText(contentText)
    .setContentIntent(pendingIntent)
    .setAutoCancel(true)
    .build();
```

Constructors and methods of the NotificationCompat.Builder class

Constructor	Description
`NotificationCompat.Builder(context)`	Creates an object from the class with the specified context.
Method	**Description**
`setSmallIcon(int)`	Set the resource ID of the icon for the notification.
`setTicker(string)`	Set the ticker text that's briefly displayed when the notification is first displayed.
`setContentTitle(string)`	Sets the title of the notification.
`setContentText(string)`	Sets the text of the notification.
`setContentIntent(pendingIntent)`	Sets the intent of the notification.
`setAutoCancel(boolean)`	If true, Android automatically removes the intent after it is selected.
`build()`	Returns a Notification object.

Description

- You can use the NotificationCompat.Builder class to create a Notification object that's compatible with older versions of Android.

Figure 11-11   How to create a notification

How to work with system services

A *system service* is a service that's provided by the Android operating system. Android provides system services that allow you to work with notifications, locations, network connections, soft keyboards, and so on.

In chapter 6, for example, you learned how to use the input method service to get an InputMethodManager object, and you learned how to use that object to hide a soft keyboard. In this chapter, you'll learn how to use two services: the notification service and the network connection service. Both of these system services can be used by the News Reader app.

How to display or remove a notification

Figure 11-12 shows how to use a system service to display or remove a notification. To start, you can call the getSystemService method to get a NotificationManager object. Since this method is available from the Context class, you can call it directly from any component such as an activity or service.

Once you have a NotificationManager object, you can call its notify method to display the notification. But first, you should declare an ID for the notification so you can pass it to the notify method. This ID should be unique within your app. That way, your app can uniquely identify that notification later. For example, if you want to remove a notification later, you can pass this ID to the cancel method of the NotificationManager object.

Some constants and methods of the Context class

Method	Description
getSystemService(name)	Returns the appropriate object for the specified system-level service.

Constant	Description
NOTIFICATION_SERVICE	Returns a NotificationManager object for displaying and removing notifications.
LOCATION_SERVICE	Returns a LocationManager object for working with GPS updates.
CONNECTIVITY_SERVICE	Returns a ConnectivityManager object for working with network connections.
WIFI_SERVICE	Returns a WifiManager object for working with Wi-Fi connectivity.
INPUT_METHOD_SERVICE	Returns an InputMethodManager object for displaying and hiding soft keyboards.

The classes used to work with notifications

```
android.app.NotificationManager
```

Some methods of the NotificationManager class

Method	Description
notify(id, notification)	Displays the specified notification and sets its ID to the specified ID. This ID should be unique within your application.
cancel(id)	Cancels the notification with the specified ID.

How to get a NotificationManager object

```
NotificationManager manager = (NotificationManager)
        getSystemService(NOTIFICATION_SERVICE);
```

How to display a notification

```
final int NOTIFICATION_ID = 1;
manager.notify(NOTIFICATION_ID, notification);
```

How to remove a notification

```
manager.cancel(NOTIFICATION_ID);
```

Description

- A *system service* is a service that's provided by the Android operating system.
- You can use the NotificationManager object to display or remove a notification.

Figure 11-12   How to display and remove notifications

How to check if a network connection is available

Figure 11-13 shows how to use a system service to check if a network connection is available to the device. This is often helpful since attempting to connect to the network when the network isn't available uses the battery unnecessarily. As a result, most apps should check if a connection is available before connecting to the network. In particular, the News Reader app should check if a connection is available before attempting to download the news feed from the Internet.

The first statement in the example calls the getSystemService method of the Context class to get a ConnectivityManager object. Then, the second statement calls the getActiveNetworkInfo method to get a NetworkInfo object that contains data about the active network for the device. If this object is not null, the example calls its isConnected method to determine whether it's possible for the device to connect to a network. If so, it downloads the news feed from the Internet.

The classes used to work with connectivity

```
android.app.ConnectivityManager
android.net.NetworkInfo
```

How to check if you have a connection to the network

```
// get NetworkInfo object
ConnectivityManager connectivityManager = (ConnectivityManager)
    getSystemService(Context.CONNECTIVITY_SERVICE);
NetworkInfo networkInfo =
    connectivityManager.getActiveNetworkInfo();

// if network is connected, download feed
if (networkInfo != null && networkInfo.isConnected()) {
    new DownloadFeed().execute();
}
```

A method of the ConnectivityManager class

Method	Description
`getActiveNetworkInfo()`	Returns a NetworkInfo object that contains info about the active network for the device.

Some methods of the NetworkInfo class

Method	Description
`isConnected()`	Returns a boolean value that indicates whether it's possible for the device to connect to the network.
`isRoaming()`	Returns a boolean value that indicates whether the device is roaming on the network and whether the connection may incur additional costs.
`getTypeString()`	Returns a string for the type of the network such as "WIFI" or "MOBILE".

The permission in the AndroidManifest.xml file

```
<uses-permission android:name="android.permission.ACCESS_NETWORK_STATE" />
```

Description

- You can use a ConnectivityManager object to check if a network connection is available to the device.

Figure 11-13   How to check if a network connection is available

The News Reader app

At this point, you've learned the skills necessary to make three improvements to the News Reader app. First, you can add a service that periodically downloads a new feed and checks for an updated feed. Second, you can have that service display a notification whenever an updated feed is available. Third, you can add a custom Application object to store the milliseconds for the current news feed and to start the service.

To make these improvements, you can add a custom Application class like the one shown in figure 11-1. In addition, you can add a NewsReaderService class like the one shown in figure 11-14. Finally, you can make some changes to the ItemsActivity and FileIO classes as shown in figures 11-15 and 11-16.

The NewsReaderService class

Figure 11-14 shows the code for the NewsReaderService class. Part 1 shows the lifecycle methods for the service. Here, the onCreate method gets references to the Application and FileIO objects and calls the startTimer method that starts the thread that performs the tasks for the service. The onStartCommand doesn't perform any tasks, but it returns the START_STICKY constant so Android can automatically restart the service if necessary. The onBind method isn't used, so it returns a null value. And the onDestroy method stops the thread for the service.

Part 2 begins by showing the startTimer method. This method creates the thread that executes the task for the service. This thread begins by downloading the file for the feed from the Internet. Then, it reads the file and parses it into an RSSFeed object.

After reading the RSSFeed object, this code checks whether the publication date of the new feed is newer than the current publication date that's stored in the Application object. If so, this code updates the Application object with the milliseconds for the publication date of the new feed. Then, it uses the sendNotification method to display a notification that indicates that an updated feed is available.

The stopTimer method stops the timer thread for the service. Since this method is called from the onDestroy method, this only occurs if the service is being destroyed. And since this app doesn't call the stopService or stopSelf methods, the service is typically only destroyed when the user turns the device off.

Part 3 shows the sendNotification method. This method creates a pending intent for the notification. It creates the variables for the notification. It creates the notification. And it uses a system service to display the notification.

This class contains many statements that display messages to the LogCat view. That provides a way for you to verify that the service is working as expected. Once you're sure the service is working correctly, you can remove some or all of these statements.

The NewsReaderService class   **Page 1**

```java
package com.murach.newsreader;

import java.util.Timer;
import java.util.TimerTask;

import android.app.Notification;
import android.app.NotificationManager;
import android.app.PendingIntent;
import android.app.Service;
import android.content.Intent;
import android.os.IBinder;
import android.support.v4.app.NotificationCompat;
import android.util.Log;

public class NewsReaderService extends Service {

    private NewsReaderApp app;
    private Timer timer;
    private FileIO io;

    @Override
    public void onCreate() {
        Log.d("News reader", "Service created");
        app = (NewsReaderApp) getApplication();
        io = new FileIO(getApplicationContext());
        startTimer();
    }

    @Override
    public int onStartCommand(Intent intent, int flags, int startId) {
        Log.d("News reader", "Service started");
        return START_STICKY;
    }

    @Override
    public IBinder onBind(Intent intent) {
        Log.d("News reader", "Service bound - not used!");
        return null;
    }

    @Override
    public void onDestroy() {
        Log.d("News reader", "Service destroyed");
        stopTimer();
    }
```

Figure 11-14   The NewsReaderService class (part 1 of 3)

The NewsReaderService class

```
    private void startTimer() {
        TimerTask task = new TimerTask() {

            @Override
            public void run() {
                Log.d("News reader", "Timer task started");

                io.downloadFile();
                Log.d("News reader", "File downloaded");

                RSSFeed newFeed = io.readFile();
                Log.d("News reader", "File read");

                // if new feed is newer than old feed
                if (newFeed.getPubDateMillis() > app.getFeedMillis()) {
                    Log.d("News reader", "Updated feed available.");

                    // update app object
                    app.setFeedMillis(newFeed.getPubDateMillis());

                    // display notification
                    sendNotification("Select to view updated feed.");
                }
                else {
                    Log.d("News reader", "Updated feed NOT available.");
                }

            }
        };

        timer = new Timer(true);
        int delay = 1000 * 60 * 60;        // 1 hour
        int interval = 1000 * 60 * 60;    // 1 hour
        timer.schedule(task, delay, interval);
    }

    private void stopTimer() {
        if (timer != null) {
            timer.cancel();
        }
    }
```

Figure 11-14   The NewsReaderService class (part 2 of 3)

The NewsReaderService class   **Page 3**

```
    private void sendNotification(String text) {
        // create the intent for the notification
        Intent notificationIntent = new Intent(this, ItemsActivity.class)
            .addFlags(Intent.FLAG_ACTIVITY_NEW_TASK);

        // create the pending intent
        int flags = PendingIntent.FLAG_UPDATE_CURRENT;
        PendingIntent pendingIntent =
                PendingIntent.getActivity(this, 0, notificationIntent, flags);

        // create the variables for the notification
        int icon = R.drawable.ic_launcher;
        CharSequence tickerText = "Updated news feed is available";
        CharSequence contentTitle = getText(R.string.app_name);
        CharSequence contentText = text;

        // create the notification and set its data
        Notification notification =
                new NotificationCompat.Builder(this)
            .setSmallIcon(icon)
            .setTicker(tickerText)
            .setContentTitle(contentTitle)
            .setContentText(contentText)
            .setContentIntent(pendingIntent)
            .setAutoCancel(true)
            .build();

        // display the notification
        NotificationManager manager = (NotificationManager)
                getSystemService(NOTIFICATION_SERVICE);
        final int NOTIFICATION_ID = 1;
        manager.notify(NOTIFICATION_ID, notification);
    }
}
```

Figure 11-14   The NewsReaderService class (part 3 of 3)

The ItemsActivity class

Figure 11-15 shows the ItemsActivity class. For the most part, this class is the same as the ItemsActivity class described in the previous chapter. As a result, you should already understand most of the code for this class.

However, this class has a few new statements that work with the Application object. These statements are highlighted. In the onResume method, the second statement gets the milliseconds for the publication date of the feed from the Application object. Then, an if statement checks whether the milliseconds for the publication date of the feed are equal to -1. If so, the app is being started for the first time. As a result, this code downloads the feed, reads the feed, and displays it. Otherwise, this code reads and displays the feed, or just displays it.

In the onPostExecute method of the ReadFeed class, the second statement sets the milliseconds for the publication date of the feed in the Application object. In other words, every time the app reads the feed, it sets the milliseconds in the Application object. As a result, the milliseconds for the feed are only equal to -1 the first time the app starts.

The FileIO class

Figure 11-16 shows the FileIO class. For the most part, this class is the same as the FileIO class described in the previous chapter. As a result, you should already understand most of the code for this class.

However, the downloadFile method of this class uses a system service to check if a network connection is available before it attempts to download the file. These statements are highlighted. Here, the first two statements get a NetworkInfo object. Then, this code uses the NetworkInfo object to check whether a network connection is available. If so, this code downloads the file from the Internet. If not, this code does not attempt to connect to the network, which is good since that conserves the battery of the device.

The ItemsActivity class

```java
package com.murach.newsreader;

import java.util.ArrayList;
import java.util.HashMap;
import java.util.List;

import android.os.AsyncTask;
import android.os.Bundle;
import android.app.Activity;
import android.content.Intent;
import android.util.Log;
import android.view.Menu;
import android.view.MenuItem;
import android.view.View;
import android.widget.AdapterView;
import android.widget.ListView;
import android.widget.SimpleAdapter;
import android.widget.TextView;
import android.widget.Toast;
import android.widget.AdapterView.OnItemClickListener;

public class ItemsActivity extends Activity
implements OnItemClickListener {

    private NewsReaderApp app;
    private RSSFeed feed;
    private long feedPubDateMillis;
    private FileIO io;

    private TextView titleTextView;
    private ListView itemsListView;

    @Override
    protected void onCreate(Bundle savedInstanceState) {
        super.onCreate(savedInstanceState);
        setContentView(R.layout.activity_items);

        titleTextView = (TextView) findViewById(R.id.titleTextView);
        itemsListView = (ListView) findViewById(R.id.itemsListView);

        itemsListView.setOnItemClickListener(this);

        // get references to Application and FileIO objects
        app = (NewsReaderApp) getApplication();
        io = new FileIO(getApplicationContext());
    }

    @Override
    public void onResume() {
        super.onResume();

        // get feed from app object
        feedPubDateMillis = app.getFeedMillis();
```

Figure 11-15   The ItemsActivity class (part 1 of 3)

The ItemsActivity class   **Page 2**

```
if (feedPubDateMillis == -1) {
    new DownloadFeed().execute();    // download, read, and display
}
else if (feed == null) {
    new ReadFeed().execute(); // read and display
}
else {
    updateDisplay();    // just display
}
}

class DownloadFeed extends AsyncTask<Void, Void, Void> {
    @Override
    protected Void doInBackground(Void... params) {
        io.downloadFile();
        return null;
    }

    @Override
    protected void onPostExecute(Void result) {
        Log.d("News reader", "Feed downloaded");
        new ReadFeed().execute();
    }
}

class ReadFeed extends AsyncTask<Void, Void, Void> {
    @Override
    protected Void doInBackground(Void... params) {
        feed = io.readFile();
        return null;
    }

    @Override
    protected void onPostExecute(Void result) {
        Log.d("News reader", "Feed read");
        app.setFeedMillis(feed.getPubDateMillis());
        ItemsActivity.this.updateDisplay();
    }
}

public void updateDisplay()
{
    if (feed == null) {
        titleTextView.setText("Unable to get RSS feed");
        return;
    }

    // set the title for the feed
    titleTextView.setText(feed.getTitle());

    // get the items for the feed
    List<RSSItem> items = feed.getAllItems();
```

Figure 11-15   The ItemsActivity class (part 2 of 3)

The ItemsActivity class   **Page 3**

```java
    // create a List of Map<String, ?> objects
    ArrayList<HashMap<String, String>> data =
            new ArrayList<HashMap<String, String>>();
    for (RSSItem item : items) {
        HashMap<String, String> map = new HashMap<String, String>();
        map.put("date", item.getPubDateFormatted());
        map.put("title", item.getTitle());
        data.add(map);
    }

    // create the resource, from, and to variables
    int resource = R.layout.listview_item;
    String[] from = {"date", "title"};
    int[] to = { R.id.pubDateTextView, R.id.titleTextView};

    // create and set the adapter
    SimpleAdapter adapter =
            new SimpleAdapter(this, data, resource, from, to);
    itemsListView.setAdapter(adapter);

    Log.d("News reader", "Feed displayed");
}

@Override
public void onItemClick(AdapterView<?> parent, View v,
        int position, long id) {
    RSSItem item = feed.getItem(position);
    Intent intent = new Intent(this, ItemActivity.class);
    intent.putExtra("pubdate", item.getPubDate());
    intent.putExtra("title", item.getTitle());
    intent.putExtra("description", item.getDescription());
    intent.putExtra("link", item.getLink());
    this.startActivity(intent);
}

@Override
public boolean onCreateOptionsMenu(Menu menu) {
    getMenuInflater().inflate(R.menu.activity_items, menu);
    return true;
}

@Override
public boolean onOptionsItemSelected(MenuItem item) {
    switch (item.getItemId()) {
        case R.id.menu_refresh:
            new DownloadFeed().execute();
            Toast.makeText(this, "Feed refreshed!",
                    Toast.LENGTH_SHORT).show();
            return true;
        default:
            return super.onOptionsItemSelected(item);
    }
}
}
```

Figure 11-15   The ItemsActivity class (part 3 of 3)

The FileIO class

```java
package com.murach.newsreader;

import java.io.FileInputStream;
import java.io.FileOutputStream;
import java.io.IOException;
import java.io.InputStream;
import java.net.URL;

import javax.xml.parsers.SAXParser;
import javax.xml.parsers.SAXParserFactory;
import org.xml.sax.InputSource;
import org.xml.sax.XMLReader;

import android.content.Context;
import android.net.ConnectivityManager;
import android.net.NetworkInfo;
import android.util.Log;

public class FileIO {

    private final String URL_STRING = "http://rss.cnn.com/rss/cnn_tech.rss";
    private final String FILENAME = "news_feed.xml";
    private Context context = null;

    public FileIO (Context context) {
        this.context = context;
    }

    public void downloadFile() {
        // get NetworkInfo object
        ConnectivityManager cm = (ConnectivityManager)
            context.getSystemService(Context.CONNECTIVITY_SERVICE);
        NetworkInfo networkInfo = cm.getActiveNetworkInfo();

        // if network is connected, download feed
        if (networkInfo != null &&  networkInfo.isConnected()) {

            try{
                // get the URL
                URL url = new URL(URL_STRING);

                // get the input stream
                InputStream in = url.openStream();

                // get the output stream
                FileOutputStream out =
                    context.openFileOutput(FILENAME, Context.MODE_PRIVATE);
```

Figure 11-16   The FileIO class (part 1 of 2)

The FileIO class   **Page 2**

```java
                // read input and write output
                byte[] buffer = new byte[1024];
                int bytesRead = in.read(buffer);
                while (bytesRead != -1)
                {
                    out.write(buffer, 0, bytesRead);
                    bytesRead = in.read(buffer);
                }
                out.close();
                in.close();
            }
            catch (IOException e) {
                Log.e("News reader", e.toString());
            }
        }
    }

    public RSSFeed readFile() {
        try {
            // get the XML reader
            SAXParserFactory factory = SAXParserFactory.newInstance();
            SAXParser parser = factory.newSAXParser();
            XMLReader xmlreader = parser.getXMLReader();

            // set content handler
            RSSFeedHandler theRssHandler = new RSSFeedHandler();
            xmlreader.setContentHandler(theRssHandler);

            // read the file from internal storage
            FileInputStream in = context.openFileInput(FILENAME);

            // parse the data
            InputSource is = new InputSource(in);
            xmlreader.parse(is);

            // set the feed in the activity
            RSSFeed feed = theRssHandler.getFeed();
            return feed;
        }
        catch (Exception e) {
            Log.e("News reader", e.toString());
            return null;
        }
    }
}
```

Figure 11-16   The FileIO class (part 2 of 2)

Perspective

The skills presented in this chapter should be enough to get you started with services and notifications. However, this chapter doesn't show how to work with bound services. As a result, if you need to use bound services, you can search the Internet for more information about working with them.

In this chapter, the service is started when the app starts, which is appropriate for some apps. However, you may also want to start a service when the device boots or when the network becomes available. In the next chapter, you'll learn how to do that.

Terms

service	pending intent
unbound service	back stack
bound service	task
notification	task history
notification area	system service
notification drawer	

Summary

- To store data and methods that apply to the entire application, you can extend the Application class and add instance variables and methods.

- A *service* performs tasks in the background, does not provide a user interface, and continues to run even if the user switches to another app.

- An *unbound service* does not interact with other components such as activities, and runs until it's stopped by another component or by itself.

- A *bound service* can interact with components such as activities. This type of service runs only as long as another component is bound to it. Multiple components can bind to the service at once, but when all of them unbind, the service is destroyed.

- To test a service, you can print messages to the LogCat view.

- You can use the Settings app to view all of the apps and services that are running on a device or emulator.

- Before you can use an Application object or a service, you must register it.

- A *notification* provides a way for a service to display a message even when another app is running.

- When it's first displayed, a notification appears as an icon in the *notification area* at the top of the screen.

- To view a notification, the user can pull down on the notification area to open the *notification drawer*.

- A *pending intent* is an intent that can be passed to other apps so that they can execute the intent at a later time.

- The *back stack* is a stack of all recently used activities. These activities are sorted in the order in which they were used.

- A *task* is a cohesive unit that can contain multiple activities. Every activity belongs to a task.

- The *task history* is the back stack for a particular task.

- A *system service* is a service that's provided by the Android operating system.

- You can use the NotificationManager object to display or remove a notification, and you can use the ConnectivityManager object to check if a network connection is available to the device.

Exercise 11-1   Work with a service

In this exercise, you'll create, register, and test a service. Then, you'll modify the service so it stops itself.

Test the app

1. Start Eclipse and import the project named ch11_ex1_Tester.

2. Run the app. This should launch an activity that displays a message that indicates that you should start this exercise by adding a service.

Create a service that runs indefinitely

3. Add a class named TesterService to the product. Modify this class so it extends the Service class.

4. Add code that implements all four methods as described in figure 11-4. Within each method, add a statement that displays a message in the LogCat view. For the onCreate method, for example, add a statement like this:

    ```
    Log.d("Test service", "Service created");
    ```

5. Open the AndroidManifest.xml file for the project and register this service as described in figure 11-5.

6. Open the TesterActivity class, and add a statement to its onCreate method that starts the service.

7. Run the app. This should launch the activity described earlier in this exercise, and it should display these messages in the LogCat view:

    ```
    Service created
    Service started
    ```

8. Change the orientation for the app one or more times. This should display additional "Service started" messages in the LogCat view.

9. Navigate to another app. Then, navigate back to the Tester app. Note that this does not display the "Service destroyed" message. As a result, this approach is useful for services that you want to continue running.

10. Use the Settings app to view the service for the Tester app. This app should be using 1 process and 1 service.

Create a service that stops itself

11. Open the TesterService class, and add two statements to the onStartCommand method. The first statement should display a message in the LogCat view that says, "Task completed". The second statement should call the stopSelf method.

12. Run the app. This should display these messages in the LogCat view:
    ```
    Service created
    Service started
    Task completed
    Service destroyed
    ```
    This approach is useful for a service that you want to stop after it completes a task. For long-running tasks, it's generally considered a best practice to use the AsyncTask class to execute the task in a separate thread.

13. Comment out the two statements that you added to the onStartCommand method in step 11.

Add a timer that performs a task at specified intervals

14. Open the TesterService class, and add the startTimer and stopTimer methods shown in figure 11-6.

15. Modify the startTimer method so it displays the first message after a delay of 1 second. Then, it should display the message every 10 seconds.

16. Modify the onCreate so it calls the startTimer method.

17. Modify the onDestroy method so it calls the stopTimer method.

18. Run the app. This should display these messages in the LogCat view:
    ```
    Service created
    Service started
    Timer task executed
    Timer task executed
    Timer task executed
    ...
    ```

19. Comment out the statements that call the startTimer and stopTimer methods.

20. Run the app. This should stop the timer from displaying a new message in the LogCat view every 10 seconds.

Exercise 11-2   Modify the notification for the News Reader app

In this exercise, you'll modify the notification used by the News Reader app presented in this chapter.

1. Start Eclipse and import the project named ch11_ex2_NewsReader.

2. Open the NewsReaderService class. Then, review the code, and note how the various methods display messages in the LogCat view.

3. In the startTimer method, modify the code so it only delays for 5 seconds (instead of 1 hour) before it runs the timer task. Then, in the if/else statement, add a statement that displays a notification even when no update is available. The text for this notification should be: "Updated feed NOT available."

4. Run the app. After a delay of 5 seconds, it should display a notification in the notification area. Pull down on this area to display the drawer and select the notification. This should display the News Reader app.

5. Press the Back button to return to remove the current News Reader activity from the stack and return to the previous activity such as the Home screen.

6. Modify the sendNotification method so it uses the "new task" flag instead of the "clear top" flag.

7. Repeat step 4.

8. Press the Back button to remove the current activity from the stack. This should exit the News Reader activity displayed by the notification, but it should leave the original News Reader activity on the stack.

9. Press the Back button again to remove that activity from the stack. This should remove the original News Reader activity from the stack and display the previous activity such as the Home screen.

12

How to work
with broadcast receivers

Some Android apps need to respond to actions that are broadcast by the system.
For example, an app may need to start a service after the user turns the device
on. To do that, you can use an Android component known as a broadcast
receiver.

Other apps need to broadcast an action. For example, the service for the
News Reader app may want to broadcast an action when a new RSS feed is
available. Then, other components within the News Reader app, such as the
Items activity, can use a broadcast receiver to respond to that action. Or, other
apps can use a broadcast receiver to respond to that action.

How to work with system broadcasts

As you use your Android device, the operating system can *broadcast* an *action* that has occurred. For example, when the device finishes starting up, Android broadcasts an action that indicates that the device has completed booting. To receive these broadcasts, an app can include a component known as a *broadcast receiver*. A broadcast receiver provides a way for the app to execute code when the broadcast is received.

A summary of the system broadcasts

Every broadcast has an *action string* that uniquely identifies the action. Figure 12-1 begins by showing five constants that define the action string for five broadcasts that are commonly made by the Android operating system. The first four constants are stored in the Intent class. These constants are general-purpose constants that apply to many different types of apps. However, the fifth constant is stored in the ConnectivityManager class. This constant is stored here because it only applies to apps that use a network connection.

This figure only shows the action strings for five system broadcasts. However, Android 2.2 (API 8) provides 98 of these actions, and each new version of Android seems to add a few more. To view a list of these action strings, you can use a text editor to open the broadcast_actions.txt file for one of the Android platforms that have been installed on your system. This figure, for example, shows the path for this text file for Android 4.2 (API 17).

Two classes used to work with broadcasts

```
android.content.Intent
android.net.ConnectivityManager
```

Some action constants of the Intent class

Constant	Action string / Description
`ACTION_BOOT_COMPLETED`	`android.intent.action.BOOT_COMPLETED` Boot completed. This broadcast is typically used to start services that should always run.
`ACTION_BATTERY_LOW`	`android.intent.action.BATTERY_LOW` Battery is low. This broadcast is typically used to pause activities that consume power.
`ACTION_BATTERY_OKAY`	`android.intent.action.BATTERY_OKAY` Battery is OK. This broadcast can be used to resume activities that consume power
`ACTION_POWER_CONNECTED`	`android.intent.action.ACTION_POWER_CONNECTED` Power source connected to the device.

An action constant of the ConnectivityManager class

Constant	Action string / Description
`CONNECTIVITY_ACTION`	`android.net.conn.CONNECTIVITY_CHANGE` A change in network connectivity has occurred.

A text file that contains the broadcast actions for the system

```
\android-sdks\platforms\android-17\data\broadcast_actions.txt
```

Description

- The Android operating system *broadcasts* certain *actions* that occur as a device is being used.
- Every broadcast has an *action string* that uniquely identifies the action.
- A *broadcast receiver* is an application component that listens for a broadcast and executes when it receives that broadcast.

Figure 12-1   A summary of system broadcasts

How to code a receiver
for the boot completed broadcast

Figure 12-2 shows how to code a receiver for the boot completed broadcast. To do that, you can code a class that extends the BroadcastReceiver class. In this figure, for example, the BootReceiver class extends the BroadcastReceiver class.

Within the class for the receiver, you can override the onReceive method. This method is executed when the broadcast is received. In this figure, this method starts the service for the News Reader app described in the previous chapter.

After you create the class for a broadcast receiver, you need to register the receiver, so Android knows what broadcasts it should receive. The easiest way to register a receiver is to add a receiver element to the project's AndroidManifest.xml file. This receiver element should specify the name of the class for the receiver, and it should specify the action or actions to receive. To do that, you can use the name attribute of the action element to specify the action string. In this figure, the second example shows how to register the BootReceiver class so it listens for the BOOT_COMPLETED action.

Some broadcast receivers require additional permissions. For example, the BOOT_COMPLETED action in this figure requires the RECEIVE_BOOT_COMPLETED permission. To request this permission, you can add a uses-permission element to the AndroidManifest.xml file. Then, when users install the app, they can decide whether or not to grant this permission.

The BootReceiver class

```
package com.murach.newsreader;

import android.content.BroadcastReceiver;
import android.content.Context;
import android.content.Intent;
import android.util.Log;

public class BootReceiver extends BroadcastReceiver {

    @Override
    public void onReceive(Context context, Intent intent) {
        Log.d("News reader", "Boot completed");

        // start service
        Intent service = new Intent(context, NewsReaderService.class);
        context.startService(service);
    }
}
```

The receiver element in the AndroidManifest.xml file

```
<receiver android:name="BootReceiver" >
    <intent-filter>
        <action android:name="android.intent.action.BOOT_COMPLETED" />
    </intent-filter>
</receiver>
```

The permission in the AndroidManifest.xml file

```
<uses-permission android:name="android.permission.RECEIVE_BOOT_COMPLETED" />
```

Description

- You can create a broadcast receiver that executes code when the device finishes starting, which is also known as booting.

- To define a broadcast receiver, code a class that inherits the BroadcastReceiver class and override its onReceive method. This method is executed when the broadcast is received.

- To register a broadcast receiver, open the project's AndroidManifest.xml file and add a receiver element. This element should specify the name of the class for the receiver, and it should specify the action or actions to receive. To do that, you can add an action element and use its name attribute to specify the action string for the action.

- If a broadcast receiver requires permissions, you can request those permissions by adding a uses-permission element to the AndroidManifest.xml file.

Figure 12-2   How to code a receiver for the boot completed broadcast

How to code a receiver
for the connectivity changed broadcast

Figure 12-3 shows how to code a receiver for the connectivity changed broadcast. This works similarly to coding a receiver for the boot completed broadcast as shown in the previous figure. As a result, you shouldn't have much trouble understanding how this works.

The ConnectivityReceiver class shown in this figure extends the BroadcastReceiver class and overrides its onReceive method. Within this method, the second and third statements get a NetworkInfo object as described in the previous chapter. Then, the fourth statement creates an intent that can be used to start or stop the service for the News Reader class. After that, this code checks whether a network connection is available. If so, this code starts the service. Otherwise, it stops the service.

Since this receiver makes sure that the service doesn't attempt to connect to the network when a connection isn't available, it conserves the battery of the device. In the previous chapter, you learned another technique for conserving battery life. Although these techniques work differently, both accomplish the same goal of conserving battery life.

Like the BootReceiver class, you can register the ConnectivityReceiver class by adding a receiver element to the AndroidManifest.xml file. Similarly, you can add a uses-permissions element to request permissions for the ACCESS_NETWORK_STATE privilege that's required to allow this class to access information about the network. This permission isn't required by the CONNECTIVITY_CHANGE action. However, it is required by the getActiveNetworkInfo method that's used by the onReceive method.

The ConnectivityReceiver class

```
package com.murach.newsreader;

import android.content.BroadcastReceiver;
import android.content.Context;
import android.content.Intent;
import android.net.ConnectivityManager;
import android.net.NetworkInfo;
import android.util.Log;

public class ConnectivityReceiver extends BroadcastReceiver {

    @Override
    public void onReceive(Context context, Intent intent) {
        Log.d("News reader", "Connectivity changed");

        ConnectivityManager connectivityManager = (ConnectivityManager)
                context.getSystemService(Context.CONNECTIVITY_SERVICE);
        NetworkInfo networkInfo =
                connectivityManager.getActiveNetworkInfo();

        Intent service = new Intent(context, NewsReaderService.class);
        if (networkInfo != null && networkInfo.isConnected()){
            Log.d("News reader", "Connected");
            context.startService(service);
        }
        else {
            Log.d("News reader", "NOT connected");
            context.stopService(service);
        }
    }
}
```

The receiver element in the AndroidManifest.xml file

```
<receiver android:name="ConnectivityReceiver" >
    <intent-filter>
        <action android:name="android.net.conn.CONNECTIVITY_CHANGE" />
    </intent-filter>
</receiver>
```

The permission in the AndroidManifest.xml file

```
<uses-permission android:name="android.permission.ACCESS_NETWORK_STATE" />
```

Description

- You can code a broadcast receiver to execute code when the connectivity for the device changes.

Figure 12-3   How to code a receiver for the connectivity change broadcast

How to work with custom broadcasts

In some cases, you may want to send a custom broadcast that other components within your app can receive. Or, you may want to send a custom broadcast that other apps can receive. Then, you can code a receiver for that broadcast.

In the News Reader app, for example, you may want to automatically update the display for the Items activity if a new feed becomes available. That way, if a user is viewing the Items activity and a new update becomes available, the News Reader app will automatically update the display. To accomplish this, the service for this app can send a custom broadcast that indicates that a new feed is available. Then, the Items activity can receive that broadcast and update the display.

How to create and send a custom broadcast

Figure 12-4 shows how to create and send a custom broadcast. To start, you can define a constant for the action string for the broadcast. By convention, this string should begin with the name of the package for your app, followed by the name of the constant. This should create an action string that's globally unique. In addition, you should typically code this constant in a class that's related to the action. In this figure, for example, the first step defines a constant named NEW_FEED in the RSSFeed class. This constant indicates that a new feed is available.

Once you create a constant for a broadcast, you can add code that creates the intent for the broadcast and sends the broadcast. To start, you can create an Intent object for the broadcast by passing the constant for the broadcast to the constructor of this class. Then, if necessary, you can use the putExtra method to store data in the Intent object. In this figure, the example puts a test string in the Intent object. However, you can use the putExtra method to store most types of data that you want to pass to the receiver. Finally, you can use the sendBroadcast method to send the broadcast.

Two classes used to work with broadcasts

```
android.content.Intent
android.content.Context
```

How to send a broadcast

Step 1: Define a constant for the broadcast (in the RSSFeed class)

```
public final static String NEW_FEED = "com.murach.newsreader.NEW_FEED";
```

Step 2: Create and send the broadcast (in the NewsReaderService class)

```
Intent intent = new Intent(RSSFeed.NEW_FEED);
intent.putExtra("test", "test1 test2");
sendBroadcast(intent);
```

A constructor of the Intent class

Constructor	Description
Intent(actionString)	Create an intent with the specified action string.

A method of the Context class

Method	Description
sendBroadcast(intent)	Broadcast the specified intent to all registered receivers.

Description

- The constant for a custom broadcast should store an action string that's globally unique. By convention, this string should begin with the name of the package for your app, followed by the name of the constant.

- To send a custom broadcast, create an Intent object for the broadcast by passing the constant for the action string to the constructor of the Intent class. Then, use the sendBroadcast method to send the broadcast.

- If you want to pass data to the receiver, you can use the putExtra method to store extra data in the Intent object.

Figure 12-4   How to create and send a custom broadcast

How to code a receiver for a custom broadcast

Figure 12-5 shows how to code a receiver for a custom broadcast. The steps for coding a receiver for a custom broadcast are similar to the steps for coding a receiver for a system broadcast. As a result, you shouldn't have much trouble understanding how to code the receiver shown in this figure.

However, if you want to update the user interface when a broadcast is received, you can code the receiver as an inner class of the activity. That way, you can easily access elements of the user interface, including any methods that update the user interface. In this figure, for example, the NewFeedReceiver class is coded as an inner class of the ItemsActivity class. As a result, it can access the updateDisplay method that's available from the ItemsActivity class.

If you want to get extra data from the Intent for the action, you can use the getXxxExtra methods of the Intent class. In this figure, the second statement in the onReceive method uses the getStringExtra method to get the test string that was set by the code that broadcast the action. Then, the third statement sends this data to the LogCat view. Although this doesn't do anything useful for the News Reader app, it illustrates how a broadcast can be used to pass data from one component to another.

When you code a receiver as an inner class of an activity, you typically use Java code to register and unregister the receiver. To do that, you create a receiver object and a filter object for the intent that has the specified action string. To create an IntentFilter object, you can pass the constant for the action string to the constructor of the IntentFilter class.

After you create these objects, you can use the methods of the Context class to register and unregister the receiver. Typically, you add these methods to the onResume and onPause methods of the activity. In this figure, the code examples register and unregister the receiver defined by the NewFeedReceiver class.

Since the NEW_FEED action that's broadcast and received by the News Reader app doesn't contain any sensitive information, I decided to not require privileges to broadcast or receive this action. As a result, it may be possible for another app to broadcast or receive this action. However, it's unlikely that another app (even a malicious one) would want to do this, and it wouldn't be too harmful if it did.

However, if your app broadcasts or receives an action that should be secure, you can use permissions to secure the broadcast. To do that, you can define the send and receive permissions in the manifest file for your project. Then, you can modify the sendBroadcast and registerReceiver methods so they enforce these permissions. For more information, you can search the Internet.

A class used to work with receivers
```
android.content.Context
```

How to receive a broadcast in the ItemsActivity class

Step 1: Define an inner class for the broadcast receiver
```
class NewFeedReceiver extends BroadcastReceiver {

    @Override
    public void onReceive(Context context, Intent intent) {
        Log.d("News reader", "New items broadcast received");

        // if necessary, get data from intent
        String test = intent.getStringExtra("test");
        Log.d("News reader", "test: " + test);

        // update the display
        updateDisplay();
    }
}
```

Step 2: Define instance variables for the broadcast receiver and intent filter
```
private NewFeedReceiver newFeedReceiver;
private IntentFilter newFeedFilter;
```

Step 3: Create the broadcast receiver and intent filter (onCreate)
```
newFeedFilter = new IntentFilter(RSSFeed.NEW_FEED);
newFeedReceiver = new NewFeedReceiver();
```

Step 4: Register the receiver (onResume)
```
registerReceiver(newFeedReceiver, newFeedFilter);
```

Step 5: Unregister the receiver (onPause)
```
unregisterReceiver(newFeedReceiver);
```

Two methods of the Context class

Methods	Description
`registerReceiver(receiver, filter)`	Register the specified broadcast receiver with the specified intent filter.
`unregisterReceiver(receiver)`	Unregister the specified broadcast receiver.

Description
- To code a receiver for a custom broadcast, you can define a class for the receiver. If you want to update the user interface when a broadcast is received, you can code the class for the receiver as an inner class of the activity.
- If you want to get extra data from the Intent for the action, you can use the getXxxExtra methods of the Intent class.
- You can use Java code to register and unregister a receiver. To do that, you can create a receiver object and a filter object for the intent with the specified action.

Figure 12-5   How to code a receiver for a custom broadcast

Perspective

The skills presented in this chapter should be enough to get you started with broadcast receivers. However, Android broadcasts actions that you can use to execute code due to changes in the battery, text messages, phone state, external hardware devices, and so on.

If you are developing an app that needs to receive and work with these types of broadcasts, you can search the Internet for more information. To do that, you can begin by using the broadcast_actions.txt described in figure 12-1 to find the action string for the broadcast. Then, you can search the Internet to find the API documentation for that broadcast action. This documentation should include a description of the action, a description of any extra data that's stored in the intent for the action, and a description of any permissions that are required by the action.

Terms

broadcast	broadcast receiver
action	action string

Summary

- The Android operating system *broadcasts* certain *actions* that occur as a device is being used.
- Every broadcast has an *action string* that uniquely identifies the action.
- A *broadcast receiver* is an application component that listens for a broadcast and executes code when it receives that broadcast.
- You can code a broadcast receiver to execute code when the connectivity for the device changes.
- To code a receiver for a custom broadcast, you can define a class for the receiver.
- If you want to get extra data from the Intent for the action, you can use the getXxxExtra methods of the Intent class.
- You can use Java code to register and unregister a receiver.

Exercise 12-1 Work with broadcast receivers

In this exercise, you'll experiment with the broadcast receivers that are available from the News Reader app.

Test the existing broadcast receivers

1. Start Eclipse and import the project named ch12_ex1_NewsReader.

2. Open the BootReceiver and ConnectivityReceiver classes. Then, review the code for these classes.

3. Open the AndroidManifest.xml file. Note that it registers both of the broadcast receivers described in the previous step and that it requests all permissions required by those receivers.

4. Run the app. If you're using an actual device, test the Connectivity receiver by turning on airplane mode. This should display messages in the LogCat view that say:

   ```
   Connectivity changed
   NOT connected
   ```

   NOTE: Due to a bug in most emulators, this might not work correctly on an emulator.

5. Turn off airplane mode. This should display messages in the LogCat view that say:

   ```
   Connectivity changed
   Connected
   ```

   NOTE: Again, this might not work correctly on an emulator.

6. Open the broadcast_actions.txt file for one of the Android SDKs that's installed on your system. Then, review the various broadcasts that your application can receive.

Add a new broadcast receiver

7. Add a broadcast receiver that displays a message on the LogCat view that says "Battery low!". This receiver should also stop the service for the app.

8. Use the AndroidManifest.xml file to register this receiver. Unfortunately, there's no easy way to test this receiver. As a result, don't test this receiver yet. Instead, review the code and make sure that it seems right.

Modify a broadcast

9. Open the RSSFeed class. Then, change the name of the constant for the broadcast from NEW_FEED to ACTION_UPDATE_AVAILABLE. Also, change the name of the action string so it matches the name of the constant.

10. Open the ItemsActivity and NewsReaderService classes. Then, make sure the statement that creates the IntentFilter object uses the new name for the constant.

Section 4

The Task List app

This section shows how to develop the Task List app. As you learn to develop this app, you'll also learn many important Android skills that can be applied to other apps. To start, chapter 13 shows how to use SQLite to create and work with the database for an app. Then, chapter 14 shows how to use tabs and custom adapters to display data on the user interface for the app. Next, chapter 15 shows how to use a content provider to make the data for an app available to other apps. Finally, chapter 16 shows how to create an app widget than can display the app's data on a device's Home screen.

13

How to work with SQLite databases

Some types of Android apps need to store data in a local database. For example, a Contacts app needs to store contact information such as names and phone numbers in a database. Fortunately, Android includes libraries that make it easy to create and use local databases. In this chapter, you'll learn how to create and use a database for a Task List app.

An introduction to databases

The skills presented in this chapter are necessary to create the Task List app. To put these skills in context, this chapter begins by showing the user interface for this app.

The user interface for the Task List app

Figure 13-1 shows the interface for the Task List app. When the app starts, the Task List activity displays a list of tasks where each task has a check box that indicates whether it has been completed, a name, and optional notes.

In addition, the Task List activity provides tabs for two lists: Personal and Business. To display lists, the user can click on the tabs that are available from the Task List activity. In this figure, for example, the Personal tab is selected and the tasks for that tab are displayed. However, the user could click on the Business tab to display the tasks for that list.

From the Task List activity, the user can mark a task as completed by selecting a check box. Then, the user can remove all tasks that have been marked as completed by clicking on the Delete Tasks icon (the small trash can). This icon is available from the task bar. However, older versions of Android don't provide for the task bar. As a result, on older devices, the user can display the options menu and select the Delete Tasks item.

The user can add a task by clicking on the Add Task icon (the + sign) that's available from the task bar. Or, if the device is too old to display a task bar, the user can display the options menu and select the Add Task item. This displays the Add/Edit activity in "add mode". At this point, the user can use the spinner to select the list. Then, the user can use the editable text views to add or edit the text for the task.

Similarly, the user can edit a task by clicking on the task in the Task List activity. This displays the Add/Edit activity in "edit mode" so the user can edit the data for the task. In this figure for example, I clicked on the "Get hair cut" task in the Task List activity. As a result, this task is displayed in the Add/Edit activity and is ready to be edited.

Take a moment to reflect on the data this app needs. First, it needs the names of the lists. Second, it needs at least four pieces of data about a task: its list, its name, its notes, and whether it has been completed. In this chapter, you'll learn how to store this type of data in a local database.

The Task List activity

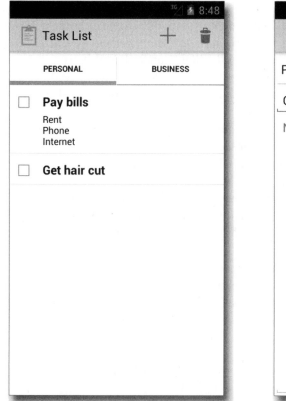

The Add/Edit activity

Description

- When the app starts, the Task List activity displays a list of tasks where each task has a check box that indicates whether it has been completed, a name, and optional notes.

- The app provides for two lists: Personal and Business.

- To display lists, the user can click on the tabs that are available in the Task List activity.

- To edit a task, the user can click on the task in the Task List activity. This displays the Add/Edit activity in "edit mode" so the user can edit the data for the task.

- To mark a task as completed, the user can select the check box in the Task List activity.

- To remove all completed tasks, the user can click the Delete Tasks button that's available from the task bar.

- To add a new task, the user can navigate to the Task List activity and click on the Add Task icon. This displays the Add/Edit activity in "add mode" so the user can add a new task.

- In the Add/Edit activity, the user can use the spinner to select the list. Then, the user can use the editable text views to add or edit the text for the task.

Figure 13-1   The user interface for the Task List app

An introduction to SQLite

Figure 13-2 describes the SQLite database that's available from Android. What is SQLite? To start, SQLite is a *relational database management system* (*RDBMS*) that implements most, but not all, of the SQL standard. SQLite is a lightweight (approximately 350KB) programming library. This library is available on every Android device and does not require any configuration or database administration.

When you create a SQLite database, that database is stored in a single file on a device. In this figure, for example, you can see where the file for the Task List app is stored on a device. Since that file is only accessible by the app that created the file, a SQLite database is relatively secure.

SQLite is a popular choice as an embedded database. It is used today by several browsers and operating systems, including the Android operating system. One of the reasons that SQLite is used so widely is because it is open-source. As a result, it's free for most purposes.

If you're familiar with other database management systems such as MySQL or Oracle, you shouldn't have much trouble learning to work with SQLite. However, SQLite is different than other database systems in a few ways.

First, SQLite does not run in a server process that's accessed by client apps, which run in separate processes. Instead, it's embedded as part of the client app and runs in the same process.

Second, SQLite only supports three data types (TEXT, INTEGER, and REAL). These data types correspond with these three Java data types: String, long, and double. As a result, other Java data types such as date/time and Boolean values must be converted into one of the SQLite types before saving them in the database. For example, a Date type can be converted to a TEXT type before it's stored in the database. Then, when it's retrieved from the database, it can be converted from a TEXT type to a Date type.

Third, SQLite is weakly-typed. In other words, a column in a SQLite database does not reject a value of an incorrect data type. For example, if you define a database column to accept the INTEGER type, you could still store a string value in this column and it would not be rejected by the database. Usually, this isn't a problem as the code for the app shouldn't attempt to store a string in an INTEGER column. However, you should be aware of this as it means that the code for your app must make sure to insert values of the correct type in each column.

SQLite is...

- **A RDBMS.** SQLite is a *relational database management system* (*DBMS*).
- **Standards-compliant.** SQLite implements most of the SQL standard.
- **Embedded.** Unlike most database management systems, SQLite does not run in a server process that's accessed by client apps, which run in separate processes. Instead, it's embedded as part of the client app and runs in the same process.
- **Zero-configuration.** SQLite is available on every Android device and does not require any database administration.
- **Lightweight**. The SQLite programming library is approximately 350 KB.
- **Secure**. A SQLite database is only accessible by the app that created the file for the database.
- **Popular.** SQLite is a popular choice as an embedded database.
- **Open-source.** The source code for SQLite is in the public domain.

Data types supported by SQLite

SQLite data type	Java data type
TEXT	String
INTEGER	long
REAL	double

The location of the SQLite database file for the Task List app

`/data/data/com.murach.tasklist/databases/tasklist.db`

Description

- SQLite only supports three data types (String, long, and double). As a result, the code for the app must convert other Java data types such as date/time and Boolean values into one of the SQLite types before saving them in the database.
- SQLite is weakly-typed. In other words, a column in a SQLite database does not reject a value of an incorrect data type. As a result, the code for the app must make sure to insert values of the correct type.

Figure 13-2   An introduction to SQLite

An introduction to the Task List database

Figure 13-3 begins by showing a diagram for the Task List database. This diagram shows that the data for the Task List database is stored in two tables: the List table and the Task table. These tables are related to each other in a one-to-many relationship. As a result, a single list can have many tasks, but a task can only have a single list.

The List table only has two columns. The _id column uniquely identifies each row, and the list_name column provides the name of the list.

The Task table has six columns. The _id column uniquely identifies each row, and the list_id column identifies the list for this task. The task_name and notes columns store the name and notes for the task.

The date_completed column stores the date and time that the task was completed. As a result, this column can be used to determine whether the task has been completed. Alternately, this column could store a Boolean value. However, for this database, I choose to store the date/time value since this allows the database to store more potentially useful information about the task.

The hidden column stores a Boolean value that indicates whether the task should be shown on a list. I choose to use this column because I'd prefer to hide tasks instead of deleting them. In other words, when a user marks a task as complete and removes it from a list, the Task List app doesn't delete it from the database. Instead, it marks it as hidden, and doesn't display it on the user interface anymore. That way, a history of completed tasks remains in the database.

When naming columns, it's a common coding convention to use _id as the name of the primary key column. This isn't required, but it makes it easier to create a content provider for a database as described in chapter 15.

The CREATE TABLE statements provide the names of the tables as well as the column names, data types, and attributes of those tables. Here, the first column in each table uses the PRIMARY KEY attribute to define a primary key column. This column requires that each row must have a value, and that value must be unique. Then, the first column uses the AUTO_INCREMENT attribute to automatically increment the value of the column when a new row is inserted.

In the List table, the list_name column uses the NOT NULL attribute to specify that this column must store a not-null value. In addition, it uses the UNIQUE attribute to specify that each value stored in the column must be unique. As a result, for a new row, the database requires the user to specify a non-null value for the list name, and it doesn't allow a list name that already exists in the database. This is usually what you want, since it doesn't make sense to use a null value for a list name or to have two lists with the same name.

In the Task table, the list_id and task_name columns use the NOT NULL attribute to specify that these columns must store a non-null value. This is usually what you want since a task should have a list and a name.

The diagram for the Task List database

The SQL statement that creates the List table

```
CREATE TABLE list (
    _id         INTEGER PRIMARY KEY AUTOINCREMENT,
    list_name   TEXT    NOT NULL    UNIQUE
)
```

The SQL statement that creates the Task table

```
CREATE TABLE task (
    _id             INTEGER PRIMARY KEY AUTOINCREMENT,
    list_id         INTEGER NOT NULL,
    task_name       TEXT    NOT NULL,
    notes           TEXT,
    date_completed  TEXT,
    hidden          TEXT
)
```

The SQL statement that drops the List table

```
DROP TABLE IF EXISTS list
```

The SQL statement that drops the Task table

```
DROP TABLE IF EXISTS task
```

Common column attributes

Attribute	Description
PRIMARY KEY	Defines a primary key column. A *primary key* column requires that each row must have a value, and that value must be unique.
AUTO_INCREMENT	Identifies a column whose value is automatically incremented by SQLite when a new row is inserted. An auto increment column is typically defined with the INTEGER data type.
NOT NULL	Specifies that the column can't store NULL values.
UNIQUE	Specifies that each value stored in the column must be unique.

Description

- The CREATE TABLE statement creates a table based on the column names, data types, and attributes that you specify.
- The DROP TABLE statement deletes the specified table.

Figure 13-3   An introduction to the Task List database

The DROP TABLE statements delete the specified table. This deletes the structure for the table as well as any data that's stored in the table. Since these statements include the IF EXISTS keywords, they don't cause an error if the table doesn't already exist.

The business objects for the Task List app

Figure 13-4 shows the classes that define the *business objects* for the Task List app. These business objects map to the tables of the database. As a result, the data for a row can be stored in its corresponding business object. For example, a row from the List table can be stored in a List object. Similarly, a row from the Task table can be stored in a Task object. Then, if necessary, the business object can provide additional methods for working with the data.

When working with databases, it's usually considered a good practice to map rows to Java objects like this. This allows you to store all database code in a class or a series of classes that's known as the database layer. Since this layer provides access to *plain old Java objects* (sometimes called *POJOs*), it separates the database layer from the user interface.

The List class defines an instance variable for each column in the List table. Then, it defines three contructors that can be used to create a List object. Next, it defines get and set methods for the instance variables. Finally, it defines a toString method that provides a way to convert a List object to a string. In this case, it does that by returning the name of the list.

The Task class works similarly. However, it provides TRUE and FALSE constants that define two strings that can be used to represent Boolean values. Here, a string of "0" represents a false value, and a string of "1" represents a true value. Within this class, the first constructor uses the FALSE constant to set default values for the completed and hidden variables.

Of the get and set methods, the getCompletedDate method returns the string for the completed_date column. This string stores the number of milliseconds for the date. However, the getCompletedDateMillis method converts the string for the date_completed column from its string value to a long value and it returns that long value. Similarly, the setCompletedDate method is overridden so it can accept a string value or a long value. If the user passes a long value to this method, it converts the long value to a string value.

The List class

```
package com.murach.tasklist;

public class List {

    private int id;
    private String name;

    public List() {}

    public List(String name) {
        this.name = name;
    }

    public List(int id, String name) {
        this.id = id;
        this.name = name;
    }

    public void setId(int id) {
        this.id = id;
    }

    public int getId() {
        return id;
    }

    public void setName(String name) {
        this.name = name;
    }

    public String getName() {
        return name;
    }

    @Override
    public String toString() {
        return name;    // used for add/edit spinner
    }
}
```

Description

- The List class defines an object that can store the data from the List table.

Figure 13-4   The business objects for the Task List app (part 1 of 3)

The Task class

```java
package com.murach.tasklist;

public class Task {

    private int taskId;
    private int listId;
    private String name;
    private String notes;
    private String completed;
    private String hidden;

    public static final String TRUE = "1";
    public static final String FALSE = "0";

    public Task() {
        name = "";
        notes = "";
        completed = FALSE;
        hidden = FALSE;
    }

    public Task(int listId, String name, String notes,
            String completed, String hidden) {
        this.listId = listId;
        this.name = name;
        this.notes = notes;
        this.completed = completed;
        this.hidden = hidden;
    }

    public Task(int taskId, int listId, String name, String notes,
            String completed, String hidden) {
        this.taskId = taskId;
        this.listId = listId;
        this.name = name;
        this.notes = notes;
        this.completed = completed;
        this.hidden = hidden;
    }

    public int getId() {
        return taskId;
    }

    public void setId(int taskId) {
        this.taskId = taskId;
    }

    public int getListId() {
        return listId;
    }

    public void setListId(int listId) {
        this.listId = listId;
    }
```

Figure 13-4   The business objects for the Task List app (part 2 of 3)

The Task class   Page 2

```
    public String getName() {
        return name;
    }

    public void setName(String name) {
        this.name = name;
    }

    public String getNotes() {
        return notes;
    }

    public void setNotes(String notes) {
        this.notes = notes;
    }

    public String getCompleted() {
        return completed;
    }

    public long getCompletedMillis() {
        return Long.valueOf(completed);
    }

    public void setCompletedDate(String completed) {
        this.completed = completed;
    }

    public void setCompletedDate(long millis) {
        this.completedDate = Long.toString(millis);
    }

    public String getHidden(){
        return hidden;
    }

    public void setHidden(String hidden) {
        this.hidden = hidden;
    }
}
```

Description

- The Task class defines an object that can store the data from the Task table.

Figure 13-4   The business objects for the Task List app (part 3 of 3)

How to create a database class

Now that you have some background about SQLite and the Task List database, you're ready to learn how to code a class that you can use to work with a database. This approach allows you to store all of the database code for the app in a single class. This is generally considered a best practice. However, if the class becomes too long to easily maintain, you can split it into multiple database classes.

How to define the constants for a database

Figure 13-5 shows the beginning of the TaskListDB class. This class contains all of the database code for the Task List app.

The TaskListDB class begins by defining the constants for the name and version number of the database. To start, the DB_NAME constant specifies a name of "tasklist.db". As a result, Android stores the file for the database in a file named tasklist.db. Although you can use any name you want, it's a common convention to use ".db" as the filename extension.

The DB_VERSION constant specifies the database version. To start, you can create version 1 of the database. Then, if you make changes to the structure of the database, you can increment this version number. When you do that, Android updates the database as specified later in this class.

This class also defines the constants for the List and Task tables. These constants include the names of the tables, their columns, and their column numbers. For example, the LIST_TABLE constant specifies the name of the List table ("list"), the LIST_ID constant specifies the name of the _id column ("_id"), and the LIST_ID_COL constant specifies the column number of the _id column (0). When specifying the column numbers, you should begin the numbering with 0. As a result, the first column is 0, the second column is 1, and so on.

The TaskListDB class   Page 1

```
package com.murach.tasklist;

import java.util.ArrayList;

import android.content.ContentValues;
import android.content.Context;
import android.database.Cursor;
import android.database.sqlite.SQLiteDatabase;
import android.database.sqlite.SQLiteDatabase.CursorFactory;
import android.database.sqlite.SQLiteOpenHelper;
import android.util.Log;

public class TaskListDB {

    // database constants
    public static final String DB_NAME = "tasklist.db";
    public static final int    DB_VERSION = 1;

    // list table constants
    public static final String LIST_TABLE = "list";

    public static final String LIST_ID = "_id";
    public static final int    LIST_ID_COL = 0;

    public static final String LIST_NAME = "list_name";
    public static final int    LIST_NAME_COL = 1;

    // task table constants
    public static final String TASK_TABLE = "task";

    public static final String TASK_ID = "_id";
    public static final int    TASK_ID_COL = 0;

    public static final String TASK_LIST_ID = "list_id";
    public static final int    TASK_LIST_ID_COL = 1;

    public static final String TASK_NAME = "task_name";
    public static final int    TASK_NAME_COL = 2;

    public static final String TASK_NOTES = "notes";
    public static final int    TASK_NOTES_COL = 3;

    public static final String TASK_COMPLETED = "date_completed";
    public static final int    TASK_COMPLETED_COL = 4;

    public static final String TASK_HIDDEN = "hidden";
    public static final int    TASK_HIDDEN_COL = 5;
```

Description

- It's generally considered a good practice to define constants for the name and version number of the database.

- It's generally considered a good practice to define constants for the table names, column names, and column numbers.

Figure 13-5   How to define the constants for the database

How to define the SQL statements
that create a database

Now that the class has defined the constants for the table and column names, it can build the CREATE TABLE statements needed to create the List and Task tables. In figure 13-6, for example, the first constant contains the CREATE TABLE statement needed to create the List table. This statement uses the LIST_TABLE constant to specify the name of the table, and it uses the LIST_ID and LIST_NAME constants to specify the names of the columns.

Similarly, the second constant contains the CREATE TABLE statement needed to create the Task table. These constants create the same CREATE TABLE statements described earlier in this chapter.

This code also defines the constants for the DROP TABLE statements that are needed to delete the List and Task tables. These constants are used to upgrade a database as shown in the next figure.

The TaskListDB class

```
// CREATE and DROP TABLE statements
public static final String CREATE_LIST_TABLE =
        "CREATE TABLE " + LIST_TABLE + " (" +
        LIST_ID   + " INTEGER PRIMARY KEY AUTOINCREMENT, " +
        LIST_NAME + " TEXT    NOT NULL    UNIQUE);";

public static final String CREATE_TASK_TABLE =
        "CREATE TABLE " + TASK_TABLE + " (" +
        TASK_ID          + " INTEGER PRIMARY KEY AUTOINCREMENT, " +
        TASK_LIST_ID     + " INTEGER NOT NULL, " +
        TASK_NAME        + " TEXT    NOT NULL, " +
        TASK_NOTES       + " TEXT, " +
        TASK_COMPLETED   + " TEXT, " +
        TASK_HIDDEN      + " TEXT);";

public static final String DROP_LIST_TABLE =
        "DROP TABLE IF EXISTS " + LIST_TABLE;

public static final String DROP_TASK_TABLE =
        "DROP TABLE IF EXISTS " + TASK_TABLE;
```

Description

- You can use the constants defined in the previous figure to build the CREATE TABLE and DROP TABLE statements needed to create and upgrade a database.

Figure 13-6   How to define the SQL statements that create a database

How to create or upgrade a database

Figure 13-7 shows the code for a static inner class named DBHelper that extends the SQLiteOpenHelper class. This class begins by defining a constructor. This constructor contains a single statement that passes four arguments to the superclass (the SQLiteOpenHelper class).

After the constructor, this class overrides the onCreate method of the superclass. If Android doesn't find the specified database on the device, it executes this method. Within this method, the first two statements call the execSQL method of the SQLiteDatabase parameter to execute the CREATE TABLE statements that create the List and Task tables. Then, the next four statements execute INSERT statements to insert rows into the List and Task tables. These rows provide some default and sample data.

After the onCreate method, this class overrides the onUpgrade method of the superclass. If Android finds the database on the device and a newer version of the database is available, it executes this method. The first statement in this method displays a message in the LogCat view. This message indicates that the database is being upgraded, and it includes the old and new version numbers. Then, the second and third statements call the execSQL method of the SQLiteDatabase parameter to execute the DROP TABLE statements that drop the List and Task tables. This deletes the structure of the database as well as all of its data. Finally, the fourth statement calls the onCreate method to recreate the database and to reinsert the starting data.

During development, it's a common practice for the onUpgrade method to drop a database and recreate it. However, this deletes all data in the database. Although this is often acceptable during testing, it's not usually what you want for a production app. In that case, the onUpgrade method typically uses an ALTER TABLE statement to alter the structure of the table without deleting all of its data. For example, you could use an ALTER TABLE statement to add a new column to a table without deleting the data that's already in the other columns of the table.

The execSQL method that's available from the SQLiteDatabase class is useful for executing SQL statements such as the CREATE TABLE and DROP TABLE statements shown earlier in this chapter. In addition, it's useful for executing INSERT statements such as the ones shown in this figure. However, if a SQL statement allows the user to insert a parameter, the execSQL method is susceptible to a security vulnerability known as a SQL injection attack. In that case, you should use the techniques specified later in this chapter. For example, you can use the insert method to allow the user to insert data into the database.

The TaskListDB class

```java
        private static class DBHelper extends SQLiteOpenHelper {

            public DBHelper(Context context, String name,
                    CursorFactory factory, int version) {
                super(context, name, factory, version);
            }

            @Override
            public void onCreate(SQLiteDatabase db) {
                db.execSQL(CREATE_LIST_TABLE);
                db.execSQL(CREATE_TASK_TABLE);

                // insert default lists
                db.execSQL("INSERT INTO list VALUES (1, 'Personal')");
                db.execSQL("INSERT INTO list VALUES (2, 'Business')");

                // insert sample tasks
                db.execSQL("INSERT INTO task VALUES (1, 1, 'Pay bills', " +
                        "'Rent\nPhone\nInternet', '0', '0')");
                db.execSQL("INSERT INTO task VALUES (2, 1, 'Get hair cut', " +
                        "'', '0', '0')");
            }

            @Override
            public void onUpgrade(SQLiteDatabase db,
                    int oldVersion, int newVersion) {

                Log.d("Task list", "Upgrading db from version "
                        + oldVersion + "to " + newVersion);

                db.execSQL(TaskListDB.DROP_LIST_TABLE);
                db.execSQL(TaskListDB.DROP_TASK_TABLE);
                onCreate(db);
            }
        }
```

A method of the SQLiteDatabase class

Method	Description
execSQL(sqlString)	Executes the specified SQL string.

Description

- If Android doesn't find the database on the device, it executes the onCreate method.
- If Android finds the database on the device and a newer version of the database is available, it executes the onUpgrade method.
- During development, it's a common practice for the onUpgrade method to use a DROP TABLE statement to drop a database and to call the onCreate method to recreate the database. This deletes all data in the database.
- For a production app, the onUpgrade method typically uses an ALTER TABLE statement to alter a table without deleting all of its data.

Figure 13-7   How to create or upgrade a database

How to open and close a database connection

Figure 13-8 shows how to code the private methods that make it easy to open and close a connection to the database. These private methods are used by the public methods shown in the next few figures.

The code in this figure begins by declaring the SQLiteDatabase and DBHelper objects as instance variables. The SQLiteDatabase object defines a connection to the database, and the DBHelper object contains the methods that can open that connection.

The constructor of the TaskListDB class creates an instance of the DBHelper object. To do that, this code passes the constants for the database name and version to the DBHelper class.

The openReadableDB method calls the getReadableDatabase method from the DBHelper object to return a SQLiteDatabase object for a read-only connection to the database. As a result, you can't use this connection to write data to the database.

The openWriteableDB method works like the openReadableDB method. However, the openWriteableDB method returns a SQLiteDatabase object for a read-write connection to the database. As a result, you can use this connection to read or write data.

The closeDB method begins by checking to make sure the SQLiteDatabase object exists. If so, it calls the close method of this object to close the connection. In general, it's considered a best practice to close the database connection after you're done using it.

The TaskListDB class   **Page 4**

```
// database object and database helper object
private SQLiteDatabase db;
private DBHelper dbHelper;

// constructor
public TaskListDB(Context context) {
    dbHelper = new DBHelper(context, DB_NAME, null, DB_VERSION);
}

// private methods
private void openReadableDB() {
    db = dbHelper.getReadableDatabase();
}

private void openWriteableDB() {
    db = dbHelper.getWritableDatabase();
}

private void closeDB() {
    if (db != null)
        db.close();
}
```

Two methods of the SQLiteOpenHelper class

Method	Description
getReadableDatabase()	Opens a read-only connection to the database.
getWriteableDatabase()	Opens a read-write connection to the database.

A method of the SQLiteDatabase class

Method	Description
close()	Closes a connection.

Description

- Within a database class, you can use private methods that make it easy to open and close a connection to the database.

Figure 13-8   How to open and close a database connection

How to add public methods to a database class

Now that you understand how to code the private methods for a database class, you're ready to learn how to code its public methods. These methods provide a way for the rest of the app to work with the database.

How to retrieve multiple rows from a table

The first method in figure 13-9 shows how to code a public method named getTasks that retrieves multiple rows from a table. To start, the declaration for this method indicates that the method accepts the name of the list as a parameter and that it returns an ArrayList of Task objects.

Within the method, the first statement defines a string named where that specifies which rows to retrieve. If you're familiar with SQL, you should recognize that this works like the condition in a WHERE clause. Here, the question mark (?) identifies a parameter that will be supplied later. As a result, the WHERE clause retrieves all rows where the list_id column is equal to the supplied listID value and where the hidden column is not equal to a true value. Since the hidden column is stored with the TEXT data type, this code uses a string of '1' for a true value.

The second statement calls the getList method that's defined later in this class to return a List object that corresponds with the specified list name. Although this method isn't explained in this chapter, it works much like the getTask method. Then, it calls the getID method to get the ID for that list.

The third statement defines an array of strings that contains the arguments for the WHERE clause. In this case, the WHERE clause contains a single parameter. As a result, this code supplies a single string for the array. However, if the WHERE clause contained multiple parameters, you could supply multiple strings, separating each string with a comma.

The fourth statement uses one of the private methods defined earlier in this class to open a read-only connection to the database.

The fifth statement uses the query method of the database object to return a Cursor object that contains the rows that are retrieved. In this case, the first argument specifies the name of the table, the third argument specifies the string for the WHERE clause, and the fourth argument specifies the array that contains the arguments for the WHERE clause.

This code specifies null values for the other arguments. As a result, the query method retrieves all columns and doesn't include GROUP BY, HAVING, or ORDER BY clauses. However, if you want to specify the columns to retrieve, you can code an array of strings that contains the names of the columns that you want to retrieve. Then, you can supply this array as the second argument. Or, if you want to group or sort the rows that are retrieved, you can supply a string for the appropriate clause. For example, to sort the rows, you can specify a string for the ORDER BY clause.

The TaskListDB class   **Page 5**

```
public ArrayList<Task> getTasks(String listName) {
    String where =
            TASK_LIST_ID + "= ? AND " +
            TASK_HIDDEN + "!='1'";
    int listID = getList(listName).getId();
    String[] whereArgs = { Integer.toString(listID) };

    this.openReadableDB();
    Cursor cursor = db.query(TASK_TABLE, null,
            where, whereArgs,
            null, null, null);
    ArrayList<Task> tasks = new ArrayList<Task>();
    while (cursor.moveToNext()) {
        tasks.add(getTaskFromCursor(cursor));
    }
    if (cursor != null)
        cursor.close();
    this.closeDB();

    return tasks;
}

public Task getTask(int id) {
    String where = TASK_ID + "= ?";
    String[] whereArgs = { Integer.toString(id) };

    this.openReadableDB();
    Cursor cursor = db.query(TASK_TABLE,
            null, where, whereArgs, null, null, null);
    cursor.moveToFirst();
    Task task = getTaskFromCursor(cursor);
    if (cursor != null)
        cursor.close();
    this.closeDB();

    return task;
}
```

A method of the SQLiteDatabase class

Method	Description
query(table, columns, where, whereArgs, groupBy, having, orderBy)	Queries the specified table and returns a cursor for the result set. All arguments are a string or an array of strings. If you don't want to use an argument, you can specify a null value.

Some methods of the Cursor class

Method	Description
moveToFirst()	Moves to the first row in the cursor.
moveToNext()	Moves to the next row in the cursor.
close()	Closes the cursor.

Figure 13-9   How to retrieve rows from a table

The sixth statement declares an ArrayList of Task objects. Then a while loop uses the moveToNext method of the Cursor object to loop through all rows in the cursor. Within the while loop, the statement uses the getTaskFromCursor method shown in figure 13-10 to create a Task object from the current row and to add that task to the array of tasks.

After the while loop, the last statement in the method cleans up resources by closing the cursor as well as the connection to the database. Then, it returns the array of tasks.

How to retrieve a single row from a table

The second method in figure 13-9 shows how to code the public getTask method. This method works similarly to the getTasks method, but it retrieves a single row. To start, the declaration for the getTask method indicates that it returns a Task object for the task with the specified ID.

Within the method, the first two statements define the condition for the WHERE clause and its arguments. Once again, the WHERE clause only contains a single parameter. This time, the parameter is for the task ID instead of the list ID.

After defining the arguments for the WHERE clause, this method opens a read-only connection and uses the query method to retrieve a cursor for a single row. Then, it uses the moveToFirst method to move the cursor to the first (and only) row in the cursor. Next, it uses the getTaskFromCursor method to get the Task object from the row. Finally, it closes the cursor and the database connection and returns the Task object.

How to get data from a cursor

Figure 13-10 shows the getTaskFromCursor method that's used by the getTasks and getTask methods shown earlier. This method accepts a Cursor object and returns a Task object.

The getTaskFromCursor method begins by checking whether the cursor is null or doesn't contain any rows. If so, it returns a null value and exits the method. Otherwise, it attempts to create a Task object from the current row.

To do that, this code creates a new Task object and uses the getXxx methods to get the data from the current row and passes that data to the constructor of the Task class. For example, this code uses the getInt method to get the value for the task ID, and it uses the getString method to get the value for the task name. Both of these methods accept a parameter for the column index, not the column name. As a result, this code uses the constants for column indexes, not the column names.

If the method successfully creates the Task object, the try clause returns that object and exits. Otherwise, the catch clause returns a null value.

The TaskListDB class   **Page 6**

```java
private static Task getTaskFromCursor(Cursor cursor) {
    if (cursor == null || cursor.getCount() == 0){
        return null;
    }
    else {
        try {
            Task task = new Task(
                cursor.getInt(TASK_ID_COL),
                cursor.getInt(TASK_LIST_ID_COL),
                cursor.getString(TASK_NAME_COL),
                cursor.getString(TASK_NOTES_COL),
                cursor.getString(TASK_COMPLETED_COL),
                cursor.getString(TASK_HIDDEN_COL));
            return task;
        }
        catch(Exception e) {
            return null;
        }
    }
}
```

More methods of the Cursor class

Method	Description
`getInt(columnIndex)`	Gets an int value from the cursor.
`getDouble(columnIndex)`	Gets a double value from the cursor.
`getString(columnIndex)`	Gets a string from the cursor.

Description

- Once you have moved to a row in a cursor, you can use the getXxx methods to get data from that row.

Figure 13-10   How to get data from a cursor

How to insert, update, and delete rows

Figure 13-11 shows how to code the public methods that insert, update, and delete rows. Although these methods work with the Task table, you can use similar techniques to work with the List table and other tables.

The insertTask method provides a way to insert a row into the database. This method accepts a Task paramter and returns a long value for the ID of the newly inserted row.

Within the method, the first statement creates a ContentValues object that's used to store the column names and their corresponding values. Then, the next five statements use the column constants to specify the column names, and they usc the methods of the Task object to get the corresponding values. Since the task_id column is defined with the AUTO_INCREMENT attribute, the database automatically generates a value for this column. As a result, you don't have to supply one.

After creating the ContentValues object, this code calls the openWriteableDB method defined earlier in this class to open a read-write connection to the database. Then, it calls the insert method from thc database object to insert the row and to store the ID of the newly inserted row. Here, the first argument specifies the name of the table, and the third argument specifies the column names and values. Since the second argument isn't needed, this code passes a null value for this argument.

After inserting the row, this method calls the closeDB method defined earlier in this class to close the connection to the database. Finally, it returns the ID of the newly inserted row.

The updateTask method provides a way to update an existing row in the database. This method accepts a Task parameter and returns an int value for the count of updated rows. If the update is successful, this should return a value of 1. Otherwise, it should return a value of 0.

Like the insertTask method, the updateTask method begins by creating a ContentValues object to store the column names and their corresponding values. Then, it creates a string for the WHERE clause and an array of strings that contains its argument. This works similarly to the WHERE clause for the getTask method.

After creating the WHERE clause and its arguments, this code opens a read-write connection to the database. Then, it calls the update method from the database connection to update the row and return the count of updated rows. Here, the first argument specifies the name of the table, the second argument specifies columns and their values, the third argument specifies the condition for the WHERE clause, and the fourth argument specifies the argument for the WHERE clause. Finally, this code closes the database connection and returns the count of updated rows.

The TaskListDB class

```java
    public long insertTask(Task task) {
        ContentValues cv = new ContentValues();
        cv.put(TASK_LIST_ID, task.getListId());
        cv.put(TASK_NAME, task.getName());
        cv.put(TASK_NOTES, task.getDescription());
        cv.put(TASK_COMPLETED, task.getCompleted());
        cv.put(TASK_HIDDEN, task.getHidden());

        this.openWriteableDB();
        long rowID = db.insert(TASK_TABLE, null, cv);
        this.closeDB();

        return rowID;
    }

    public int updateTask(Task task) {
        ContentValues cv = new ContentValues();
        cv.put(TASK_LIST_ID, task.getListId());
        cv.put(TASK_NAME, task.getName());
        cv.put(TASK_NOTES, task.getDescription());
        cv.put(TASK_COMPLETED, task.getCompleted());
        cv.put(TASK_HIDDEN, task.getHidden());

        String where = TASK_ID + "= ?";
        String[] whereArgs = { String.valueOf(task.getId()) };

        this.openWriteableDB();
        int rowCount = db.update(TASK_TABLE, cv, where, whereArgs);
        this.closeDB();

        return rowCount;
    }
```

Two methods of the SQLiteDatabase class

Method	Description
insert(table, columns, values)	Inserts a row into the table and returns a long value for the ID of the newly inserted row.
update(table, values, where, whereArgs)	Updates one or more rows in a table and returns an int value for the number of rows that were successfully updated.

One method of the ContentValues class

Method	Description
put(column, value)	Specifies the column name and value.

Description

- You can use the methods of a SQLiteDatabase object to insert or update a row. For these operations, you can store the column names and their corresponding values in a ContentValues object.

Figure 13-11   How to insert, update, and delete rows (part 1 of 2)

The deleteTask method provides a way to delete a row from the database. This method accepts the ID of the task to delete and returns an int value for the count of deleted rows. If the deletion is successful, this should return a value of 1. Otherwise, it should return a value of 0.

Within the method, the first two statements define the condition and arguments for the WHERE clause. Then, the third statement opens a read-write connection to the database.

The fourth statement calls the delete method from the database connection to delete the row and return the count of deleted rows. Here, the first argument specifies the name of the table, the second argument specifies the condition for the WHERE clause, and the third argument specifies the argument for the WHERE clause. Finally, this code closes the database connection and returns the count of deleted rows.

The TaskListDB class

```
    public int deleteTask(long id) {
        String where = TASK_ID + "= ?";
        String[] whereArgs = { String.valueOf(id) };

        this.openWriteableDB();
        int rowCount = db.delete(TASK_TABLE, where, whereArgs);
        this.closeDB();

        return rowCount;
    }
}
```

Another method of the SQLiteDatabase class

Method	Description
delete(table, where, whereArgs)	Deletes one or more rows in a table and returns an int value for the number of rows that were deleted.

Description

- You can use the delete method of a SQLiteDatabase object to delete one or more rows.

Figure 13-11   How to insert, update, and delete rows (part 2 of 2)

How to test the database class and clear its data

Once you write a database class, you need to test it to make sure it works correctly. To do that, you can write code that creates an object from the database class. Then, you can call public methods from that object to make sure they work correctly. As you test and modify the database class, you may want to clear data from a device, or you may want to use a graphical tool such as the SQLite Database Browser to view a database.

How to test the database class

Figure 13-12 shows how to test the TaskListDB class described in the previous figures. You can add code like this to an activity. Then, you can run that activity to display the results on the user interface. Or, if you prefer, you can add code that uses the Log class to display similar messages in the LogCat view.

The code in this figure begins by creating a TaskListDB object named db. Then, it creates a StringBuilder object named sb. This object is used to build the string that's displayed on the user interface.

The code that tests the insertTask method begins by creating a Task object that has some test data in it. This test data includes a list ID of 1. Then, this code calls the insertTask method from the db object and stores the long value that's returned in a variable named insertID. Next, this code checks whether the value of the insert ID is greater than 0. If so, the insert was successful, and the code appends an appropriate message to the StringBuilder object.

The code that tests the updateTask method begins by setting the task ID in the Task object. That way, the Task object contains all of the columns necessary for an update operation. Next, this code changes the name that's stored in the Task object, calls the updateTask method to update the row, and stores the count of updated rows in a variable named updateCount. Finally, this code checks whether the count of updated rows is equal to 1. If so, the update was successful, and the code appends an appropriate message to the StringBuilder object.

The code that tests the deleteTask method begins by calling the deleteTask method from the db object. This code passes the ID of the newly inserted row to this method and stores the count of deleted rows in a variable named deleteCount. Then, this code checks whether the count of deleted rows is equal to 1. If so, the deletion was successful, and the code appends an appropriate message to the StringBuilder object.

The code that tests the getTasks method begins by getting all tasks for the list with a name of "Personal". Then, this code loops through all of the tasks and appends the task ID and name for each task to the StringBuilder object.

Code that tests the database class

```
// get db and StringBuilder objects
TaskListDB db = new TaskListDB(this);
StringBuilder sb = new StringBuilder();

// insert a task
Task task = new Task(1, "Make dentist appointment", "", "0", "0");
long insertId = db.insertTask(task);
if (insertId > 0) {
    sb.append("Row inserted! Insert Id: " + insertId + "\n");
}

// update a task
task.setId((int) insertId);
task.setName("Update test");
int updateCount = db.updateTask(task);
if (updateCount == 1) {
    sb.append("Task updated! Update count: " + updateCount + "\n");
}

// delete a task
int deleteCount = db.deleteTask(insertId);
if (deleteCount == 1) {
    sb.append("Task deleted! Delete count: " + deleteCount + "\n\n");
}

// display all tasks (id + name)
ArrayList<Task> tasks = db.getTasks("Personal");
for (Task t : tasks) {
    sb.append(t.getId() + "|" + t.getName() + "\n");
}

// display string on UI
TextView taskListTextView = (TextView)
        findViewById (R.id.taskListTextView);
taskListTextView.setText(sb.toString());
```

Sample text displayed on the user interface

```
Task inserted! Insert Id: 3
Task updated! Update count: 1
Task deleted! Delete count: 1

1|Pay bills
2|Get hair cut
```

Description

- To test the database class, you can write code that creates an instance of the database class and uses its public methods. Then, you can display data on the user interface. Alternately, you can display the data in the LogCat view.

- If you make changes to the structure of the database, you can increment the version number for the database.

Figure 13-12   How to test the database class

After this code finishes building the string, it displays that string on a TextView widget. To do that, this code gets a reference to the widget. Then, it calls the setText method of that widget to display the string that's stored in the StringBuilder object.

As you test an app, you may decide that you need to change the structure of the database. To do that, you can modify the code in the onUpgrade method of the DBHelper class. However, to get Android to execute this method on each device, you need to increment the version number for the database. Then, the next time you run the app, Android calls the onUpgrade method of the helper class and upgrades the database to the most current version.

How to clear test data from a device

If you don't want to continue to increment the version number for a database, you can delete all data for the app from your test devices as shown in figure 13-13. Then, when you run the app, Android executes the onCreate method of the DBHelper class since no database exists on the device. This allows you to keep the database version at 1 or to reset the database version to 1.

When you use this technique, keep in mind that it also deletes all other data for the app including any shared preferences. As a result, if you want to keep shared preferences or any other test data that has been stored on the device, you shouldn't use this technique.

The Settings app displaying info about the Task List app

Procedure

1. Navigate to the Home screen.
2. Start the Settings app. This app may be available on the Home screen, or you may need to display all apps.
3. Click on the Apps item.
4. Display the Downloaded tab. To do that, you may need to scroll left or right.
5. Click the app.
6. Click the Clear Data button.

Description

- You can use the Settings app to delete the database for an app that's running on a device. This also deletes all other data for the app including any shared preferences.

Figure 13-13   How to clear test data from a device

How to use the DDMS perspective to work with database files

As you test a database, you may want to view the database file that's created on a device. To do that, you can use Eclipse's DDMS perspective to connect to a device (or emulator) as shown figure 13-14. Then, you can use the File Explorer view to navigate to the databases directory for your app. In this figure, for example, the File Explorer shows the databases directory for the Task List app. This directory shows the tasklist.db file that's used by the TaskListDB class.

If you want to copy the database file from the device to your computer, you can select the file and click the "Pull a file" button. Then, you can use the resulting dialog box to specify the directory on your computer where you'd like to store the file. This can be useful if you want to use an app like the SQLite Database Browser app to work with a database.

If you want to copy a file from your computer to a device, you can select the directory on the device and click the "Push a file" button. Then, you can use the resulting dialog box to specify the file that you want to put on the device. This can be useful if you have a SQLite database on your computer that you want to use for testing and you need to copy it onto a device.

The database files

How to view a database

1. Switch to the DDMS perspective by selecting Window→Perspective→Other→DDMS.
2. Open the File Explorer view by selecting Windows→Show View→File Explorer.
3. In the Devices view, select the device.
4. In the File Explorer view, navigate to the databases directory for a particular app.

The database directory and file for the Task List app

```
/data/data/com.murach.tasklist/databases/tasklist.db
```

Description

- To copy a file from the device to your computer, select the file and click on the "Pull a file" button. Then, use the resulting dialog box to locate the file that you want to get from the device.
- To copy a file from your computer to a device, select the directory on the device and click the "Push a file" button. Then, use the resulting dialog box to specify the file that you want to put on the device.

Figure 13-14 How to use the DDMS perspective to work with databases

How to use the SQLite Database Browser

Figure 13-15 shows how to install and use an app known as the SQLite Database Browser. This often makes it easier to test the SQL statements for your app.

For a new database, for example, you can use the New Database button to create the new database. Then, you can use the Execute SQL tab to test the CREATE TABLE statements that create the tables for the database. After that, you can use the Database Structure tab to verify that each table has been created.

Similarly, you can use this tab to test the INSERT statements that insert test data into the database. Then, you can use the Browse Data tab to view this data. Once you're sure that your SQL statements are working correctly, you can use Java statements to build the SQL statements in the database class.

If you want to work with an existing database that's on a device or emulator, you need to start by pulling the file from the device or emulator onto your computer as described in the previous figure. Then, you can use the Open Database button to open that database. After that, you can view the structure or data for the database, and you can quickly and easily test SQL statements against that database.

The SQLite Database Browser

The URL for downloading the SQLite Database Browser

`http://sourceforge.net/projects/sqlitebrowser/`

Description

- You can use a tool like the SQL Database Browser to work with a database that's on your computer (not on a device or emulator).

- To install the SQLite Database Browser, you can search the Internet and follow the instructions to download and install it.

- To create a new database, click the New Database button in the toolbar and respond to the resulting dialog boxes.

- To open an existing database, click the Open Database button in the toolbar and respond to the resulting dialog boxes.

- To view the structure of the database, use the Database Structure tab.

- To view or edit the data for a table, use the Browse Data tab.

- To execute SQL statements, use the Execute SQL tab.

Figure 13-15   How to use the SQLite Database Browser

Perspective

The skills presented in this chapter should be enough to get you started with SQLite databases. However, there is much more to learn. For example, the SQLiteDatabase class provides many more methods for querying a database and for inserting, updating, and deleting data. In addition, it provides methods for other features such as working with transactions. To get an idea of the methods that are available from this class, you can review the documentation for the SQLiteDatabase class.

Another issue to consider is that it often makes sense to store the data for a mobile app in the cloud. That way, if you have multiple devices, they can all work with the same data and that data should always be current. If you store the data for the Tasks List application in the cloud, for example, you should be able to update tasks from any of your devices (computer, tablet, phone, and so on) and the most current tasks should always be available to these devices.

To get this to work correctly, you usually need to store data on the local device. That way, the data is available even if a connection to the cloud isn't available. However, when you are able to connect to the cloud, the app should be able to synch the data on your devices with the data that's stored in the cloud. There are several Task List apps available from the Android market, including GTasks, that use this approach.

Although you learned how to work with a SQLite database in this chapter, you haven't yet learned how to display that data on a user interface. In the next chapter, you'll learn how to use tabs and a custom adapter to do that.

Terms

relational database management system (RDBMS)	business objects
	plain old Java objects (POJOs)

Summary

- SQLite only supports three data types (String, long, and double), and is weakly-typed, meaning a column in a SQLite database does not reject a value of an incorrect data type

- It's generally considered a good practice to define constants for the name and version number of the database, and to define constants for the table names, column names, and column numbers.

- The SQLiteOpenHelper class provides the onCreate and onUpgrade methods that are used to create and upgrade the database.

- Within a database class, you can use private methods that make it easy to open and close a connection to the database.

- You can use the methods of a SQLiteDatabase object to insert, update, and delete rows.

- To test the database class, you can write code that creates an instance of the database class and uses its public methods. Then, you can display data on the user interface. Alternately, you can display the data in the LogCat view.

- If you make changes to the structure of the database, you can increment the version number for the database.

- You can use the Settings app to delete the database for an app that's running on a device. This also deletes all other data for the app including any shared preferences.

- You can use a tool like the SQL Database Browser to work with a database that's on your computer (not on a device or emulator).

Exercise 13-1   Review the Task List app and use its database class

In this exercise, you'll review the code that creates and uses the Task List database. Then, you'll modify the code that uses the Task List app.

1. Start Eclipse and import the project named ch13_ex1_TaskList.

2. Open the Task, List, TaskListDB, and TaskListActivity classes and review their code.

3. Run the app. This should display a series of messages on the user interface for the activity.

4. Switch to the TaskListActivity class and add code that inserts a second task with a list ID of 1 and a name of "Take car in for oil change".

5. Run the app. This should display a similar series of messages as before, but it should also display the task that you inserted in step 4.

6. Change the orientation of the activity one or more times. This should run the test code again, which should insert the task from step 4 again. This shows that the database doesn't use the UNIQUE attribute to require that all names are unique.

7. Add code that deletes the rows that you added in steps 4-6. To do that, you can get the IDs for the tasks by viewing the messages on the Task List activity. Then, you can pass these IDs to the deleteTask method.

8. Run the app. This time the tasks inserted in steps 4-6 should be deleted, but the task inserted by this step should still be in the database.

Exercise 13-2   Use the SQLite Database Browser

In this exercise, you'll use the SQLite Database Browser app to work with an existing Task List database.

1. Start Eclipse and import the project named ch13_ex2_TaskList.

2. Run the app in an emulator. This should display a series of message on the user interface for the activity.

3.   If necessary, install the SQLite Database Browser on your computer.

4.   Start the SQLite Database Browser.

5.   In Eclipse, switch to the DDMS perspective, display the File Explorer, and pull the tasklist.db file for the database from the emulator to the ch13_ex2_TaskList directory on your computer.

6.   Use the SQLite Database Browser to open the tasklist.db file.

7.   View the database structure for the List and Task tables. You should be able to view all of the column names and their data types.

8.   Browse the data that's stored in the List and Task tables. This should help you understand how these tables work.

9.   Execute the following SQL statement:
```
INSERT INTO task VALUES (1, 'Test!', '', '0', '0')
```
This should display an error that indicates that the Task table requires 6 columns.

10. Execute the following SQL statement:
```
INSERT INTO task VALUES (100, 1, 'Test!', '', '0', '0')
```
This should not display any errors.

11. View the data for the Task table to make sure the "Test!" row was inserted.

Exercise 13-3   Modify the database class for the Task List app

In this exercise, you'll modify the database class for the Task List app.

Modify the data types that are used by the database

1.   Start Eclipse and import the project named ch13_ex3_TaskList.

2.   Run the app. This should display a series of messages on the user interface for the activity.

3.   Open the TaskListDB class.

4.   Modify the CREATE TABLE statement for the Task table so it stores the date_completed and hidden columns using the INTEGER types instead of using the TEXT type.

5.   Modify the INSERT statements for the Task table so they don't include single quotes around the last two columns. (You don't need to use quotes for INTEGER values, only for TEXT values.)

6.   In the getTasks method, modify the variable named where so it doesn't include quotes around the 1. (They are no longer needed.)

7.   Open the Task class and modify it so it uses the long type for the completedDate instance variable and the int type for the hidden variable. To get this to work, you can change the data type for the TRUE and FALSE constants to the int type. In addition, you need to modify the constructors, and some of the methods of this class, so they use the correct data types.

8. Switch to the TaskListDB class. Then, modify the getTaskFromCursor method so it uses the getInt method to get values from the date_completed and hidden columns.

9. Open the TaskListActivity class. Then, modify the code that creates that Task object so that it doesn't include quotes around the completedDate and hidden values.

10. Increment the database version number to 2.

11. Run the app. It should work the same as before. However, the app is now storing its data with different types. To verify this, you can check the LogCat view. It should display a message that indicates that the database has upgraded from version 1 to 2.

12. Set the database version number to 1 and run the app again. The app should crash because the database can't be downgraded from 2 to 1.

13. Delete all data for the app on the device.

14. Run the app again. This time, the app should run, and the database version number should be reset to 1.

Add a private method to the TaskListDB class

15. Switch to the TaskListDB class and scroll down to the closeDB method. After this method, add a private method named closeCursor method that accepts a Cursor object. This method should check whether the Cursor object exists (is not null). If so, this method should close the Cursor object.

16. Modify the getLists, getList, getTasks, and getTask methods so they use the closeCursor method.

17. Run the app again. It should work the same as before, but you have made a minor improvement to the database class.

14

How to work with tabs and custom adapters

In the last chapter, you learned how to write the database and business classes for the Task List app. Now, you'll learn how to use those classes to display data on the user interface for the Task List app. To do that, you'll learn how to use tabs and custom adapters.

How to use tabs

The Task List app displays the name of each list on a tab. To do that, this app uses the TabManager class that was originally made available by Google. This class provides a way to use a fragment to display the content for each tab.

Prior to Android 3.0, developers often used the TabActivity class to display tabs and tab content. However, that class has been deprecated. As a result, it's generally considered a best practice to use the FragmentActivity class to display tabs and tab content as shown in the next few figures.

How to add the TabManager class to your project

The TabManager class isn't available as part of the Android API. As a result, before you can use this class, you must add its library to your project. To do that, you can get the TabManager.jar file for the library by downloading the source code for this book. Then, you can copy the TabManager.jar file into the libs directory for your project. In figure 14-1, for example, the libs directory of the project shows that it includes both the android-support-v4.jar and TabManager.jar files.

The layout for an activity that displays tabs

Figure 14-1 also shows the layout for an activity that displays tabs. If you want to display tabs across the top of an activity, you can add code like the code shown in this figure. This code is boilerplate code that you can copy and paste from the downloadable source code for this book into your activity.

The TabHost widget defines a container for tabs. This container has two sets of children: (1) a TabWidget object that can contain one or more tabs (2) a FrameLayout object that can display the content for the selected tab. The code in this figure provides a TabHost object named tabhost, a TabWidget object named tabs, and two FrameLayout objects named tabcontent and realtabcontent. Here, the first FrameLayout object is a placeholder that's required by the TabHost object. However, it isn't used by the app. That's why both its height and width have been set to 0dp. The second FrameLayout object, on the other hand, has a height of 0dp, but a weight of 1. As a result, it takes all of the vertical space between the TabWidget object and the bottom of the screen.

Since the tabs aren't added to the TabWidget object until runtime, the Graphical Layout editor doesn't show the tabs. Instead, it shows an area for the tabs with a label that says "TAB LABEL". Below that, the Graphical Layout editor displays a white area for the content of the selected tab.

A tab widget displayed in the Graphical Layout editor

The XML for the layout

```
<TabHost
    xmlns:android="http://schemas.android.com/apk/res/android"
    android:id="@android:id/tabhost"
    android:layout_width="match_parent"
    android:layout_height="match_parent">

    <LinearLayout
        android:orientation="vertical"
        android:layout_width="match_parent"
        android:layout_height="match_parent">

        <TabWidget
            android:id="@android:id/tabs"
            android:orientation="horizontal"
            android:layout_width="match_parent"
            android:layout_height="wrap_content"
            android:layout_weight="0"/>

        <FrameLayout
            android:id="@android:id/tabcontent"
            android:layout_width="0dp"
            android:layout_height="0dp"
            android:layout_weight="0"/>

        <FrameLayout
            android:id="@+android:id/realtabcontent"
            android:layout_width="match_parent"
            android:layout_height="0dp"
            android:layout_weight="1"/>

    </LinearLayout>
</TabHost>
```

Description

- To get the TabManager.jar file, you can download the source code for this book. This .jar file is stored in the book_apps\ch14_TaskList\libs directory.
- To add the TabManager library to your project, copy the TabManager.jar file into the libs directory for your project.

Figure 14-1   The layout for an activity that displays tabs

The class for an activity that displays tabs

Figure 14-2 shows the class for an activity that displays tabs. This class begins by importing the TabManager class that's stored in the TabManager library. Then, it imports the FragmentActivity class that's included with the android-support-v4 library. In addition, it imports the TabHost and TabSpec classes that are used to work with tabs.

The TaskListActivity class begins by extending the FragmentActivity class. As mentioned earlier, this is generally considered the best way to display tabs since it allows you to display the content for a tab as a fragment.

Within the TaskListActivity class, the onCreate method gets a reference to the TabHost object. Then, it calls the setup method from that object. When you load the TabHost object by calling the findViewById method, you must call this method before you add tabs to the TabHost object.

After setting up the TabHost object, this code creates a new TabManager object. To do that, this code passes three arguments to the constructor of the TabManager class: (1) the context, (2) the TabHost object, and (3) the ID of the FrameLayout widget that's used to display the content for the current tab. This ID corresponds with the second FrameLayout widget shown in the previous figure.

After creating the TabManager object, this code creates a new object from the TaskListDB class described in the previous chapter. Then, it calls the getLists method from this object to get an ArrayList of List objects.

If this ArrayList object contains one or more List objects, this code loops through all of the List objects. Then, it displays the name of each list on a tab. To do that, this code creates a TabSpec object that represents the specification for the tab, and it adds that tab to the tabs.

Within the loop, the first statement uses the newTabSpec method of the TabHost class to create a TabSpec object and set its tag to the name of the list. Then, the second statement uses the setIndicator method to set the label that's used for the tab to the name of the list. Next, the third statement uses the addTab method of the TabManager class to add the tab and its content to the TabHost widget. To do that, this statement passes the TabSpec object and the name of the class for the fragment that displays the content of the tab. In the next figure, you'll see how this class works.

After the while loop, the onCreate method finishes by checking whether the Bundle object that's passed to it is not null. If so, this code gets the name of the current tab's tag from the Bundle object. Then, it sets the current tab to that tag.

The onSaveInstanceState method begins by passing the Bundle object to the onSaveInstanceState superclass. Then, it saves the name of the current tab's tag in the Bundle object. This makes the tag for the current tab available from the Bundle object of the onCreate method.

The class for an activity that displays tabs

```
package com.murach.tasklist;

import java.util.ArrayList;

import com.google.tabmanager.TabManager;

import android.os.Bundle;
import android.support.v4.app.FragmentActivity;
import android.widget.TabHost;
import android.widget.TabHost.TabSpec;

public class TaskListActivity extends FragmentActivity {
    TabHost tabHost;
    TabManager tabManager;

    @Override
    protected void onCreate(Bundle savedInstanceState) {
        super.onCreate(savedInstanceState);
        setContentView(R.layout.activity_task_list);

        // get tab manager
        tabHost = (TabHost) findViewById(android.R.id.tabhost);
        tabHost.setup();
        tabManager = new TabManager(this, tabHost, R.id.realtabcontent);

        // get the lists from the database
        TaskListDB db = new TaskListDB(this);
        ArrayList<List> lists = db.getLists();

        // add a tab for each list
        if (lists != null && ! lists.isEmpty()) {
            for (List list : lists) {
                TabSpec tabSpec = tabHost.newTabSpec(list.getName());
                tabSpec.setIndicator(list.getName());
                tabManager.addTab(tabSpec, TaskListFragment.class, null);
            }
        }

        // set current tab to the last tab opened
        if (savedInstanceState != null) {
            tabHost.setCurrentTabByTag(savedInstanceState.getString("tab"));
        }
    }

    @Override
    protected void onSaveInstanceState(Bundle outState) {
        super.onSaveInstanceState(outState);
        outState.putString("tab", tabHost.getCurrentTabTag());
    }
}
```

Description

- Within the class for an activity, you can use a TabManager object to add one or more tabs. For example, you can add one tab for each list stored in a database.

Figure 14-2   The class for an activity that displays tabs

The class for a fragment that displays tab content

Figure 14-3 shows the class for a fragment that displays the content for a tab. This class begins by extending the Fragment class.

The onCreate method begins by inflating the layout for the fragment. Although it isn't shown here, the layout for this class uses a vertical linear layout that contains a single TextView widget named taskTextView. The second statement gets a reference to this TextView widget.

The third statement gets a reference the TabHost object. To do that, this statement calls the getParent method of the container object twice. The first call gets the LinearLayout widget that's used by the fragment, and the second call gets the TabHost widget that's used by the activity.

The fourth statement uses the getCurrentTabTag method of the TabHost object to get the name of the current tab. Then, the fifth statement calls the refreshTaskList method. This method displays a message on the TextView widget that includes the name of the current tab. For example, if the user clicks on the Personal tab, the content for that tab displays a message that says, "This is the Personal tab."

The onResume method begins by calling the onResume method of the superclass. Then, it calls the refreshTaskList method to display a message on the TextView widget.

Now that you can display a simple message as the content for a tab, you're just a step or two away from displaying more complex data. For example, you might want to look up all of the tasks for the current tab and display those tasks as the content for that tab. In the next few figures, you'll learn how to do that.

The class for a fragment that displays tab content

```
package com.murach.tasklist;

import android.os.Bundle;
import android.support.v4.app.Fragment;
import android.view.LayoutInflater;
import android.view.View;
import android.view.ViewGroup;
import android.widget.TextView;
import android.widget.TabHost;

public class TaskListFragment extends Fragment {

    private TextView taskTextView;
    private String currentTabTag;

    @Override
    public View onCreateView(LayoutInflater inflater, ViewGroup container,
            Bundle savedInstanceState) {

        // inflate the layout for this fragment
        View view = inflater.inflate(R.layout.fragment_task_list,
            container, false);

        // get references to widgets
        taskTextView = (TextView) view.findViewById (R.id.taskTextView);

        // get the current tab
        TabHost tabHost = (TabHost) container.getParent().getParent();
        currentTabTag = tabHost.getCurrentTabTag();

        // refresh the task list view
        refreshTaskList();

        // return the view
        return view;
    }

    public void refreshTaskList() {
        String text = "This is the " + currentTabTag + " list.";
        taskTextView.setText(text);
    }

    @Override
    public void onResume() {
        super.onResume();
        refreshTaskList();
    }
}
```

Description

- Within the class for a fragment, you can use a TabHost class to get the tag for the current tab. You can use this tag to display the appropriate data for the current tab.

Figure 14-3   The class for a fragment that displays tab content

How to use a custom adapter

After the Task List app reads a list of tasks from the database, it needs to display those tasks on the user interface. To do that, you could use a simple adapter to display those tasks in a ListView widget as shown in the previous chapters. However, the Task List app includes a check box in the ListView widget. To get this check box to work correctly, you can use a custom adapter as shown in the next few figures.

A layout for a list view item

Figure 14-4 shows the layout for an item in the ListView widget that's used to display a list of tasks. This layout displays a CheckBox widget that indicates whether the task has been completed, a TextView widget for the name of the task, and a TextView widget for any notes for the task. If you've read chapter 5, you shouldn't have any trouble understanding how this layout works. However, there are a couple of attributes that haven't been presented yet.

For the CheckBox widget, the button attribute specifies a custom layout for the button. This layout is stored in a file named btn_check.xml in the res\drawable directory of the project, and it specifies the .png files for the checked and unchecked versions of the check box. Although this code isn't necessary, it provides a way to control the appearance of the check box. By default, the CheckBox widget was too light to see on some devices, so I specified some slightly darker .png files for these check boxes. These .png files are stored in the appropriate res\drawable-xxx directories for the project.

For the TextView widgets, the textColor attribute specifies black for the color of the text. Again, this code isn't necessary. However, the default color was light gray on some devices, and I think black is easier to read.

The listview_task layout

The XML for the layout

```xml
<?xml version="1.0" encoding="utf-8"?>
<RelativeLayout xmlns:android="http://schemas.android.com/apk/res/android"
    android:id="@+id/relativeLayoutTaskItem"
    android:layout_width="match_parent"
    android:layout_height="match_parent" >

    <CheckBox
        android:id="@+id/completedCheckBox"
        android:layout_width="wrap_content"
        android:layout_height="wrap_content"
        android:layout_alignParentLeft="true"
        android:layout_alignParentTop="true"
        android:layout_margin="5dp"
        android:button="@drawable/btn_check" />

    <TextView
        android:id="@+id/nameTextView"
        android:layout_width="wrap_content"
        android:layout_height="wrap_content"
        android:layout_alignBaseline="@+id/completedCheckBox"
        android:layout_alignBottom="@+id/completedCheckBox"
        android:layout_toRightOf="@+id/completedCheckBox"
        android:text="@string/task_name"
        android:textColor="@android:color/black"
        android:textSize="18sp"
        android:textStyle="bold" />

    <TextView
        android:id="@+id/notesTextView"
        android:layout_width="wrap_content"
        android:layout_height="wrap_content"
        android:layout_below="@+id/nameTextView"
        android:layout_marginTop="-5dp"
        android:layout_toRightOf="@+id/completedCheckBox"
        android:paddingBottom="7dp"
        android:text="@string/task_notes"
        android:textColor="@android:color/black"/>

</RelativeLayout>
```

Figure 14-4   The layout for a list view item

A class that extends the layout for a list view item

When the layout for a ListView widget contains complex widgets such as a check box, you code a class that extends the layout as shown in figure 14-5. This allows you to display the widgets correctly and to handle events that occur upon them.

The TaskLayout class begins by extending the RelativeLayout class and implementing the OnClickListener interface. The first constructor of this class supports some Android tools. It isn't required for this app to work, but it's considered a good practice to include it.

The second constructor accepts two parameters: a Context object and a Task object. Within this constructor, the first statement passes the Context object to the constructor of the superclass. Then, the next two statements set the context for the current object and create a TaskListDB object.

The fourth statement uses the getSystemService method of the Context object to get a LayoutInflater object. Then, the fifth statement uses that object to inflate the layout defined in the previous figure. After that, the next three statements get references to the widgets on the layout.

The eighth and ninth statements set the listeners for the widgets. Here, the eighth statement sets the listener for the check box, and the ninth statement sets the listener for the entire layout.

A class that extends the layout   **Page 1**

```
package com.murach.tasklist;

import android.content.Context;
import android.content.Intent;
import android.view.LayoutInflater;
import android.view.View;
import android.view.View.OnClickListener;
import android.widget.CheckBox;
import android.widget.RelativeLayout;
import android.widget.TextView;

public class TaskLayout extends RelativeLayout implements OnClickListener {

    private CheckBox completedCheckBox;
    private TextView nameTextView;
    private TextView notesTextView;

    private Task task;
    private TaskListDB db;
    private Context context;

    public TaskLayout(Context context) {    // used by Android tools
        super(context);
    }

    public TaskLayout(Context context, Task t) {
        super(context);

        // set context and get db object
        this.context = context;
        db = new TaskListDB(context);

        // inflate the layout
        LayoutInflater inflater = (LayoutInflater)
                context.getSystemService(Context.LAYOUT_INFLATER_SERVICE);
        inflater.inflate(R.layout.listview_task, this, true);

        // get references to widgets
        completedCheckBox = (CheckBox) findViewById(R.id.completedCheckBox);
        nameTextView = (TextView) findViewById(R.id.nameTextView);
        notesTextView = (TextView) findViewById(R.id.notesTextView);

        // set listeners
        completedCheckBox.setOnClickListener(this);
        this.setOnClickListener(this);
```

Figure 14-5   A class that extends the layout (part 1 of 2)

The last statement in the constructor calls the setTask method that's defined later in this class. This method sets the Task object as an instance variable, and it displays the data in the Task object on the user interface.

The setTask method begins by setting the instance variable for the Task object equal to the Task parameter. Then, it updates the user interface. First, it sets the name of the task on the appropriate TextView widget. Then, it checks whether the notes for the task are equal to an empty string. If so, it hides the TextView widget for the notes. Otherwise, it sets the notes for the task on the appropriate TextView widget.

This code continues by checking whether the milliseconds for the completed date are greater than zero. If so, the task has been marked as completed. As a result, this code checks the CheckBox widget. Otherwise, this code removes the check from the CheckBox widget.

The onClick method begins by using a switch statement to check whether the CheckBox widget was clicked. If so, this code checks whether the CheckBox widget was checked. If so, this code sets the completed date for the task to the number of milliseconds for the current date. Otherwise, this code sets the completed date for the task to zero. This indicates that the task has not been completed. Either way, this code uses the database object to update the current task in the database.

If the CheckBox widget wasn't clicked, some other area of the layout for the item was clicked. For example, the user may have clicked the name or notes for the task. If so, this code executes the code for the default case. This code begins by creating an intent for the Add/Edit activity. Then, it adds a "new task" flag to the intent and stores two pieces of extra data for the intent: (1) the ID of the task and (2) a Boolean variable that indicates that the layout is in edit mode. Finally, this code uses the intent to start the Add/Edit activity.

A class that extends the layout   **Page 2**

```
        // set task data on widgets
        setTask(t);
    }

    public void setTask(Task t) {
        task = t;
        nameTextView.setText(task.getName());

        // Remove the notes if empty
        if (task.getNotes().equalsIgnoreCase("")) {
            notesTextView.setVisibility(GONE);
        }
        else {
            notesTextView.setText(task.getNotes());
        }

        if (task.getCompletedDateMillis() > 0){
            completedCheckBox.setChecked(true);
        }
        else{
            completedCheckBox.setChecked(false);
        }
    }

    @Override
    public void onClick(View v) {
        switch (v.getId()) {
            case R.id.completedCheckBox:
                if (completedCheckBox.isChecked()){
                    task.setCompletedDate(System.currentTimeMillis());
                }
                else {
                    task.setCompletedDate(0);
                }
                db.updateTask(task);
                break;
            default:
                Intent intent = new Intent(context, AddEditActivity.class);
                intent.addFlags(Intent.FLAG_ACTIVITY_NEW_TASK);
                intent.putExtra("taskId", task.getId());
                intent.putExtra("editMode", true);
                context.startActivity(intent);
                break;
        }
    }
}
```

Description

- When the layout for a ListView widget contains complex widgets such as a check box, you code a class that extends the layout. This allows you to display the widgets correctly and to handle events that occur upon them.

Figure 14-5   A class that extends the layout (part 2 of 2)

A class for a custom adapter

Figure 14-6 shows a class for a custom adapter. This class begins by extending the BaseAdapter class. Then, it declares two instance variables: (1) a Content object and (2) an ArrayList of Task objects. Both of these instance variables are assigned values in the constructor for this class.

After the constructor, the TaskListAdapter class overrides four methods of the Adapter class: (1) getCount, (2) getItem, (3) getItemId, and (4) getView. These methods are needed to display the tasks in the adapter as items on a ListView widget.

The first three methods are easy to understand. The getCount method returns a count of the number of items in the ArrayList object. The getItem method returns the Task object that's at the specified position in the ArrayList object. And the getItemId method returns the position parameter. As a result, this method uses the same value for position and item ID.

The fourth method, the getView method, is more complicated. This method returns a TaskLayout object like the one defined by the class shown in the previous figure. This object corresponds with the data for the task at the specified position. To start, the getView method defines a variable for a TaskLayout object. Then, it gets the Task object for the specified position parameter. Next, it checks whether the View parameter is null. If so, it creates a new TaskLayout object for the current task. Otherwise, it converts the View parameter to a TaskLayout object and uses the setTask method of that object to set the current task for the layout. Either way, this method returns the TaskLayout object.

A class for a custom adapter

```java
package com.murach.tasklist;

import java.util.ArrayList;

import android.content.Context;
import android.view.View;
import android.view.ViewGroup;
import android.widget.BaseAdapter;

public class TaskListAdapter extends BaseAdapter {

    private Context context;
    private ArrayList<Task> tasks;

    public TaskListAdapter(Context context, ArrayList<Task> tasks){
        this.context = context;
        this.tasks = tasks;
    }

    @Override
    public int getCount() {
        return tasks.size();
    }

    @Override
    public Object getItem(int position) {
        return tasks.get(position);
    }

    @Override
    public long getItemId(int position) {
        return position;
    }

    @Override
    public View getView(int position, View convertView, ViewGroup parent) {
        TaskLayout taskLayout = null;
        Task task = tasks.get(position);

        if (convertView == null) {
            taskLayout = new TaskLayout(context, task);
        }
        else {
            taskLayout = (TaskLayout) convertView;
            taskLayout.setTask(task);
        }
        return taskLayout;
    }
}
```

Description

- The class for a custom adapter that extends the BaseAdapter class can supply the
  data for ListView and Spinner widgets.

Figure 14-6   A class for a custom adapter

A class for a fragment
that uses a custom adapter

Figure 14-7 shows a class for a fragment that uses a custom adapter like the one shown in the previous figure. This class works much like the class shown in figure 14-3. However, the refreshTaskList method in figure 14-7 gets the tasks for a list from the database and uses a custom adapter to display them.

The refreshTaskList method begins by getting the context for the application. Then, it creates an object from the TaskListDB class and calls the getTasks method from that object to get the tasks for the current tab.

The refreshTaskList method finishes by creating a TaskListAdapter object and passing it the application context as well as the tasks. Then, it sets that object as the adapter for the ListView control for the fragment.

A class for a fragment that uses a custom adapter

```
package com.murach.tasklist;

import java.util.ArrayList;

import android.content.Context;
import android.os.Bundle;
import android.support.v4.app.Fragment;
import android.view.LayoutInflater;
import android.view.View;
import android.view.ViewGroup;
import android.widget.ListView;
import android.widget.TabHost;

public class TaskListFragment extends Fragment {
    private ListView taskListView;
    private String currentTabTag;

    @Override
    public View onCreateView(LayoutInflater inflater, ViewGroup container,
            Bundle savedInstanceState) {
        View view = inflater.inflate(R.layout.fragment_task_list,
            container, false);
        taskListView = (ListView) view.findViewById (R.id.taskListView);
        TabHost tabHost = (TabHost) container.getParent().getParent();
        currentTabTag = tabHost.getCurrentTabTag();
        refreshTaskList();
        return view;
    }

    public void refreshTaskList() {
        // get task list for current tab from database
        Context context = getActivity().getApplicationContext();
        TaskListDB db = new TaskListDB(context);
        ArrayList<Task> tasks = db.getTasks(currentTabTag);

        // create adapter and set it in the ListView widget
        TaskListAdapter adapter = new TaskListAdapter(context, tasks);
        taskListView.setAdapter(adapter);
    }

    @Override
    public void onResume() {
        super.onResume();
        refreshTaskList();
    }
}
```

Description

- Within the class for a fragment, you can use a custom adapter to display appropriate data for the current tab.

Figure 14-7   A class for a fragment that uses a custom adapter

The Task List app

So far, this chapter has presented enough of the code for the Task List app that you should have a general idea of how it works. Now, this chapter presents some more of the code to fill in the gaps about how this app works. But first, to put this code in context, this chapter reviews the user interface for the app, which was originally presented at the beginning of the previous chapter.

The user interface

Figure 14-8 reviews the interface for the Task List app. When the app starts, the Task List activity displays a list of tasks where each task has a check box that indicates whether it has been completed, a name, and optional notes.

In addition, the Task List activity provides tabs for two lists: Personal and Business. To display lists, the user can click on the tabs that are available from the Task List activity. In this figure, for example, the Personal tab is selected and the tasks for that tab are displayed. However, the user could click on the Business tab to display the tasks for that list.

From the Task List activity, the user can mark a task as completed by selecting a check box. Then, the user can remove all tasks that have been marked as completed by clicking on the Delete Tasks icon (the small trash can). This icon is available from the task bar. However, older versions of Android don't provide for the task bar. As a result, on older devices, the user can display the options menu and select the Delete Tasks item.

The user can add a task by clicking on the Add Task icon (the + sign) that's available from the task bar. Or, if the device is too old to display a task bar, the user can display the options menu and select the Add Task item. This displays the Add/Edit activity in add mode. At this point, the user can use the spinner to select the list. Then, the user can use the editable text views to add or edit the text for the task.

Similarly, the user can edit a task by clicking on the task in the Task List activity. This displays the Add/Edit activity in edit mode so the user can edit the data for the task. In this figure for example, I clicked on the "Get hair cut" task in the Task List activity. As a result, the Add/Edit activity displays this task in edit mode.

The Task List activity   **The Add/Edit activity**

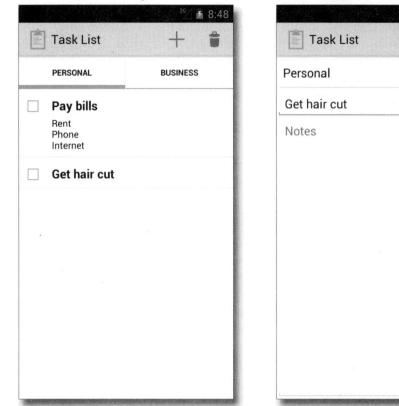

Description

- When the app starts, the Task List activity displays a list of tasks where each task has a check box that indicates whether it has been completed, a name, and optional notes.

- The app provides for two lists: Personal and Business.

- To display lists, the user can click on the tabs that are available in the Task List activity.

- To edit a task, the user can click on the task in the Task List activity. This displays the Add/Edit activity in edit mode so the user can edit the data for the task.

- To mark a task as completed, the user can select the check box in the Task List activity.

- To remove all completed tasks, the user can click the Delete Tasks button that's available from the task bar.

- To add a new task, the user can navigate to the Task List activity and click on the Add Task icon. This displays the Add/Edit activity in add mode so the user can add a new task.

- In the Add/Edit activity, the user can use the spinner to select the list. Then, the user can use the editable text views to add or edit the text for the task.

Figure 14-8   The user interface for the Task List app

The activity_task_list menu

Figure 14-9 shows the code that defines the menu for the TaskListActivity class. This code displays two menu items: (1) an Add Task item and (2) a Delete item.

Both of these menu items display icons in the task bar if there is enough room. These icons are stored in the res\drawable-xxx directory that's appropriate for the resolution of the icon. Otherwise, they display text in the options menu. Also, if the device is old and doesn't support the task bar, these menu items are available from the options menu.

The TaskListActivity class

Figure 14-10 shows the complete code for the TaskListActivity class. Part 1 of this figure works like the class described in figure 14-2. As a result, you shouldn't have much trouble understanding how it works. However, part 2 of this figure includes code that handles the events that occur when the user selects an item from the options menu.

To start, the onOptionsItemSelected method checks which menu item was selected. If the Add Task item was selected, the first statement creates an intent for the Add/Edit activity. Then, the second statement stores the current tab in the extra data for the intent. Finally, the third statement uses the intent to start the Add/Edit activity. Since this intent doesn't include extra data that specifies edit mode, this displays the Add/Edit activity in add mode.

If the Delete item was selected, this code hides all tasks that have been marked as complete. To do that, the first statement gets all tasks for the current tab. Then, the code loops through all of these tasks. If the milliseconds for the completed date are greater than 0, the task is marked as hidden, and the database is updated.

After updating the database, this code refreshes the task list. To do that, this code gets a reference to the fragment that corresponds with the current tab. Then, it calls the refreshTaskList method from that fragment to refresh the list of tasks. As a result, any tasks marked as hidden will no longer be displayed on the list.

The activity_task_list menu

```xml
<menu xmlns:android="http://schemas.android.com/apk/res/android" >
    <item
        android:id="@+id/menuAddTask"
        android:icon="@drawable/ic_action_add"
        android:orderInCategory="1"
        android:showAsAction="ifRoom"
        android:title="@string/add_task"/>
    <item
        android:id="@+id/menuDelete"
        android:icon="@drawable/ic_action_delete"
        android:orderInCategory="2"
        android:showAsAction="ifRoom"
        android:title="@string/delete"/>
</menu>
```

Figure 14-9   The activity_task_list menu

The TaskListActivity class   Page 1

```java
package com.murach.tasklist;

import java.util.ArrayList;

import com.google.tabmanager.TabManager;

import android.content.Intent;
import android.os.Bundle;
import android.support.v4.app.FragmentActivity;
import android.view.Menu;
import android.view.MenuItem;
import android.widget.TabHost;
import android.widget.TabHost.TabSpec;

public class TaskListActivity extends FragmentActivity {
    TabHost tabHost;
    TabManager tabManager;
    TaskListDB db;

    @Override
    protected void onCreate(Bundle savedInstanceState) {
        super.onCreate(savedInstanceState);
        setContentView(R.layout.activity_task_list);

        // get tab manager
        tabHost = (TabHost) findViewById(android.R.id.tabhost);
        tabHost.setup();
        tabManager = new TabManager(this, tabHost, R.id.realtabcontent);

        // get database
        db = new TaskListDB(getApplicationContext());

        // add a tab for each list in the database
        ArrayList<List> lists = db.getLists();
        if (lists != null && lists.size() > 0) {
            for (List list : lists) {
                TabSpec tabSpec = tabHost.newTabSpec(list.getName());
                tabSpec.setIndicator(list.getName());
                tabManager.addTab(tabSpec, TaskListFragment.class, null);
            }
        }

        // sets current tab to the last tab opened
        if (savedInstanceState != null) {
            tabHost.setCurrentTabByTag(savedInstanceState.getString("tab"));
        }

    }

    @Override
    protected void onSaveInstanceState(Bundle outState) {
        super.onSaveInstanceState(outState);
        outState.putString("tab", tabHost.getCurrentTabTag());
    }
```

Figure 14-10   The TaskListActivity class (part 1 of 2)

The TaskListActivity class   **Page 2**

```java
    @Override
    public boolean onCreateOptionsMenu(Menu menu) {
        getMenuInflater().inflate(R.menu.activity_task_list, menu);
        return true;
    }

    @Override
    public boolean onOptionsItemSelected(MenuItem item) {
        switch (item.getItemId()){
            case R.id.menuAddTask:
                Intent intent = new Intent(this, AddEditActivity.class);
                intent.putExtra("tab", tabHost.getCurrentTabTag());
                startActivity(intent);
                break;
            case R.id.menuDelete:
                // Hide all tasks marked as complete
                ArrayList<Task> tasks =
                    db.getTasks(tabHost.getCurrentTabTag());
                for (Task task : tasks){
                    if (task.getCompletedDateMillis() > 0){
                        task.setHidden(Task.TRUE);
                        db.updateTask(task);
                    }
                }

                // Refresh list
                TaskListFragment currentFragment = (TaskListFragment)
                        getSupportFragmentManager().
                        findFragmentByTag(tabHost.getCurrentTabTag());
                currentFragment.refreshTaskList();

                break;
        }
        return super.onOptionsItemSelected(item);
    }
}
```

Figure 14-10   The TaskListActivity class (part 2 of 2)

The activity_add_edit and spinner_list layout

Figure 14-11 begins by showing the layout for the Add/Edit activity. This activity allows the user to use a spinner to select the name of the list. Then, the user can use the two editable text views to edit the name and notes of the task. Both of the editable text views use the hint attribute to specify a hint to the user. For example, the editable text view for notes specifies a hint of "Notes". That way, the user knows to enter notes in this editable text view.

Since the layout for the activity uses a custom layout for the spinner, this figure also shows the spinner_list layout. This layout is used to display the names of the lists in the spinner. To do that, it uses a TextView widget with black text color and a text size of 18sp.

The activity_add_edit menu

Figure 14-11 also shows the code that defines the menus for the Add/Edit activity. This displays two items: (1) the Save item and (2) the Cancel item. Like the menu items for the Task List activity, both of these menu items display icons in the task bar if there is enough room. Otherwise, they display text in the options menu.

The activity_add_edit layout

```xml
<?xml version="1.0" encoding="utf-8"?>
<LinearLayout xmlns:android="http://schemas.android.com/apk/res/android"
    android:layout_width="match_parent"
    android:layout_height="match_parent"
    android:orientation="vertical" >

    <Spinner
        android:id="@+id/listSpinner"
        android:layout_width="match_parent"
        android:layout_height="wrap_content" />

    <EditText
        android:id="@+id/nameEditText"
        android:layout_width="match_parent"
        android:layout_height="wrap_content"
        android:hint="@string/task_name_hint" >

        <requestFocus />
    </EditText>

    <EditText
        android:id="@+id/notesEditText"
        android:layout_width="match_parent"
        android:layout_height="0dp"
        android:layout_weight="1"
        android:gravity="top"
        android:hint="@string/task_notes_hint"
        android:inputType="textMultiLine" />

</LinearLayout>
```

The spinner_list layout

```xml
<?xml version="1.0" encoding="utf-8"?>
<TextView xmlns:android="http://schemas.android.com/apk/res/android"
    android:layout_width="fill_parent"
    android:layout_height="wrap_content"
    android:padding="5dp"
    android:textColor="@android:color/black"
    android:textSize="18sp" />
```

The activity_add_edit menu

```xml
<?xml version="1.0" encoding="utf-8"?>
<menu xmlns:android="http://schemas.android.com/apk/res/android" >
    <item android:id="@+id/menuSave"
        android:showAsAction="ifRoom"
        android:title="@string/save"
        android:icon="@drawable/ic_action_save"
        android:orderInCategory="2"></item>
    <item android:id="@+id/menuCancel"
        android:showAsAction="ifRoom"
        android:title="@string/cancel"
        android:icon="@drawable/ic_action_cancel"
        android:orderInCategory="1"></item>
</menu>
```

Figure 14-11   The activity_add_edit layout and menu

The AddEditActivity class

Figure 14-12 shows the code for the AddEditActivity class. In part 1, the code works like most of the other activities you've seen throughout this book. However, it also creates a TaskListDB object. Then, it uses that database object to get an array of lists that are used as the data source for the adapter of the spinner.

Part 2 begins by getting the edit mode from the intent. Then, this code checks if the activity is in edit mode. If so, it uses the ID of the task to retrieve that task from the database. Then, it updates the task name and notes on the user interface. To do that, this code uses the setText methods of the editable text views for the name and notes of the task.

After setting the name and notes for the task, this code sets the appropriate list in the spinner. To do that, it checks if the activity is in edit mode. If so, it gets the ID for the same list as the selected task. Otherwise, it gets the ID for the list that was current when the user clicked the Add Task item. Either way, this code subtracts 1 from the database ID to get the correct position. This is necessary because the database numbering starts with 1, and the list item positions start with 0. Then, this code uses the current list item position to select the correct list from the spinner.

The onOptionsItemSelected method checks which item was selected. If the Save item was selected, this code saves all data on the activity to the database. To do that, it calls the saveToDB method shown in part 3 of this figure. Then, it calls the finish method of the activity to remove the Add/Edit activity from the stack. This displays the Task List activity.

If the Cancel item was selected, this code calls the finish method of the activity. This removes the Add/Edit activity from the stack without saving any of its data to the database.

The saveToDB method shown in part 3 begins by getting data from the three widgets displayed on the Add/Edit activity. Then, this code checks to make sure that the user has entered a task name. If the user hasn't entered a task name, this code exits the method without saving the task. That's because it doesn't make sense to save a task that doesn't have a name.

After making sure the user has entered a name, this code checks if the activity is in add mode. If so, this code creates a new Task object. This isn't necessary if the activity is in edit mode as the Task object should already exist.

After creating the Task object, this code stores the data in the Task object. Then, it checks if the activity is in edit mode. If so, this updates the task in the database. Otherwise, the activity is in add mode. As a result, this code adds the task to the database.

The onKey method hides the soft keyboard if the user presses the Center key on a D-Pad. Also, if the user presses the Back key, the onKey method saves the data to the database before it allows the system to process this key, which causes the Add/Edit activity to be removed from the stack. That's usually what you want since it's usually nice to save the user's data even if he or she presses the Back key.

The AddEditActivity class

```
package com.murach.tasklist;

import java.util.ArrayList;

import android.app.Activity;
import android.content.Context;
import android.content.Intent;
import android.os.Bundle;
import android.view.KeyEvent;
import android.view.Menu;
import android.view.MenuItem;
import android.view.View;
import android.view.View.OnKeyListener;
import android.view.inputmethod.InputMethodManager;
import android.widget.ArrayAdapter;
import android.widget.EditText;
import android.widget.Spinner;

public class AddEditActivity extends Activity
implements OnKeyListener {

    private EditText nameEditText;
    private EditText notesEditText;
    private Spinner listSpinner;

    private TaskListDB db;
    private boolean editMode;
    private String currentTabName = "";
    private Task task;

    public void onCreate(Bundle savedInstanceState) {
        super.onCreate(savedInstanceState);
        setContentView(R.layout.activity_add_edit);

        // get references to widgets
        listSpinner = (Spinner) findViewById(R.id.listSpinner);
        nameEditText = (EditText) findViewById(R.id.nameEditText);
        notesEditText = (EditText) findViewById(R.id.notesEditText);

        // set listeners
        nameEditText.setOnKeyListener(this);
        notesEditText.setOnKeyListener(this);

        // get the database object
        db = new TaskListDB(this);

        // set the adapter for the spinner
        ArrayList<List> lists = db.getLists();
        ArrayAdapter<List> adapter = new ArrayAdapter<List>(
                this, R.layout.spinner_list, lists);
        listSpinner.setAdapter(adapter);
```

Figure 14-12   The AddEditActivity class (part 1 of 3)

The AddEditActivity class

```java
        // get edit mode from intent
        Intent intent = getIntent();
        editMode = intent.getBooleanExtra("editMode", false);

        // if editing
        if (editMode) {
            // get task
            long taskId = intent.getLongExtra("taskId", -1);
            task = db.getTask(taskId);

            // update UI with task
            nameEditText.setText(task.getName());
            notesEditText.setText(task.getNotes());
        }

        // set the correct list for the spinner
        long listID;
        if (editMode) {    // edit mode - use same list as selected task
            listID = (int) task.getListId();
        }
        else {                // add mode - use the list for the current tab
            currentTabName = intent.getStringExtra("tab");
            listID = (int) db.getList(currentTabName).getId();
        }
        // subtract 1 from database ID to get correct list position
        int listPosition = (int) listID - 1;
        listSpinner.setSelection(listPosition);
    }

    @Override
    public boolean onCreateOptionsMenu(Menu menu) {
        getMenuInflater().inflate(R.menu.activity_add_edit, menu);
        return true;
    }

    @Override
    public boolean onOptionsItemSelected(MenuItem item) {
        switch (item.getItemId()){
            case R.id.menuSave:
                saveToDB();
                this.finish();
                break;
            case R.id.menuCancel:
                this.finish();
                break;
        }
        return super.onOptionsItemSelected(item);
    }
```

Figure 14-12   The AddEditActivity class (part 2 of 3)

The AddEditActivity class   **Page 3**

```java
        private void saveToDB() {
            // get data from widgets
            int listID = listSpinner.getSelectedItemPosition() + 1;
            String name = nameEditText.getText().toString();
            String notes = notesEditText.getText().toString();

            // if no task name, exit method
            if (name == null || name.equals("")) {
                return;
            }

            // if add mode, create new task
            if (!editMode) {
                task = new Task();
            }

            // put data in task
            task.setListId(listID);
            task.setName(name);
            task.setNotes(notes);

            // update or insert task
            if (editMode) {
                db.updateTask(task);
            }
            else {
                db.insertTask(task);
            }
        }

        @Override
        public boolean onKey(View view, int keyCode, KeyEvent event) {
            if (keyCode == KeyEvent.KEYCODE_DPAD_CENTER) {
                // hide the soft Keyboard
                InputMethodManager imm = (InputMethodManager)
                        getSystemService(Context.INPUT_METHOD_SERVICE);
                imm.hideSoftInputFromWindow(view.getWindowToken(), 0);
                return true;
            }
            else if (keyCode == KeyEvent.KEYCODE_BACK) {
                saveToDB();
                return false;
            }
            return false;
        }
    }
```

Figure 14-12   The AddEditActivity class (part 3 of 3)

Perspective

The skills presented in this chapter should be enough to get you started with tabs and custom adapters. This is a complicated subject. Fortunately, you can begin by using the Task List app presented in this chapter as a starting point. Then, you can copy and paste the necessary code into your app and modify it as necessary to get your tabs and custom adapters to work correctly.

So far, the database that's used by the Task List app is only available to the Task List app. In the next chapter, you'll learn how make the data in this database available to other apps. In addition, you'll learn how to work with other system databases that are available to your apps.

Summary

- To add the TabManager library to your project, copy the TabManager.jar file into the libs directory for your project. To get the TabManager.jar file, you can download the source code for this book.

- Within the class for an activity, you can use a TabManager object to add one or more tabs. For example, you can add one tab for each list stored in a database.

- Within the class for a fragment, you can use a TabHost class to get the tag for the current tab. You can use this tag to display the appropriate data for the current tab.

- When the layout for a ListView widget contains complex widgets such as a check box, you code a class that extends the layout, which allows you to display the widgets correctly and to handle events that occur upon them.

- The class for a custom adapter that extends the BaseAdapter class can supply the data for ListView and Spinner widgets.

- Within the class for a fragment, you can use a custom adapter to display appropriate data for the current tab.

Exercise 14-1   Work with tabs

In this exercise, you'll work with the tabs that are available from a simplified version of the TaskList app.

Review the app

1. Start Eclipse and open the project named ch14_ex1_TaskList.

2. Open the TaskListActivity and TaskListFragment classes and review the code.

3. Run the app. It should display a Task List app that displays two tabs. The Personal tab should be selected, and it should display a message that says, "This is the Personal list."

4. Click on the Business tab. It should display a message that says, "This is the Business list".

Use a string to display the content for a tab

5. Switch to the TaskListFragment class.

6. Modify the refreshTaskList method so it displays a string that contains all task names for the current list. To do that, create a TaskListDB object and use it to get the tasks for the current list from the database. Then, loop through the tasks and build a string that contains each task on a separate line. Finally, display the string on the fragment.

7. Run the app. This should display the task names for the selected list. These names should reflect the data that's in the current Task List database. Since this data is stored on the device (or emulator), the data that's displayed may vary from device to device.

Use a simple adapter to display the content for a tab

8. Open the fragment_task_list layout and add a ListView widget below the TextView widget.

9. Switch to the TaskListFragment class and modify the refreshTaskList method so it uses a simple adapter to display each task as a ListView item. To get this to work, you can use the listview_item layout that's already in the project. In addition, you may be able to reuse some code from the News Reader app presented earlier in this book.

10. Run the app. This should display the task names for the selected list in the TextView and ListView widgets.

Exercise 14-2   Work with a custom adapter

In this exercise, you'll modify the News Reader app presented earlier in this book so it uses a custom adapter instead of a simple adapter.

Review the app

1. Start Eclipse and open the project named ch14_ex2_NewsReader.

2. Open the listview_item layout. This layout contains the TextView widgets for displaying the publication date and title of a news item.

3. Run the app. It should display a list of news items. Then, click on a news item. It should display more information about the news item.

Add a custom adapter

4. Copy the TaskLayout class from the ch14_ex1_TaskList project to this project.

5. Rename this class from TaskLayout to ItemLayout.

6. Modify the newly renamed ItemLayout class so it works with an RSSItem object and a listview_item layout instead of a Task object and a listview_task layout. This class should map the data from an RSSItem to the two widgets in the listview_item layout.

7. Modify the onClick method of the ItemLayout class so it creates an intent for the ItemActivity class, stores all necessary extra data in that intent, and uses the intent to start the activity. To do that, you can open the ItemsActivity class and copy the code that stores the extra data for the intent into this class.

8. Switch to the ItemsActivity class. Then, delete the onItemClick method and all other code that implements the OnItemClickListener.

9. Copy the TaskListAdapter class from the ch14_ex1_TaskList project to this project.

10. Rename this class from TaskListAdapter to ItemsAdapter.

11. Modify the newly renamed ItemsAdapter class so it works with a List of RSSItem objects instead of an ArrayList of Task objects.

12. Switch to the ItemsActivity class. Then, delete all code that's necessary to create the SimpleAdapter object and replace it with a single statement that creates the ItemsAdapter object.

13. Run the app. It should work the same as before. However, the code in the ItemsActivity class should be shorter and easier to read since the adapter code has been moved from the ItemsActivity class into the ItemsAdapter and ItemLayout classes.

15

How to work with content providers

In Android, a content provider allows multiple apps to share the same data. Android uses built-in content providers to make certain types of data available to multiple apps. For example, Android includes a content provider for your contacts. That way, built-in Android apps such as the phone dialer and text messaging apps can share this data. Similarly, third-party apps such as a custom contacts app can share this data.

In this chapter, you'll learn how to add a content provider to the Task List app from the previous two chapters. This makes the data in the Task List database available to all apps. Then, you'll learn how to create a Task History app that uses this content provider. This illustrates most of the concepts you need to work with content providers. In addition, you'll learn a little about using Android's built-in content provider for contacts.

An introduction to content providers

As mentioned earlier, a *content provider* allows multiple apps to share the same data. This data can be stored in a database or in files. Android includes content providers for your contacts, calendar, settings, bookmarks, and media including images, music, and video. That way, built-in and third-party apps can share this data.

URIs for content providers

Figure 15-1 begins by showing the syntax for a *URI* (*Uniform Resource Identifier*) for a content provider. This syntax begins with "content://", followed by the name of the provider (the *authority*), followed by a *path* to the content. Typically, a path begins with a name that corresponds with a table or file. In addition, a path can include an optional ID that points to an individual row in a table. In this figure, the first URI specifies all rows in the Task table of the Task List database. The second URI specifies all rows in the List table of the Task List database. And the third URI specifies the row in the Task table that has an ID of 2. In other words, the row with a primary key value of 2.

When you work with URIs, you should know that each front slash in the path begins a new *segment*. For the third URI, for example, the first segment is "tasks" and the second segment is "2". Typically, the URI for a content provider only has one or two segments. However, a URI for a content provider can have more segments if that helps organize and structure the data.

MIME types for content providers

When you create a URI for a content provider, you need to specify the type of content that's returned by that URI. To do that, you use a *MIME type*, which is an Internet standard for defining types of content.

This figure shows how to specify the MIME type for a URI. Here, the first syntax is for a URI that can return multiple rows, and the second syntax is for a URI that returns a single row. The only difference between these syntaxes is that the first syntax uses "dir" to indicate that it returns a directory of items, and the second syntax uses "item" to indicate that it returns a single item. For example, the first MIME type indicates that the URI can return multiple tasks, and the second MIME type indicates that the URI only returns a single task.

Built-in Android content providers

- Contacts
- Calendar
- Settings
- Bookmarks
- Media (images, music, video)

URIs for content providers

The syntax
```
content://authority/path[/id]
```

A URI for all rows of the Task table
```
content://com.murach.tasklist.provider/tasks
```

A URI for all rows of the List table
```
content://com.murach.tasklist.provider/lists
```

A URI for a single row in the Task table with an ID of 2
```
content://com.murach.tasklist.provider/tasks/2
```

MIME types for content providers

The syntax for multiple rows
```
vnd.android.cursor.dir/vnd.company_name.content_type
```

The syntax for a single row
```
vnd.android.cursor.item/vnd.company_name.content_type
```

A MIME type for multiple rows
```
vnd.android.cursor.dir/vnd.murach.tasklist.tasks
```

A MIME type for a single row
```
vnd.android.cursor.item/vnd.murach.tasklist.tasks
```

Description

- A *content provider* allows multiple apps to share the same data. This data can be stored in a database or in files. Android uses built-in content providers to make certain types of data available to all apps.
- A *URI* (*Uniform Resource Identifier*) for a content provider includes the name of the provider (the *authority*) and a *path* to the content. This path typically begins with a name that corresponds with a table or file, and it can include an optional ID that points to an individual row in a table.
- Within a path, each front slash begins a new *segment*.
- A *MIME type* is an Internet standard for defining types of content.

Figure 15-1   An introduction to content providers

How to add supporting methods to the database class

In chapter 13, you learned how to create a database class named TaskListDB. This class includes methods like getTasks and insertTask that are used by the app to work with the database. Now, figure 15-2 shows some methods that you can add to this database class that can be used by the content provider class. These supporting methods provide basic query, insert, update, and delete operations for the Task table. In the next few figures, you'll see how these methods are used by the content provider.

These supporting methods open connections to the database, but they don't close those connections. That's because the database connections are closed automatically when the process that hosts the content provider is destroyed.

Four supporting methods in the TaskListDB class

```
public Cursor queryTasks(String[] columns, String where,
        String[] whereArgs, String orderBy) {
    this.openReadableDB();
    return db.query(TASK_TABLE, columns, where, whereArgs,
            null, null, orderBy);
}

public long insertTask(ContentValues values) {
    this.openWriteableDB();
    return db.insert(TASK_TABLE, null, values);
}

public int updateTask(ContentValues values, String where,
        String[] whereArgs) {
    this.openWriteableDB();
    return db.update(TASK_TABLE, values, where, whereArgs);
}

public int deleteTask(String where, String[] whereArgs) {
    this.openWriteableDB();
    return db.delete(TASK_TABLE, where, whereArgs);
}
```

Description

- To make it easy to create a content provider, you can include some supporting methods in your database class that provide for the query, insert, update, and delete operations.

- These methods should not close the database connection after performing the operation.

Figure 15-2   How to add supporting methods to the database class

How to create a content provider

Now that you understand some basic concepts about content providers, you're ready to learn how to create a content provider. In particular, you're ready to learn how to add a content provider to the Task List app described in the previous chapter. This allows other apps to read and write data from the database that's used by the Task List app.

How to start a content provider class

Figure 15-3 shows how to start a class for a content provider. To begin, you declare a class like the TaskListProvider class shown in this figure that extends the ContentProvider class.

Within the TaskListProvider class, the first four statements declare the constants for the class. The first statement declares the constant for the authority that's used within the URI. Then, the next three statements declare constants that are used to determine whether the URI matches one of the URIs that are supported by this class.

After declaring the constants, this class declares two instance variables. The first instance variable is a TaskListDB object for working with the Task List database. The second instance variable is a UriMatcher object for checking whether a URI matches one of the supported URIs.

The onCreate method contains the code that initializes these variables. This method overrides the onCreate method of the ContentProvider class. As a result, Android calls this method when it creates the content provider.

Within the onCreate method, the first statement creates the TaskListDB object. Then, the next three statements create the UriMatcher object. Of these statements, the first creates the UriMatcher object and uses the NO_MATCH variable to specify the value that's returned if no matching URI is found. The second statement adds this URI:

```
content://com.murach.tasklist.provider/tasks
```

to the UriMatcher object and uses the ALL_TASKS_URI constant to specify the value that's returned if a URI matches this URI. The third statement adds this URI:

```
content://com.murach.tasklist.provider/tasks/#
```

and uses the SINGLE_TASK_URI constant to specify the value that's returned if a URI matches this URI. Here, the last segment uses the # wildcard to specify any number. As a result, this URI pattern matches a URI that specifies a single row like this:

```
content://com.murach.tasklist.provider/tasks/5
```

The TaskListProvider class   **Page 1**

```
package com.murach.tasklist;

import android.content.ContentProvider;
import android.content.ContentValues;
import android.content.UriMatcher;
import android.database.Cursor;
import android.net.Uri;

public class TaskListProvider extends ContentProvider {

    public static final String AUTHORITY = "com.murach.tasklist.provider";

    public static final int NO_MATCH = -1;
    public static final int ALL_TASKS_URI = 0;
    public static final int SINGLE_TASK_URI = 1;

    private TaskListDB db;
    private UriMatcher uriMatcher;

    @Override
    public boolean onCreate() {
        db = new TaskListDB(getContext());

        uriMatcher = new UriMatcher(NO_MATCH);
        uriMatcher.addURI(AUTHORITY, "tasks", ALL_TASKS_URI);
        uriMatcher.addURI(AUTHORITY, "tasks/#", SINGLE_TASK_URI);

        return true;
    }
```

A constructor and method of the UriMatcher class

Constructor/Method	Description
`UriMatcher(code)`	Creates a new UriMatcher object and specifies the code to return when no matching URI is found.
`addURI(authority, path, code)`	Adds a URI to the UriMatcher object and specifies the code to return when a matching URI is found for this URI.

Description

- The class for a content provider must inherit the ContentProvider class.
- To initialize variables, your content provider class can override the onCreate method of the ContentProvider class.

Figure 15-3   How to start a content provider class

How to provide for querying

Figure 15-4 shows how to code the methods of a content provider that provides for querying. To start, you can override the query method of the ContentProvider class. This method accepts five arguments (the URI, an array of column names, a WHERE clause, an array of arguments for the WHERE clause, and an ORDER BY clause), and it returns a Cursor object.

Within the query method, this code uses a switch statement to check the URI argument. Then, you can execute appropriate code depending on the URI argument. In this figure, for example, the switch statement checks whether the URI argument matches the URI that corresponds with the ALL_TASKS_URI constant. If so, this code calls the queryTasks method from the TaskListDB object and passes it the other four arguments of the query method. This returns a Cursor object that contains the columns and rows specified by those arguments.

If the URI argument doesn't match the URI that corresponds with the ALL_TASKS_URI constant, this code throws an exception that indicates that this type of operation isn't supported. In this figure, for example, if you used a URI that corresponded with the SINGLE_TASK_URI constant, this code would throw an exception. In other words, this query method doesn't support the URI for a single task. As a result, if you attempted to use a URI for a single row, it would throw an exception.

This exception sends an error message to the developers of other apps that are trying to use this content provider. This message should help the developer figure out how to use the content provider. In this case, the exception clearly indicates that the operation being requested by the URI isn't supported. If you want, you can use a different type of exception object such as an IllegalArgumentException. However, the goal is to send an error message to other developers that clearly identifies why the URI that they sent didn't work.

For querying to work, the content provider class must override the getType method of the ContentProvider class. This method accepts one argument (the URI) and returns a string that uniquely identifies the content type.

Within the getType method, this code uses a switch statement to check the URI argument. Then, you can execute the appropriate code depending on the URI argument. In this figure, for example, the switch statement returns the appropriate MIME type if the URI argument matches one of the URIs that correspond with the ALL_TASKS_URI or SINGLE_TASK_URI constants. However, if the URI argument doesn't match either of these URIs, this code throws an exception to indicate that the operation is not supported for the specified URI.

The TaskListProvider class   **Page 2**

```
@Override
public Cursor query(Uri uri, String[] columns, String where,
        String[] whereArgs, String orderBy) {
    switch(uriMatcher.match(uri)) {
        case ALL_TASKS_URI:
            return db.queryTasks(columns, where,
                    whereArgs, orderBy);
        default:
            throw new UnsupportedOperationException (
                    "URI " + uri + " is not supported.");
    }
}

@Override
public String getType(Uri uri) {
    switch(uriMatcher.match(uri)) {
        case ALL_TASKS_URI:
            return "vnd.android.cursor.dir/vnd.murach.tasklist.tasks";
        case SINGLE_TASK_URI:
            return "vnd.android.cursor.item/vnd.murach.tasklist.tasks";
        default:
            throw new UnsupportedOperationException(
                    "URI " + uri + " is not supported.");
    }
}
```

A method of the UriMatcher class

Method	Description
`match(uri)`	If the specified Uri object matches any of the URIs stored in the UriMatcher object, this method returns the code for that URI.

Description

- To allow other apps to query your data, your content provider class can override the query and getType methods of the ContentProvider class.

- Every data access method of the ContentProvider class has a Uri object as its first argument. This allows the method to determine what table, row, or file to access.

- In a URI, you can use the # wildcard to match any number.

Figure 15-4   How to provide for querying

How to provide for inserting rows

Figure 15-5 shows how to code the method of a content provider that provides for inserting rows. To do that, you can override the insert method of the ContentProvider class. This method accepts two arguments (the URI, and a ConventValues object that contains the column names and values for the row). This method returns a URI for the row that's inserted.

Within the insert method, this code uses a switch statement to check the URI argument. This works similarly to the query method described in the previous figure. If the URI argument matches the URI that corresponds with the ALL_TASKS_URI constant, the first statement calls the insertTask method from the TaskListDB object and passes it the second argument of the insert method. This returns the ID for the row that was inserted.

The second statement gets a ContentResolver object and calls the notifyChange method from it to notify registered observers that a row was changed. That way, the content provider notifies registered observers when the data changes so those observers can re-requery the provider to keep their data current. Without this statement, apps that use this provider would have to re-query the provider manually to make sure that their content is current. By default, CursorAdapter objects get this notification. As a result, you don't need to register them.

The third statement uses several methods of the Uri.Builder class to build the Uri object for the row that was inserted. To do that, it appends the ID for the row that was inserted to the URI argument. For example, if you insert a row and the database assigns that row a primary key value of 14, this statement returns this URI:

```
content://com.murach.tasklist.provider/tasks/14
```

The TaskListProvider class   **Page 3**

```
@Override
public Uri insert(Uri uri, ContentValues values) {
    switch(uriMatcher.match(uri)) {
        case ALL_TASKS_URI:
            long insertId = db.insertTask(values);
            getContext().getContentResolver().notifyChange(uri, null);
            return uri.buildUpon().appendPath(
                    Long.toString(insertId)).build();
        default:
            throw new UnsupportedOperationException(
                    "URI: " + uri + " is not supported.");
    }
}
```

A method of the Context class

Method	Description
`getContentResolver()`	Returns a ContentResolver object for your app.

A method of the ContentResolver class

Method	Description
`notifyChange(uri, observer)`	Notify registered observers that a row was updated. By default, CursorAdapter objects get this notification.

Some methods of the Uri.Builder class

Method	Description
`buildUpon()`	Creates a Uri.Builder object for building a new Uri object.
`appendPath(segment)`	Encodes the specified segment and appends it to the path. Returns a Uri.Builder object to provide for method chaining.
`build()`	Returns a Uri object for the current Uri.Builder object.

Description

- To allow other apps to insert rows into your database, you can override the insert method of the ContentProvider class.
- The insert method should return a URI for the inserted row.

Figure 15-5   How to provide for inserting rows

How to provide for updating rows

Figure 15-6 shows how to code the method of a content provider that provides for updating rows. To do that, you can override the update method of the ContentProvider class. This method accepts four arguments (the URI, a ContentValues object, a WHERE clause, and an array that contains the arguments for the WHERE clause). This method returns an int value for the count of rows that were updated.

Within the update method, this code uses a switch statement to check the URI argument. This works similarly to the query method described earlier in this chapter. However, unlike the query method described earlier in this chapter, the update method supports two URIs: (1) the URI that corresponds with the ALL_TASKS_URI constant and (2) the URI that corresponds with the SINGLE_TASK_URI constant.

If the URI argument matches the SINGLE_TASK_URI constant, this code updates the specified row. To do that, the first statement uses the getLastPathSegment method to get the ID for the row to update. The next two statements use that ID to build a WHERE clause that selects the row for the specified update. The fourth statement calls the updateTask method of the TaskListDB object to update the row and get the count of updated rows. The fifth statement notifies any registered observers that a row was changed. And the sixth statement returns the count of rows that were updated. If the update is successful, this should return a count of 1.

If the URI argument matches the ALL_TASKS_URI constant, this code updates the specified rows. To do that, the first statement passes the second, third, and fourth arguments to the updateTask method of the TaskListDB object. This may update one or more rows. Then, the second statement notifies any registered observers that a row was changed, and the third statement returns the count of rows that were updated.

How to provide for deleting rows

This figure also shows how to code the method of a content provider that provides for deleting rows. Since this method works much like the update method described in this figure, you shouldn't have much trouble understanding how it works.

The TaskListProvider class
Page 4

```java
@Override
public int update(Uri uri, ContentValues values, String where,
        String[] whereArgs) {
    int updateCount;
    switch(uriMatcher.match(uri)) {
        case SINGLE_TASK_URI:
            String taskId = uri.getLastPathSegment();
            String where2 = "_id = ?";
            String[] whereArgs2 = { taskId };
            updateCount = db.updateTask(values, where2, whereArgs2);
            getContext().getContentResolver().notifyChange(uri, null);
            return updateCount;
        case ALL_TASKS_URI:
            updateCount = db.updateTask(values, where, whereArgs);
            getContext().getContentResolver().notifyChange(uri, null);
            return updateCount;
        default:
            throw new UnsupportedOperationException (
                    "URI " + uri + " is not supported.");
    }
}

@Override
public int delete(Uri uri, String where, String[] whereArgs) {
    int deleteCount;
    switch(uriMatcher.match(uri)) {
        case SINGLE_TASK_URI:
            String taskId = uri.getLastPathSegment();
            String where2 = "_id = ?";
            String[] whereArgs2 = { taskId };
            deleteCount = db.deleteTask(where2, whereArgs2);
            getContext().getContentResolver().notifyChange(uri, null);
            return deleteCount;
        case ALL_TASKS_URI:
            deleteCount = db.deleteTask(where, whereArgs);
            getContext().getContentResolver().notifyChange(uri, null);
            return deleteCount;
        default:
            throw new UnsupportedOperationException (
                    "URI " + uri + " is not supported.");
    }
}
}
```

A method of the Uri class

Method	Description
getLastPathSegment()	Returns the value for the last path segment. For a URI that specifies a single row, this gets the ID for the row.

Description
- To allow other apps to update or delete rows in your database, you can override the update and delete methods of the ContentProvider class.

Figure 15-6   How to provide for updating and deleting rows

How to register a content provider

After you create the class for a content provider, you need to register the provider as described in figure 15-7. The easiest way to register a provider is to add a provider element to the project's AndroidManifest.xml file. When you code a provider element, you can code it at the same level as an activity element.

A provider element should use the name attribute to specify the name of the class for the provider, and it should use the authorities attribute to specify the authority for the provider. In this figure, for example, the provider element specifies the name and authority for the content provider for the Task List app.

In addition, a provider element should use the exported attribute to specify whether other apps can use the content provider. Since content providers are designed to be used by other apps, you typically set the exported attribute to a value of "true". However, in some cases, you may only want the current app to be able to use the content provider. For example, you may want to use a content provider to allow an app widget to work with the data for an app. For more information about app widgets, please see the next chapter.

The provider element in the AndroidManifest.xml file

```
<provider
    android:name="com.murach.tasklist.TaskListProvider"
    android:authorities="com.murach.tasklist.provider"
    android:exported="true" >
</provider>
```

The attributes of the provider element

Attribute	Description
name	Specifies the package and name of the content provider class.
authorities	Specifies the authority for the content provider.
exported	Specifies whether the content provider can be used by other apps.

Description

- Before other apps can use your content provider class, you must register it by adding a provider element to the AndroidManifest.xml file.

Figure 15-7   How to register a content provider

How to use a content provider

Now that you know how to create a content provider, you're ready to learn how to use a content provider. To start, you'll learn how to use the content provider for the Task List app that's described in the previous figures. Then, you'll learn how to use one of the built-in content providers that's available from Android, the Contacts Provider. This should give you a good idea of how content providers work.

How to use a custom content provider

Figure 15-8 shows how to use the content provider for the Task List app that's presented earlier in this chapter. To start, the first code example defines the constants that this code needs to refer to the columns names and indexes.

To keep things simple, I copied these constants from the TaskListDB class presented in chapter 13. However, a more elegant approach would be to store these constants in a library to make it easy to share them between apps. For this content provider, for example, you could store the constants in a class named TaskListContract. Then, you could store that class in a library named TaskListContract.jar. That way, both the Task List and Task History apps could import this library and use the same set of constants.

The second code example defines the constants for the authority and the URI for multiple tasks. Again, to keep things simple, I copied the code for the AUTHORITY constant from content provider for the Task History app. However, to reduce code duplication, these constants could be stored in a library as described in the previous paragraph.

Once you're done defining the constants that are needed by the code, you can use the getContentResolver method to get a ContentResolver object. Then, you can call any of the methods that are available from the content provider from that object as shown by the rest of the code examples.

The third, fourth, and fifth examples call the query method and pass the URI for all tasks as the first argument. The third example passes null values for all other arguments. As a result, it returns a Cursor object that contains all columns and rows.

The fourth example defines a WHERE clause that only displays tasks that have been hidden. To do that, the where variable only returns rows where the TASK_HIDDEN column is equal to value of '1', which is how the database defines a true value. Then, this example passes this variable to the query method. As a result, this example only returns the rows that have been marked as hidden. In other words, it returns all old tasks that the Task List app no longer displays.

The fifth example defines an array of three columns, the TASK_ID, TASK_NAME, and TASK_NOTES columns. Then, it passes this variable to the query method. As a result, this example only returns these three columns.

How to define constants for columns

```
public static final String TASK_ID = "_id";
public static final int    TASK_ID_COL = 0;
public static final String TASK_NAME = "task_name";
public static final int    TASK_NAME_COL = 2;
...
```

How to create the base Uri object

```
public static final String AUTHORITY = "com.murach.tasklist.provider";
public static final Uri TASKS_URI =
        Uri.parse("content://" + AUTHORITY + "/tasks");
```

How to query a content provider

```
Cursor cursor = getContentResolver()
        .query(TASKS_URI, null, null, null, null);
```

A second way to query a content provider

```
String where = TASK_HIDDEN + " = '1' ";
String orderBy = TASK_COMPLETED + " DESC";
Cursor cursor = getContentResolver()
        .query(TASKS_URI, null, where, null, orderBy);
```

A third way to query a content provider

```
String[] columns = {TASK_ID, TASK_NAME, TASK_NOTES}
Cursor cursor = getContentResolver()
        .query(TASKS_URI, columns, null, null, null);
```

How to delete data from the content provider

```
String where = TASK_ID + " = ?";
String[] whereArgs = { Integer.toString(taskId) };
int deleteCount = getContentResolver()
        .delete(TASKS_URI, where, whereArgs);
```

Another way to delete data from the content provider

```
Uri taskUri = ContentUris.withAppendedId(TASKS_URI, taskId);
int deleteCount = getContentResolver()
        .delete(taskUri, null, null);
```

A method of the ContentUris class

Method	Description
`withAppendedId(Uri, id)`	Returns the specified Uri object with the specified ID appended to the path.

Description

- You can define constants for the columns and URI for the content provider.
- You can call any methods of the content provider from the ContentResolver object.

Figure 15-8   How to use a custom content provider

The sixth and seventh examples call the delete method of the ContentResolver object. The sixth example defines a variable for a WHERE clause that selects the task with the specified ID. Then, this example passes this variable and the URI for all tasks to the delete method. As a result, this example deletes the row that matches the specified ID.

The seventh example appends the task ID to the URI for all tasks. Then, this example passes this URI followed by two null values to the delete method. This is another way to delete a row for the specified ID.

How to use a built-in content provider

Figure 15-9 shows some code that you can use to work with the Contacts Provider that comes with Android. This content provider makes it possible for you to work with your address book: names, phone numbers, email addresses, and so on.

Since the Contacts Provider is a part of the Android API, you can import the classes that contain the constants for working with this provider. In this figure, the first example imports three classes that contain constants. All three of these classes are inner classes of the ContactsContract class.

Once you import these classes, you can use the constants available from them to work with the provider. For example, the second example shows how to use the CONTENT_URI constant that's available from the ContractsContract.Data class to get the URI for the provider.

The third example shows how to query a content provider. To start, this example defines four columns for the cursor: _ID, DISPLAY_NAME, DATA1, and DATA2. To understand the details of these constants, you should view the documentation for the Contacts Provider. In brief, _ID is the primary key that's needed by the adapter, DISPLAY_NAME is the name of the person, DATA1 is the phone number, and DATA2 is the type of phone number (home, mobile, work, etc.).

This example defines a WHERE clause that selects phone number rows for contacts that have been marked as favorites. In other words, it selects contacts where the STARRED constant is true (equal to a value of '1').

This example defines an ORDER BY clause that sorts the rows in descending order by the number of times that they have been contacted. To do that, this code uses the TIMES_CONTACTED constant. This orders the rows so the most frequently contacted numbers are at the top of the cursor.

This example finishes by calling the query method from the ContentResolver object and passing it the DATA_URI constant as well as the columns, where, and orderBy variables. This returns a Cursor object with the specified columns and rows. Although it isn't shown in this figure, you can also call the insert, update, and delete methods to modify the data that's available from the Contacts Provider.

How to import the classes that contain the constants

```
import android.provider.ContactsContract.Contacts;
import android.provider.ContactsContract.Data;
import android.provider.ContactsContract.CommonDataKinds.Phone;
```

How to get a base Uri object

```
private final Uri DATA_URI = Data.CONTENT_URI;
```

How to query a content provider

```
String[] columns = {
        Data._ID,                 // primary key
        Contacts.DISPLAY_NAME,    // person's name
        Data.DATA1,               // phone number
        Data.DATA2                // phone type (mobile, home, work, etc.)
};
String where =
        "(" + Data.MIMETYPE + "='" + Phone.CONTENT_ITEM_TYPE + "' AND "
            + Contacts.STARRED + "='1' )";
String orderBy = Contacts.TIMES_CONTACTED + " DESC";

Cursor cursor = getContentResolver().query(
        DATA_URI, columns, where, null, orderBy);
```

Description

- When you work with a content provider that's included with Android, you can import the classes that contain the constants for working with the content provider.

- To learn more about a content provider that's included with Android, you can read the documentation for the content provider.

Figure 15-9   How to use a built-in content provider

How to work with a dialog box

At this point, you have all the skills you need to work with a content provider. However, the Task History app that's presented in the next few figures requires one more skill: how to work with a dialog box.

How to import the dialog class and interface

When creating apps, you often need to display a dialog box like the one shown in figure 15-10. To do that, you can begin by importing the AlertDialog.Builder class and the DialogInterface.OnClickListener interface. Then, you can use this class and interface to build and show a dialog box.

How to build and show the dialog box

To build a dialog box, you can begin by creating a new AlertDialog.Builder object. Since the AlertDialog.Builder object supports method chaining, you don't need to provide a name for this object. Instead, you can call a series of methods from it.

To start, you can call the setMessage method to set the message that's displayed across the top of the dialog box. In this figure, for example, this message says, "Are you sure you want to delete this task?"

Then, you can set the buttons on the dialog by calling the setPositiveButton and setNegativeButton methods. Android displays the positive button on the right side of the dialog, and it displays the negative button on the left. In this figure, for example, the Cancel button is the negative button, and the Yes button is the positive button. If necessary, you can specify one or more neutral buttons. In that case, Android displays the neutral buttons between the positive and negative buttons.

When you set a button, you can also specify the event handler that's executed when the user clicks the button. The easiest way to do that is to create an anonymous inner class for the listener. In this figure, for instance, the code that sets the positive button creates an anonymous inner class that implements the DialogInterface.OnClickListener interface. As a result, if the user clicks the Yes button, this code deletes the task.

The code that sets the negative button also creates an anonymous inner class that implements the DialogInterface.OnClickListener interface. However, this event handler cancels the dialog by calling the cancel method of the first parameter of the onClick method. As a result, if the user clicks the Cancel button, this code cancels the dialog box without deleting the task.

Once this code finishes setting the buttons for this dialog, it calls the show method to display the dialog. This displays a dialog like the one shown at the top of this figure.

A dialog box

How to use a dialog box

Import the dialog class and interface

```
import android.app.AlertDialog;
import android.content.DialogInterface;
```

Create and show the dialog box

```
new AlertDialog.Builder(this)
.setMessage("Are you sure you want to delete this task?")
.setPositiveButton("Yes", new DialogInterface.OnClickListener() {
    @Override
    public void onClick(DialogInterface dialog, int id) {
        delete(item);
    }
})
.setNegativeButton("Cancel", new DialogInterface.OnClickListener() {
    @Override
    public void onClick(DialogInterface dialog, int id) {
        dialog.cancel();
    }
})
.show();
```

Constructors and methods of the AlertDialog.Builder class

Constructor/Method	Description
`AlertDialog.Builder(context)`	Creates a new AlertDialog.Builder object.
`setMessage(text)`	Sets the message that's displayed above the buttons.
`setPositiveButton(text, listener)`	Sets the text and listener for the positive button.
`setNegativeButton(text, listener)`	Sets the text and listener for the negative button.
`setNeutralButton(text, listener)`	Sets the text and listener for a neutral button.
`show()`	Displays the dialog.

Description

- You can use the AlertDialog.Builder class and the DialogInterface.OnClickListener interface to build and display a dialog box that includes event handlers.

Figure 15-10   How use a dialog box

The Task History app

The Task History app uses the same database that's used by the Task List app described in the previous two chapters. This app only uses the query and delete methods that are available from the content provider, not the insert or update methods. However, this app clearly shows how one app can use a content provider to work with the same data as another app.

The user interface

Figure 15-11 shows the user interface for the Task History app. When the app starts, the Task History activity displays a list of tasks that have been completed and marked as hidden. In other words, it displays all tasks the Task List app no longer displays.

The Task History activity displays the task name, notes, and completion date for each task. If there are no notes for a task, this app displays a message that says, "No notes". In this figure, for example, none of the tasks have notes.

If you want to delete a task permanently, you can click the task. Then, the app displays a dialog box that prompts you to confirm the deletion. If you click the Yes button, this app deletes the task from the database. Otherwise, this app cancels the dialog box and does not delete the task.

The Task History activity

The dialog box for deleting a task

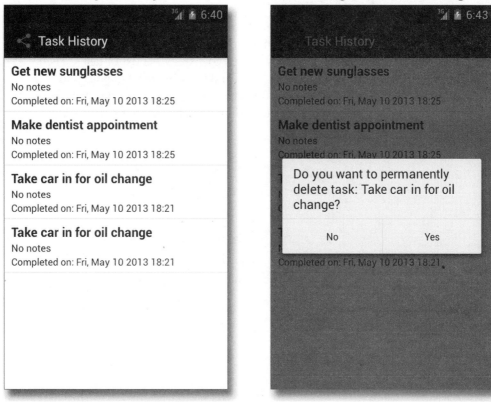

Description

- When the app starts, the Task History activity displays a list of tasks that have been completed and hidden by the Task List app. Each task includes the task name, notes, and completion date.

- To delete a task, the user can click the task. Then, the user can respond to the resulting dialog box to confirm or cancel the deletion.

Figure 15-11   The user interface for the Task History app

The XML for the layouts

Figure 15-12 shows the XML for the layouts that are used by the Task History activity. To start, the layout named activity_task_history uses a LinearLayout container and a ListView widget to display the list of tasks. Then, the listview_task layout uses a LinearLayout and three TextView widgets to format each task. Since you've seen layouts like these in the previous chapter, you shouldn't have much trouble understanding how they work by now.

The activity_task_history layout

```xml
<?xml version="1.0" encoding="utf-8"?>
<LinearLayout android:layout_width="match_parent"
    android:layout_height="match_parent"
    android:orientation="vertical"
    xmlns:android="http://schemas.android.com/apk/res/android">

    <ListView
        android:id="@+id/taskListView"
        android:layout_width="match_parent"
        android:layout_height="match_parent" >
    </ListView>

</LinearLayout>
```

The listview_task layout

```xml
<?xml version="1.0" encoding="utf-8"?>
<LinearLayout xmlns:android="http://schemas.android.com/apk/res/android"
    android:layout_width="match_parent"
    android:layout_height="match_parent"
    android:orientation="vertical" >

    <TextView
        android:id="@+id/nameTextView"
        android:layout_width="wrap_content"
        android:layout_height="wrap_content"
        android:layout_marginTop="5dp"
        android:layout_marginLeft="10dp"
        android:textSize="18sp"
        android:textStyle="bold"
        android:text="@string/name" />

    <TextView
        android:id="@+id/notesTextView"
        android:layout_width="wrap_content"
        android:layout_height="wrap_content"
        android:layout_marginLeft="10dp"
        android:text="@string/notes" />

    <TextView
        android:id="@+id/dateTextView"
        android:layout_width="wrap_content"
        android:layout_height="wrap_content"
        android:layout_marginBottom="5dp"
        android:layout_marginLeft="10dp"
        android:text="@string/date" />

</LinearLayout>
```

Figure 15-12   The layouts for the Task History app

The Java code for the activity

Figure 15-13 shows the Java code for the Task List activity. This class begins by defining the constants for the columns that it uses. These constants were copied in from the TaskListDB class presented in chapter 13. Then, this class defines a constant for the authority. This constant was copied in from the TaskListProvider class. Next, this class uses the AUTHORITY constant to define a constant for the URI for all tasks. As mentioned earlier, a more elegant solution would be to create a library for these constants so they can be shared between the Task List and Task History apps.

After defining its constants, this class defines four instance variables that are used throughout the class. The first instance variable is for the list view that stores the tasks. The second is for the cursor adapter. The third is for the cursor. And the fourth is for the application's context.

The onCreate method begins by displaying the activity_task_history layout. Then, it sets the context variable, and gets a reference to the list view.

The TaskHistoryActivity class   **Page 1**

```
package com.murach.taskhistory;

import java.text.SimpleDateFormat;
import java.util.Date;

import android.annotation.SuppressLint;
import android.app.Activity;
import android.app.AlertDialog;
import android.content.Context;
import android.content.DialogInterface;
import android.database.Cursor;
import android.net.Uri;
import android.os.Bundle;
import android.support.v4.widget.SimpleCursorAdapter;
import android.support.v4.widget.SimpleCursorAdapter.ViewBinder;
import android.view.View;
import android.widget.AdapterView;
import android.widget.AdapterView.OnItemClickListener;
import android.widget.ListView;
import android.widget.TextView;
import android.widget.Toast;

@SuppressLint("SimpleDateFormat")
public class TaskHistoryActivity extends Activity
implements OnItemClickListener {

    public static final String TASK_ID = "_id";
    public static final int    TASK_ID_COL = 0;
    public static final String TASK_NAME = "task_name";
    public static final int    TASK_NAME_COL = 2;
    public static final String TASK_NOTES = "notes";
    public static final int    TASK_NOTES_COL = 3;
    public static final String TASK_COMPLETED = "date_completed";
    public static final int    TASK_COMPLETED_COL = 4;
    public static final String TASK_HIDDEN = "hidden";
    public static final int    TASK_HIDDEN_COL = 5;

    public static final String AUTHORITY = "com.murach.tasklist.provider";
    public static final Uri TASKS_URI =
            Uri.parse("content://" + AUTHORITY + "/tasks");

    private ListView taskListView;
    private SimpleCursorAdapter adapter;
    private Cursor cursor;
    private Context context;

    @Override
    protected void onCreate(Bundle savedInstanceState) {
        super.onCreate(savedInstanceState);
        setContentView(R.layout.activity_task_history);

        context = this;
        taskListView = (ListView) findViewById(R.id.taskListView);
    }
```

Figure 15-13   The TaskHistoryActivity class (part 1 of 3)

The onResume method begins by using the query method of the ContextResolver object to get a cursor for all tasks that have been marked as hidden. This code also sorts these tasks in descending order by the completed date of the task. As a result, the most recent tasks are at the top of the cursor.

After getting the cursor, the onResume method defines the variables needed to create the adapter. More specifically, it gets the ID for the listview_task layout, it specifies the three columns to get the data from, and it specifies the three TextView widgets to send the data to. Then, this code uses these variables to create the adapter. Finally, this code sets the adapter on the ListView widget, and it sets the listener for the adapter.

After setting the listener for the adapter, this code uses the setViewBinder method to convert column data into readable values. To do that, this code creates an anonymous class that implements the ViewBinder interface and overrides its setViewValue method. Within that method, an if/else statement checks whether the column is the date completed column for the task. If so, it gets the long value for the date, converts it to a readable date value, sets the date on the appropriate TextView widget, and returns a true value to indicate that it has bound data to the view.

Then, the if/else statement checks whether the column is the notes column for the task. If so, it checks whether the string contained in this column is equal to a null value or an empty string. If so, it sets a value of "No notes" on the appropriate TextView widget and returns a true value to indicate that it has bound data to the view. Otherwise, this code returns a false value, which allows the default view binding to occur. As a result, the string contained in the name column is bound to the view.

The TaskHistoryActivity class

```
@Override
public void onResume() {
    super.onResume();

    // get cursor
    String where = TASK_HIDDEN + "= '1' ";
    String orderBy = TASK_COMPLETED + " DESC";
    cursor = getContentResolver()
            .query(TASKS_URI, null, where, null, orderBy);

    // define variables for adapter
    int layout_id = R.layout.listview_task;
    String[] fromColumns = {TASK_NAME, TASK_NOTES,
        TASK_COMPLETED};
    int[] toViews = {R.id.nameTextView, R.id.notesTextView,
        R.id.dateTextView};

    // create and set adapter
    adapter = new SimpleCursorAdapter(this, layout_id,
            cursor, fromColumns, toViews, 0);
    taskListView.setAdapter(adapter);
    taskListView.setOnItemClickListener(this);

    // convert column data to readable values
    adapter.setViewBinder(new ViewBinder() {
        @Override
        public boolean setViewValue(View view, Cursor cursor,
                int colIndex) {
            if (colIndex == TASK_COMPLETED_COL) {
                long dateMillis = cursor.getLong(colIndex);
                TextView tv = (TextView) view;
                SimpleDateFormat date =
                        new SimpleDateFormat("EEE, MMM d yyyy HH:mm");
                tv.setText("Completed on: " +
                        date.format(new Date(dateMillis)));
                return true;
            }
            else if (colIndex == TASK_NOTES_COL) {
                String notes = cursor.getString(colIndex);
                if (notes == null || notes.equals("")) {
                    TextView tv = (TextView) view;
                    tv.setText("No notes");
                    return true;
                }
            }
            return false;
        }
    });
}
```

Figure 15-13   The TaskHistoryActivity class (part 2 of 3)

The onPause method closes the cursor that's held by the adapter. Then, it closes the cursor that's held by the activity. This works because the onResume method always gets a new cursor.

The onItemClick method is executed when the user clicks on a task. This method begins by getting the ID and name for the task that was clicked. Then, it displays a dialog box that confirms the deletion.

If the user clicks the Yes button, this code uses the URI defined earlier in the class to delete the specified task. Then, it uses a toast to display a message that indicates whether the task was successfully deleted. In most cases, the task should be deleted successfully, and the Toast should display a message that says, "Delete operation successful".

If the user clicks the No button, this code calls the cancel method of the DialogInterface object, which is the first argument of the onClick method. This closes the dialog and returns to the Task History activity without deleting the task.

The TaskHistoryActivity class   **Page 3**

```java
@Override
protected void onPause() {
    adapter.changeCursor(null);    // close cursor for the adapter
    cursor.close();                // close cursor for the activity
    super.onPause();
}


@Override
public void onItemClick(AdapterView<?> adapter, View view,
        int position, long id) {

    // get data from cursor
    cursor.moveToPosition(position);
    final int taskId = cursor.getInt(TASK_ID_COL);
    final String taskName = cursor.getString(TASK_NAME_COL);

    // display a dialog to confirm the delete
    new AlertDialog.Builder(this)
    .setMessage("Do you want to permanently delete task: " +
            taskName + "?")
    .setPositiveButton("Yes", new DialogInterface.OnClickListener() {
        @Override
        public void onClick(DialogInterface dialog, int id) {
            // delete the specified task
            String where = TASK_ID + " = ?";
            String[] whereArgs = { Integer.toString(taskId) };
            int deleteCount = getContentResolver()
                    .delete(TASKS_URI, where, whereArgs);
            if (deleteCount <= 0) {
                Toast.makeText(context, "Delete operation failed",
                    Toast.LENGTH_SHORT).show();
            }
            else {
                Toast.makeText(context, "Delete operation successful",
                    Toast.LENGTH_SHORT).show();
                onResume();
            }
        }
    })
    .setNegativeButton("No", new DialogInterface.OnClickListener() {
        @Override
        public void onClick(DialogInterface dialog, int id) {
            dialog.cancel();
        }
    })
    .show();
}
}
```

Figure 15-13   The TaskHistoryActivity class (part 3 of 3)

Perspective

The skills presented in this chapter should be enough to get you started with content providers. However, there is much more to learn about working with content providers. For example, you may want to provide access to data that's stored in files instead of data that's stored in a database. If so, you can search the Android documentation for more information about how to do that.

Or, you may want to create an app that uses an existing content provider. To do that, you can begin by searching the Internet or the Android documentation for more information about the content provider that you want to use. For example, to learn more about the Contacts Provider, you can search for "android contacts provider". Then, you can use the documentation for that content provider to find the URIs and column names that you need to use the content provider.

Terms

content provider	path
URI (Uniform Resource Identifier)	segment
authority	MIME type

Summary

- A *content provider* allows multiple apps to share the same data. This data can be stored in a database or in files. Android uses built-in content providers to make certain types of data available to all apps.

- A *URI* (*Uniform Resource Identifier*) for a content provider includes the name of the provider (the *authority*) and a *path* to the content. Within a path, each front slash begins a new *segment*.

- A *MIME type* is an Internet standard for defining types of content.

- Every data access method of the ContentProvider class has a Uri object as its first argument. This allows the method to determine what table, row, or file to access.

- To allow other apps to query your database, you can override the query and getType methods of the ContentProvider class.

- To allow other apps to modify the data in your database, you can override the insert, update, and delete methods of the ContentProvider class.

- Before other apps can use your content provider class, you must register it by adding a provider element to the AndroidManifest.xml file.

- You can call any methods of a content provider from the ContentResolver object.

- When you work with a content provider that's included with Android, you can import the classes that contain the constants for working with the content provider.

- You can use the AlertDialog.Builder class and the DialogInterface.OnClickListener interface to build and display a dialog box that includes event handlers.

Exercise 15-1   Review and modify the Task List and Task History apps

In this exercise, you'll review the Task List and Task History apps to see how they share the same data. Then, you'll modify the code for the Task History app.

Review the Task List app

1. Start Eclipse and import the project named ch15_TaskList that's in the book_apps directory (not the ex_starts directory).

2. Open the TaskListDB class and scroll down to the end of it to view the last four methods that support the content provider.

3. Open the TaskListProvider class and review its code. Note that the delete method supports a URI for all tasks and a URI for a single task.

4. Run the app. Add at least 4 tasks to the list and then mark at least 3 of these tasks as completed, and delete them from the list. This may look like you are deleting these tasks. However, this code only sets the completed date for these tasks and hides them for this app. That way, these tasks are stored for future reference and can be used by other apps such as the Task History app.

Review the Task History app

5. Import the project named ch15_ex1_TaskHistory that's in the ex_starts directory.

6. Open the TaskHistoryActivity class and review its code.

7. Run the app. This should display all tasks that you removed from the Task List app earlier in this exercise.

8. Click one of the tasks to delete it.

Modify the Task History app

9. Modify the TASKS_URI constant so the path is "task" instead of "tasks".

10. Run the app. The app should crash, and the LogCat view should display an error that indicates that the URI is not supported.

11. Fix the TASKS_URI constant so the path is "tasks" again.

12. Run the app again. It should run correctly this time.

13. Modify the code that deletes the task so it uses a URI for a single row.

14. Test this change to make sure it works correctly by deleting another task.

Exercise 15-2   Review and modify the Favorites app

In this exercise, you'll review and modify the Favorites app. This app uses the Contacts Provider that's introduced in this chapter.

Review the Favorites app

1. Start Eclipse and import the project named ch15_ex2_Favorites.

2. Open the FavoritesListActivity class and review its code.

3. Run this app. This should display all phone numbers for all contacts that have been marked as a favorite. It should look something like this:

4. If the app doesn't display any names and numbers, you may need to use the built-in Contacts app to mark at least one contact as a favorite by clicking the star icon. Before you can do that on an emulator, you may need to add a contact.

5. Click a phone number for a contact. Then, test the resulting dialog box to make sure it works to start a phone call or a text. Of course, you can only complete a text message successfully for a mobile number.

Modify the Favorites app

6. Modify the app so it only displays mobile phone numbers. To do that, you can modify the WHERE clause so it only returns rows where the DATA2 column is equal to 2.

7. Run this app to make sure it works correctly.

Learn more about the Contacts Provider

8. Search the Internet to find the official documentation for the Android Contacts Provider and read more about it.

16

How to work with app widgets

An app widget can display an app's most important information on a device's Home screen. This can allow a user to work with an app without even starting it. Android typically includes several built-in widgets such as the Analog Clock, Power Control, and Music widgets. In this chapter, you'll learn how to build an app widget for the Task List app. This app widget will display the three oldest tasks from the Task List app.

An introduction to app widgets

Before you learn how to create an app widget for your app, you need to learn some basic skills for working with app widgets. For example, you need to learn how to add an app widget to the Home screen.

A Home screen that has app widgets

Figure 16-1 shows a Home screen that has three *app widgets* and seven *app icons*. Here, each icon occupies a single cell. However, each app widget can occupy more than one cell. For example, the Search widget is 4 cells wide by 1 cell tall. As a result, it occupies the entire first row of the screen. Similarly, the second app widget, the Task List widget, is 4 cells wide by 1 cell tall. As a result, it occupies the entire third row of the screen.

The third app widget is the Power Control widget that comes with most versions of Android. This widget makes it easy to extend your phone's battery life by managing power consuming features such as Wi-Fi, bluetooth, GPS, synching, and screen brightness. Of course, the user can also use the Settings app to perform the same functions, but the Power Control widget makes it easier to access these functions.

The user can click any of the app icons to start the related app. For example, the user can click the Task List icon to start the Task List app. However, these icons don't display any data or provide any other functionality.

The user can also click the Task List widget to start the Task List app. However, the user can also get the most relevant data for the app without even starting it. This can potentially save the user the time and effort.

If you want to remove an app widget from your Home screen, you can do that by long-clicking on it. Then, on Android 3.0 and later, you can drag it to the "Remove" area that appears at the top of the screen. Prior to that, you can drag it to the Trash Can icon that appears at the bottom of the screen. Similarly, if you want to move an app widget, you can long-click on it, and drag it to its new position.

A Home screen with app icons and widgets

Description

- *App icons* occupy a single cell on the Home screen and provide a way for users to start an app.

- *App widgets* occupy one or more cells on the Home screen. They can display data and receive periodic updates.

- Android typically includes several app widgets such as the Search and Power Control widgets.

- You can create custom app widgets for an app.

- From the Home screen, you can typically start the related app by clicking the app widget.

- To remove an app widget from the Home screen, long-click it and drag it to the "Remove" area that appears at the top of the screen. Or, long-click it and drag it to the Trash Can icon that appears at the bottom of the screen.

- To move an app widget, long-click it and drag it to its new position.

Figure 16-1   An introduction to app widgets

How to add app widgets to the Home screen

When you install an app that includes app widgets, those app widgets are installed on your device, but they aren't usually displayed on the Home screen. Figure 16-2 shows how to add app widgets to the Home screen. Android 3.0 introduced a new way to add apps to the Home screen. Since most devices use Android 3.0 or later, you can use this technique to add widgets to most devices. However, this figure also shows the old way of adding an app widget to the Home screen for devices prior to Android 3.0.

With Android 3.0 or later, you use the Widgets tab to display all app widgets that have been installed on a device. In this figure, for example, the Widgets tab shows the Power Control, Search, and Task List widgets that were shown in the previous figure. This tab includes a preview of the widget as well as an indication of how many cells the widget occupies on the Home screen. For example, the Music Playlist widget is 1 cell wide by 1 cell tall, the Picture Frame widget is 2 cells wide by 2 cells tall, and so on.

The tab that displays the app widgets for a device

Procedure (Android 3.0 and later)

1. From the Home screen, click the Launcher icon to display all apps.
2. Click the Widgets tab to display app widgets.
3. Swipe to the right or left to find the app widget you want to add.
4. Long-click the app widget and drag it to the desired position on the Home screen.

Procedure (prior to Android 3.0)

1. Long-click a blank spot on the Home screen.
2. Select the Widgets item from the resulting menu.
3. Select the app widget you want to add.

Figure 16-2   How to add an app widget to your Home screen

How to create app widgets

Now that you know how to use app widgets, you're ready to create one. In particular, you're ready to create a Task List app widget like the one shown in the previous two figures.

How to create the layout

Figure 16-3 shows how to create the layout for an app widget. To start, you can use the Graphical Layout editor just as you would use it for any other type of layout. However, an app widget can only use widgets that are supported by the RemoteViews class. This class supports many common widgets such as the TextView and Button widgets. However, it doesn't support other common widgets such as the EditText or CheckBox widgets. For more information about this class, see the next figure.

The layout for the Task List app widget uses a vertical linear layout to display four TextView widgets. Here, the first TextView widget displays the title of the app and the next three TextView widgets display the three oldest tasks.

When creating the layout for a widget, it's generally considered a good design guideline to add approximately 8dp of padding to an app widget. That way, there is some padding between the widget, other widgets, and the edge of the screen. However, Android 4.0 and later automatically adds padding to app widgets. As a result, you need to add padding prior to Android 4.0, but you don't need to add padding for Android 4.0 and later.

To handle this situation, you can create a *dimension resource*, which is an XML file that stores margin or padding values for different versions of Android. In this figure, for example, the LinearLayout element uses a dimension resource to specify the padding for the widget.

The Task List app widget in the Graphical Layout editor

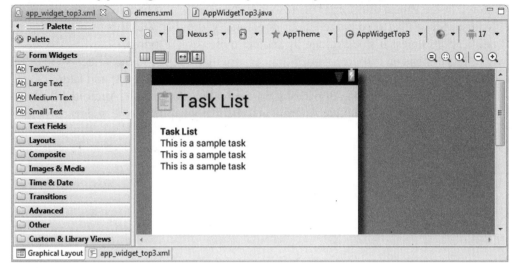

The app_widget_top3 layout

```xml
<?xml version="1.0" encoding="utf-8"?>
<LinearLayout xmlns:android="http://schemas.android.com/apk/res/android"
    android:id="@+id/appwidget_top3"
    android:layout_width="match_parent"
    android:layout_height="match_parent"
    android:orientation="vertical"
    android:padding="@dimen/widget_padding"
    android:background="@android:color/background_light" >

    <TextView
        android:id="@+id/titleTextView"
        android:layout_width="wrap_content"
        android:layout_height="wrap_content"
        android:layout_marginLeft="5dp"
        android:layout_marginTop="5dp"
        android:text="@string/app_name"
        android:textColor="@android:color/black"
        android:textStyle="bold" />

    <TextView
        android:id="@+id/task1TextView"
        android:layout_width="wrap_content"
        android:layout_height="wrap_content"
        android:layout_marginLeft="5dp"
        android:text="@string/sample_task"
        android:textColor="@android:color/black" />

    <TextView
        android:id="@+id/task2TextView"
        android:layout_width="wrap_content"
        android:layout_height="wrap_content"
        android:layout_marginLeft="5dp"
        android:text="@string/sample_task"
        android:textColor="@android:color/black" />
```

Figure 16-3   The layout for the app widget (part 1 of 2)

Part 2 of this figure shows a dimension resource named dimens.xml that's stored in the values directory. This resource specifies 8dp of padding for the widget. Since it's stored in the values directory, Android applies this resource to all versions unless it's overridden by another dimension resource.

The dimension resource named dimens.xml that's stored in the values-v14 directory overrides the first dimension file for Android 4.0 (API 14) and later. This resource specifies 0dp of padding for the widget. As a result, Android 4.0 and later uses the padding that Android automatically applies to the widget.

The app_widget_top3 layout (continued)

```
    <TextView
        android:id="@+id/task3TextView"
        android:layout_width="wrap_content"
        android:layout_height="wrap_content"
        android:layout_marginLeft="5dp"
        android:text="@string/sample_task"
        android:textColor="@android:color/black" />
</LinearLayout>
```

The values/dimens.xml file

```
<?xml version="1.0" encoding="utf-8"?>
<resources>
    <dimen name="widget_padding">8dp</dimen>
</resources>
```

The values-v14/dimens.xml file

```
<?xml version="1.0" encoding="utf-8"?>
<resources>
    <dimen name="widget_ padding">0dp</dimen>
</resources>
```

Description

- You can use the Graphical Layout editor to work with the layout for an app widget.

- The layout for an app widget can only use widgets that are supported by the RemoteViews class. For a list of supported layouts and widgets, view the Android documentation for "App Widgets".

- It's generally considered a good practice to add padding to an app widget so there is some padding between the widget, other widgets, and the edge of the screen.

- Android 4.0 and later automatically adds padding to app widgets.

- A *dimension resource* is an XML file typically named dimens.xml that contains dimension settings for different versions of Android.

How to add padding to the layout for an app widget

1. Open the AndroidManifest.xml file and make sure its targetSdkVersion attribute is set to 14 or later.

2. Create a dimension resource, store it in the values directory, and specify a widget padding value for versions of Android prior to 4.0.

3. Create a dimension resource, store it in the values-v14 directory, and specify a widget padding value of 0 for Android 4.0 and later.

4. In your layout file, refer to the padding value in the dimension resource files.

Figure 16-3   The layout for the app widget (part 2 of 2)

How to modify the database class

Figure 16-4 shows how to modify the database class (the TaskListDB class) to support the app widget. To do that, the database class sends a broadcast whenever it modifies the data in the Task table. That way, the class for the app widget can receive the broadcast and update the widget accordingly.

To do that, the database class begins by defining the TASK_MODIFIED constant that uniquely identifies the broadcast action. Then, this class defines the broadcastTaskModified method that sends the broadcast. Finally, this class calls the broadcastTaskModified method from each method that modifies data in the Task table. For example, this figure shows a deleteTask method that deletes a task and then calls the broadcastTaskModified method. However, all methods that insert, update, or delete a task should call the broadcastTaskModified method.

The database class also contains a method named getTopTaskNames that makes it easy to get the task names that are displayed on the widget. This method accepts an argument that specifies the number of names to get. Then, the code within this method gets those names, stores them in an array of strings, and returns the array.

The constant for the action

```
public static final String TASK_MODIFIED =
        "com.murach.tasklist.TASK_MODIFIED";
```

A helper method for broadcasting the action

```
private void broadcastTaskModified() {
    Intent intent = new Intent(TASK_MODIFIED);
    context.sendBroadcast(intent);
}
```

A method that modifies a task and broadcasts the action

```
public int deleteTask(long id) {
    String where = TASK_ID + "= ?";
    String[] whereArgs = { String.valueOf(id) };

    this.openWriteableDB();
    int rowCount = db.delete(TASK_TABLE, where, whereArgs);
    this.closeDB();

    broadcastTaskModified();

    return rowCount;
}
```

A method that gets the data for an app widget

```
public String[] getTopTaskNames(int taskCount) {
    String where = TASK_COMPLETED + "= '0'";
    String orderBy = TASK_COMPLETED + " DESC";
    this.openReadableDB();
    Cursor cursor = db.query(TASK_TABLE, null,
            where, null, null, null, orderBy);

    String[] taskNames = new String[taskCount];
    for (int i = 0; i < taskCount; i++) {
        if (cursor.moveToNext()) {
            Task task = getTaskFromCursor(cursor);
            taskNames[i] = task.getName();
        }
    }

    if (cursor != null)
        cursor.close();
    db.close();

    return taskNames;
}
```

Description

- To keep the data on an app widget current, you can broadcast an action when the app modifies its data. Then, the app widget can receive the broadcast and update itself.

- To get data for an app widget, you can add a method to a database class for the app.

Figure 16-4   How to modify the database class

How to create the provider class

Figure 16-5 shows how to code a provider class for an app widget. This class extends the AppWidgetProvider class.

Within this class, the onUpdate method contains the code that sets up the app widget and displays data on its user interface. Android calls this method when the user adds the app widget to the Home screen. In addition, Android calls this method at the interval specified by the widget's info file, which you'll learn about later in this chapter.

The onUpdate method accepts three parameters: a Context object, an AppWidgetManager object, and an array of IDs for all app widgets for this provider. Within this method, the code begins by looping through all app widgets for this provider. This is necessary because a user can add multiple instances of a widget to the Home screen.

Within this loop, the first two statements create a pending intent for the Task List activity. The third statement creates the RemoteViews object that stores the layout. The fourth statement sets the listener for the widget so clicking on it starts the Task List activity. The fifth and sixth statements use the TaskListDB class to get an array of task names. The next three statements display these names on the corresponding TextView widgets of the layout. Or, if the array contains a null value for one or more of these names, these statements set an empty string on the corresponding TextView widget. Finally, the last two statements use the RemoteViews object to update the app widget with the specified ID. Up until these statements, the changes have only been made to the RemoteViews object but haven't been applied to the app widget.

The RemoteViews object is necessary because it allows Android to display the layout for the app widget in another process. More specifically, it allows Android to display the layout for the app widget in the process for the Home screen. With app widgets, this is the typical use of the RemoteViews object. However, it's possible to use a RemoteViews object to display the layout for an app widget in any process that's designed as a container for an app widget. As a result, other uses of this object are possible.

The AppWidgetTop3 class   **Page 1**

```java
package com.murach.tasklist;

import android.app.PendingIntent;
import android.appwidget.AppWidgetManager;
import android.appwidget.AppWidgetProvider;
import android.content.ComponentName;
import android.content.Context;
import android.content.Intent;
import android.widget.RemoteViews;

public class AppWidgetTop3 extends AppWidgetProvider {

    @Override
    public void onUpdate(Context context,
            AppWidgetManager appWidgetManager, int[] appWidgetIds) {

        // loop through all app widgets for this provider
        for (int i = 0; i < appWidgetIds.length; i++) {

            // create a pending intent for the Task List activity
            Intent intent = new Intent(context, TaskListActivity.class);
            PendingIntent pendingIntent =
                    PendingIntent.getActivity(context, 0, intent, 0);

            // get the layout and set the listener for the app widget
            RemoteViews views = new RemoteViews(
                    context.getPackageName(), R.layout.app_widget_top3);
            views.setOnClickPendingIntent(
                    R.id.app_widget_top3, pendingIntent);

            // get the names to display on the app widget
            TaskListDB db = new TaskListDB(context);
            String[] names = db.getTopTaskNames(3);

            // update the user interface
            views.setTextViewText(R.id.task1TextView,
                    names[0] == null ? "" : names[0]);
            views.setTextViewText(R.id.task2TextView,
                    names[1] == null ? "" : names[1]);
            views.setTextViewText(R.id.task3TextView,
                    names[2] == null ? "" : names[2]);

            // update the current app widget
            int appWidgetId = appWidgetIds[i];
            appWidgetManager.updateAppWidget(appWidgetId, views);
        }
    }
```

Figure 16-5   The class for the app widget provider (part 1 of 2)

The onReceive method contains code that's executed when the app widget receives a broadcast. This method accepts two parameters: the application context and the intent that was broadcast.

Within this method, the code begins by checking whether the intent is the TASK_MODIFIED intent that's broadcast by the database class. If so, this code updates the app widget. To do that, the first statement creates an AppWidgetManager object. The second statement creates a ComponentName object for the current class, the app widget provider class. The third statement uses these objects to get an array of IDs for all app widgets for this provider. And the fourth statement passes these objects to the onUpdate method shown in part 1 of this figure. This updates the data that's displayed on the app widget. That way, the data that's displayed on the app widget is always synchronized with the data that's displayed by the app.

For most app widgets, you only need to override the onUpdate and onReceive methods of the AppWidgetProvider class to get the widget to work the way you want. However, if you want to execute code when a widget is added to or removed from the Home screen, you may want to override the onEnabled, onDeleted, or onDisabled method. Android calls the onEnabled method when an app widget is added to the Home screen, it calls the onDeleted method when an instance of an app widget is removed from the Home screen, and it calls the onDisabled method when the last instance of an app widget is removed from the Home screen.

The AppWidgetTop3 class   **Page 2**

```
    @Override
    public void onReceive(Context context, Intent intent) {
        super.onReceive(context, intent);

        if (intent.getAction().equals(TaskListDB.TASK_MODIFIED)) {
            AppWidgetManager manager =
                    AppWidgetManager.getInstance(context);
            ComponentName provider =
                    new ComponentName(context, AppWidgetTop3.class);
            int[] appWidgetIds = manager.getAppWidgetIds(provider);
            onUpdate(context, manager, appWidgetIds);
        }
    }
}
```

A constructor and some methods of the RemoteViews class

Constructor/Method	Description
`RemoteViews(package, layoutId)`	Create a new RemoteViews object that provides access to the views in the specified layout file.
`setOnClickPendingIntent(` `    appWidgetId, intent)`	Sets the pending intent that's executed when a user clicks on the layout.
`setTextViewText(textViewId, text)`	Sets the specified text on the specified TextView widget in the layout.

Some methods of the AppWidgetManager class

Method	Description
`updateAppWidget(widgetId, views)`	Updates the specified widget with the specified RemoteViews object.
`getInstance()`	Gets an AppWidgetManager object.
`getAppWidgetIds(provider)`	Gets the IDs of the installed app widgets for the specified provider.

Description

- The class for an app widget must extend the AppWidgetProvider class.
- Android calls the onUpdate method when the user adds the app widget to the Home screen and at the interval specified by the widget's info file.
- Android calls the onReceive method when an action for the app widget is broadcast.
- Android uses a RemoteViews object to display a layout in another process such as the process for the Home screen. The RemoteViews class provides some methods that you can use to modify the layout that's stored in the RemoteViews object.
- The AppWidgetProvider class also provides the onEnabled, onDeleted, and onDisabled methods. If necessary, you can override these methods to execute code when an app widget is added to or removed from the Home screen.

Figure 16-5   The class for the app widget provider (part 2 of 2)

How to configure an app widget

Figure 16-6 shows how to configure an app widget by adding an info file to the res\xml directory. The initialLayout attribute specifies the layout for the app widget.

The minHeight and minWidth attributes specify the minimum height and width for the app widget. In this figure, the minimum height has been set to 40dp, which is approximately 1 cell, and the minimum width has been set to 250dp, which is approximately 4 cells. As a result, the app widget should be 4 cells wide by 1 cell tall.

The updatePeriodMillis attribute specifies the number of milliseconds for the update interval. However, if the device is asleep, these types of updates wake the device to perform the update. As a result, frequent updates of this type can cause significant problems for the battery life. To minimize this issue, Android won't perform this type of update more than once every 30 minutes (1800000 milliseconds), and it's generally recommended to not perform this type of update more than once per hour.

So, if you want to update an app widget more often than once per hour, you need to use a different approach. On one hand, if you want to update the app widget when a particular action occurs in the app, you can broadcast an action from the app that the app widget can receive. Then, you can delete the updatePeriodMillis attribute or set it to a value of 0. This approach is described in the previous figures.

On the other hand, if you need to update an app widget at a specified time interval, you can use an alarm of the ELAPSED_REALTIME or RTC type to broadcast an update action at a specified time interval. Then, the app widget can receive the broadcast and update itself. This approach works similarly to the approach described in this chapter. However, this approach does not wake the device if the device is sleeping. As a result, it conserves battery life. For more details about working with alarms, you can view the documentation for Android's AlarmManager class.

The res\xml\app_widget_top3_info.xml

```
<appwidget-provider
    xmlns:android="http://schemas.android.com/apk/res/android"
    android:initialLayout="@layout/app_widget_top3"
    android:minHeight="40dp"
    android:minWidth="250dp"
    android:updatePeriodMillis="21600000" > <!-- every 6 hours -->
</appwidget-provider>
```

The appwidget-provider element

Attribute	Description
initialLayout	Specifies the layout file for the app widget.
minHeight	Specifies the minimum height for the app widget.
minWidth	Specifies the minimum width for the app widget.
updatePeriodMillis	Specifies the number of milliseconds for the update interval. These updates are not delivered more than once every 30 minutes (1800000 milliseconds). In addition, the actual update is not guaranteed to occur exactly on time.

Approximate sizes for minimum height and width

Cell count	Approximate size
1	40dp
2	110dp
3	180dp
4	250dp

Description

- To configure an app widget, you can add an info file for the app widget to the res\xml directory.

Figure 16-6   How to configure an app widget

How to register an app widget

Before you can use an app widget, you must register it. To do that, you can add a receiver element to the AndroidManifest.xml file as shown in figure 16-7.

When you register an app widget, you need to define a label attribute that specifies the name that Android uses for the app widget in its list of app widgets. For example, this specifies a value of "Task List (4 x 1)" that's stored in the strings.xml file for the app.

Since the AppWidgetProvider class extends the BroadcastReceiver class, registering an app widget is similar to registering a broadcast receiver. In particular, you must add an intent-filter element for every broadcast action that your app listens for. Here, the APPWIDGET_UPDATE action is necessary to update the app widget when it's added to the Home screen and at the interval specified in the info file. Similarly, the TASK_MODIFIED action is necessary to update the app widget when the user inserts, updates, or deletes a task.

To finish registering an app widget, you need to use the meta-data element to specify the type of broadcast receiver and its location. Here, the broadcast receiver is an app widget provider, and its info file is stored in the directory shown in the previous figure.

How to test an app widget

After you register an app widget, you should test it. To do that, you can run the app to install the app widget on a device. Then, you can add the app widget to your Home screen as described in figure 16-2. At this point, you can test the app widget to make sure it works correctly. For the Task List app widget, for example, you can use the Task List app to delete a task. Then, you can make sure the app widget is updated correctly.

The AndroidManifest.xml file

```xml
<receiver
    android:name="com.murach.tasklist.AppWidgetTop3"
    android:label="@string/appwidget_top3_label" >
    <intent-filter>
        <action
            android:name="android.appwidget.action.APPWIDGET_UPDATE" />
    </intent-filter>
    <intent-filter>
        <action android:name="com.murach.tasklist.TASK_MODIFIED" />
    </intent-filter>
    <meta-data
        android:name="android.appwidget.provider"
        android:resource="@xml/app_widget_top3_info" />
</receiver>
```

The receiver element

Attribute	Description
name	The class for the app widget provider.
label	The name that Android uses for the app widget in its list of app widgets. For example, "Task List (4X1)".

The meta-data element

Attribute	Description
name	The type of broadcast receiver.
resource	The name of the info file for the app widget.

Description

- The AppWidgetProvider class extends the BroadcastReceiver class. As a result, an app widget is also a broadcast receiver.

- To register an app widget, add a receiver element to the AndroidManifest.xml file. For all app widgets, you should add an intent-filter element that listens for the APPWIDGET_UPDATE action. In addition, you may want to add intent-filter elements to listen for other broadcast actions.

- To test an app widget, you can run the app and add the app widget to your Home screen as described in figure 16-2.

Figure 16-7   How to register and test an app widget

Perspective

The skills presented in this chapter should be enough to get you started with app widgets. However, Android provides many more features for working with app widgets. For example, you can create app widgets that can be added to the Lock screen. This provides a way for the user to use your app widget without even unlocking the device. You can add an activity that provides a way for users to configure an app widget. You can create app widgets that the user can resize. And so on. To learn more about these features, you can begin by searching the Android documentation for more information about app widgets.

Terms

app icons
app widgets
dimension resource

Summary

- *App icons* occupy a single cell on the Home screen and provide a way for users to start an app.

- *App widgets* occupy one or more cells on the Home screen and can display data and receive periodic updates. You can create custom app widgets for an app.

- A *dimension resource* is an XML file typically named dimens.xml that contains dimension settings for different versions of Android.

- Android calls the onUpdate method of an app widget when the user adds the widget to the Home screen and at the interval specified by the widget's info file.

- Android calls the onReceive method of an app widget when an action for the widget is broadcast.

- Android uses a RemoteViews object to display a layout in another process such as the process for the Home screen.

- To register an app widget, add a receiver element to the AndroidManifest.xml file.

Exercise 16-1   Review and modify the Task List app widget

In this exercise, you'll review the code for the Task List app widget. Then, you'll test this app widget to see how it works. Finally, you'll modify this app widget to improve how it works.

Review the app widget
1.   Start Eclipse and import the project named ch16_ex1_TaskList.

2. Open the app_widget_top3 layout. Then, view it in the Graphical Layout editor, and view it in the XML editor. This file should use a dimension resource to specify the padding for the linear layout.

3. Open the dimens.xml files that are stored in the values and values-v14 directories. These files should specify 8dp of padding prior API 14 and 0dp of padding for API 14 and later.

4. Open the AppWidgetTop3 class and view its code. This code should call the getTopTaskNames method from the TaskListDB, class, and it should receive the TASK_MODIFIED action.

5. Open the TaskListDB class. Then, review the code for the getTopTaskNames method. Also, review the methods of the TaskListDB class that modify the data in the Task table. Note that some of these methods broadcast the TASK_MODIFIED action.

6. Open the app_widget_top3_info.xml file that's in the res\xml folder. This file should specify that the app widget should be 4 cells wide by 1 cell tall and that it should be updated every 6 hours.

7. Open the AndroidManifest.xml file and review the receiver element. This file should specify the actions that the app widget receives, and it should specify the info file for the app.

Test the app widget

8. Run the app. This should start the Task List app. If this app displays less than three tasks, add tasks until this app displays at least five tasks.

9. Add the app widget to the Home screen. To do that, you may need to remove other app icons or widgets to make room for the app widget. Note the three tasks that the Task List app widget displays.

10. Switch to the Task List app and delete one of the three tasks that's currently displayed by the app widget.

11. Switch back to the Home screen to make sure the app widget has been updated to reflect the changed data.

Modify the app widget

12. Open the TaskListDB class. Make sure that all of the methods that modify the Task table broadcast the TASK_MODIFIED action. This should include the methods that support the content provider.

13. Open the app_widget_top3_info.xml file that's in the res\xml folder.

14. Modify the info file so it specifies that the app widget should be 3 cells wide by 1 cell tall.

15. Modify the info file so Android does not update the app widget at a specified interval. This should help to conserve battery life.

16. Test the app widget to make sure it still works correctly. To do that, run the app again. Then, remove the old app widget from the Home screen and add the modified app widget to the Home screen.

Exercise 16-2 Add an app widget that displays a count of tasks

In this exercise, you'll add another app widget to the Task List app presented in this chapter. This app widget should be 1 cell wide by 1 cell tall, and it should display a count of the total number of incomplete tasks.

1. Start Eclipse and import the project named ch16_ex2_TaskList.

2. Open the TaskListDB class and note that it includes a method named getTaskCount that gets a count of incomplete tasks.

3. Navigate to the res\drawable-xhdpi directory and view its contents. Note that it includes an icon file named ic_task_count.png.

4. Create a layout named app_widget_task_count that uses a TextView widget to display the count of incomplete tasks on the icon named ic_task_count. To accomplish this, you can set the background property of the TextView to the icon resource.

5. Make sure the layout uses a dimension resource to specify the padding for the widget. That way, the padding should work correctly for different versions of Android.

6. Create a class named AppWidgetTaskCount that displays the count of tasks. This class should update the data on the layout when the widget is first added to the Home screen and when the TASK_MODIFIED action is received.

7. Create an info file for the app widget that specifies that the icon is 1 cell wide by 1 cell tall.

8. Open the Android manifest file and add a receiver element that registers the app widget.

9. Run the app on an emulator or device. Note the number of tasks that are displayed by the app.

10. Add the app widget to the Home screen and check the task count to make sure the app widget is working correctly. If the app displays 3 tasks, the app widget should look like this:

Section 5

Advanced Android skills

In sections 1 through 4, you learned the skills that you need to develop many types of Android apps. In this section, chapter 17 shows several ways to deploy most apps, including how to publish an app on the Google Play store.

Chapter 18 shows how to develop the Run Tracker app. This app tracks a device's location and displays it on a map. In this chapter, you'll learn how to use Android to work with the newest version of Google Maps. Since this chapter requires some understanding of deployment, I recommend reading chapter 17 before chapter 18.

How to deploy an app

So far in this book, you've learned how to develop, test, and debug an Android app. Once you finish thoroughly testing and debugging your app, you're ready to deploy your app. One way to deploy an app is to distribute it directly to your users by making it available from a web site. Another way to deploy an app is to publish it on an application marketplace. In this chapter, you'll learn how to publish your app on the most popular application marketplace, Google Play.

An introduction to distribution and monetization

Before you distribute an app, you should examine the various distribution options. Similarly, if you want to make money from your app, you need to find a way to get paid for it. This is known as *monetization*.

How distribution works

Figure 17-1 begins by showing two ways to distribute an app. Typically, you distribute an app through an *application marketplace*. Today, Google Play (formerly known as the Android Market) is the most popular application marketplace. This figure, for example, shows the Google Play listing for a free Tip Calculator app. This listing includes descriptive text and images for the app.

If you want to use another application marketplace instead of or in addition to Google Play, that's possible too. Many other application marketplaces exist, and they all have their advantages. For example, the Amazon Appstore has some promotions and features that aren't available from Google Play.

Another distribution option is to make an app available directly to users via a web site download or an email attachment. This option is especially appropriate for a proprietary app that's only used within an organization. However, it can also be a good way to promote an app, especially a free version of an app.

A free app displayed in Google Play

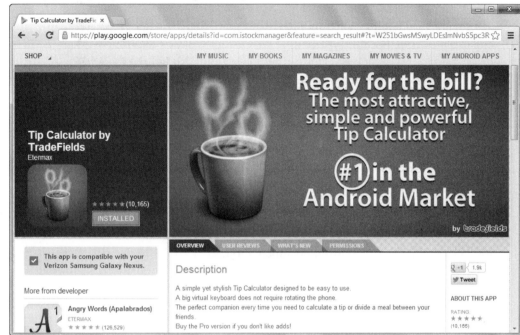

Two ways to distribute an app

- Through an application marketplace
- Directly to the user via a web site, email, etc.

Two application marketplaces

- Google Play
- Amazon Appstore

Description

- You can distribute an app through an *application marketplace* such as Google Play, or you can distribute it directly to users via a web site or email.
- Google Play, formerly known as the Android Market, is a digital marketplace for Android apps that's developed and maintained by Google.

Figure 17-1   How distribution works

How monetization works

Figure 17-2 begins by listing three ways to monetize an app. First, you can create a *paid app*. For this type of app, the customer pays a one-time fee before he or she can install the app. This figure, for example, shows the Google Play listing for a Sports Tracker app that costs $4.99.

Second, you can use *in-app billing*. This provides for *in-app products* that have a one-time billing or *subscriptions* that have a recurring, automated billing. To learn how to add in-app billing to your app, you can start by searching for "in-app billing" in the Android documentation.

Third, you can add advertising to your app. For example, you can add a banner ad to your app. To do that, you need to work with a *mobile advertising network* that serves ads to mobile devices. AdMob is a popular mobile advertising network that's owned by Google, but other advertising networks are also available. To learn how to add advertising to your app, you can start by searching for "advertising" in the Android documentation.

If you publish a free app on Google Play, you can't change it to being a paid app. However, you can add in-app products or subscriptions, and you can add advertising. On the other hand, if you publish a paid app, you can change it at any time to being a free app.

A common strategy for monetizing an app is to provide a free version of an app that includes advertising. Then, you can provide a paid version of that app that removes the advertising. Or, you can provide a free version of an app that doesn't include all features. Then, the paid version can provide additional features. If one of your goals is to make money from the apps that you develop, you should create a strategy for monetizing your app before you publish it.

A paid app displayed in Google Play

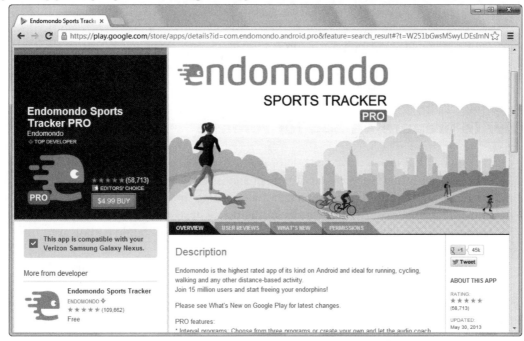

How to monetize an app

- Paid apps
- In-app billing
- Advertising

Description

- Getting paid for an app is referred to as *monetizing* an app.
- If you publish a free app, you can't change it to being a paid app. However, you can sell in-app products or subscriptions, and you can add advertising.
- If you publish a *paid app*, you can change it at any time to being a free app.
- *In-app billing* provides for *in-app products* that have a one-time billing or *subscriptions* that have a recurring, automated billing.

Figure 17-2   How monetization works

How to create a release build

Now that you understand the options for distributing and monetizing an app, you're ready to create a release build for your app. Then, you can distribute this release build via a web site or email. Or, you can publish this release build to a marketplace such as Google Play.

How to prepare an app for release

Figure 17-3 shows a Tip Calculator app that has been thoroughly tested and debugged in the development environment and is ready to be released. It also lists some tasks you should complete before creating the release build. Most of the time, you complete these tasks as you create, test, and debug the app. However, if you haven't already completed these tasks, you should complete them before releasing the app.

To start, you should make sure the package name for the app is unique and appropriate. That's because Android uses the package name to identify the app. As a result, no two apps can use the same package. If you plan to create free and paid versions of an app, you may want to adjust the package names accordingly.

You should make sure to set the final icon for the app. Most of the time, you set this icon when you create the app, but you may want to switch to a more professional app icon just before releasing the app.

You should make sure the version numbers are correct for the app and its components. To set the correct version for the app, you can edit the Android manifest file for the app. If the app uses another component such as a database, you can edit the version number for that component too. For example, you can edit the database class to set the version number for the database.

You should add any copyright info or *End User License Agreement* (*EULA*) you want. To do that, you can add copyright info to an About screen. Similarly, you can add a starting screen that displays the EULA and requires the user to accept it or decline it. If the user accepts the EULA, this screen should not be displayed on subsequent starts.

If you haven't already done it, you should finish localizing the app. This involves creating resource directories for various countries and adding strings.xml files to those directories that contain translations for other languages as described in chapter 3.

You should remove any old files from the project. During development, it's common to add files such as icon files that don't end up being used by the release version of that app. To keep the file size of the release build small, you should delete all unused files.

Finally, you should remove logging statements from the source code. These statements are useful during development, but you should remove them from a release build.

A Tip Calculator app

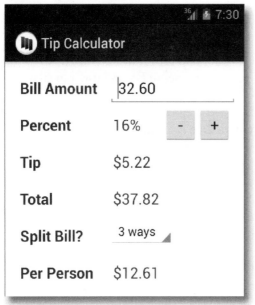

Final checklist

1. Set the final package name.
2. Set the final application icon.
3. Set the version numbers for the app and all of its components.
4. Add copyright info.
5. Add an *End User License Agreement* (*EULA*).
6. Finish localizing the app.
7. Remove any old files from the project.
8. Remove logging statements from the source code.

Description

- After you have thoroughly tested and debugged your app for all devices and emulators in your development environment, you can prepare an app for release by performing some final checks on the app.

Figure 17-3   How to prepare an app for release

How to create the signed APK file

To run an app on a device or emulator, the app must be stored in an *APK file* (*Android package file*) that's *signed* with a *digital certificate*. During development, Eclipse automatically generates a *debug key* and uses it to sign the APK file. Before distributing an app, you must generate a *release key* and use it to sign the APK file.

The JDK includes command-line tools that you can use to generate a release key and sign an app. However, Eclipse provides an easier way to generate the release key and APK file for an app. When you use the procedure shown in figure 17-4, you respond to several Export dialog boxes that are displayed by Eclipse. This figure shows three of these dialog boxes.

The first dialog box specifies the location and passwords for the *keystore*, which is a file that contains one or more digital certificates. Here, the keystore is stored in a file named murach_release_keystore. This file doesn't have an extension, but it is a file, not a directory.

Android uses the keystore file to verify any updates to an app. If you lose this file, you won't be able to update your app. Conversely, if a hacker gains access to this file, he or she can update your app. For example, a hacker could replace your app with another app that contains malicious code. As a result, it's important to store this file in a secure location. And, of course, it's important to store this file in a location that you can remember.

The Keystore dialog box

Procedure

1. Right-click the project in the Package Explorer and select the Export command. Then, select the Android→Export Android Application item.

2. Specify the location and password for a new keystore file, or select an existing keystore file.

3. If necessary, specify the information for the new release key.

4. Specify the location and filename for the APK file.

Description

- To run an app on a device or emulator, the app must be stored in an *APK file* (*Android package file*) that's *signed* with a *digital certificate*.

- During development, the IDE automatically generates a *debug key* and uses it to sign the APK file. Before distributing an app, you must generate a *release key* and use it to sign the APK file.

- A *keystore* is a file that contains one or more digital certificates. If you lose this file, you won't be able to update your app. Conversely, if a hacker gains access to this file, he or she can update your app. As a result, it's important to store this file in a secure location.

Figure 17-4   How to create the signed APK file (part 1 of 2)

The second dialog box specifies the information for a digitial certificate for a release key. When you use this dialog box, you typcially want to specify a value in the Validity text box so the digital certificate is valid for the entire life of the app. Since it's hard to predict how long an app might live, you should specify a large value such as 50 years. Another issue to consider is that Google Play requires the release key to be valid until October 22, 2033.

The third dialog box specifies the location of the signed APK file. Here, the APK is stored in a file named ch17_TipCalculator.apk. At this point, you can distribute this APK file directly to your users as described in the next figure, or you can publish it on Google Play as described after that.

The Key dialog box

The Destination dialog box

Figure 17-4   How to create the signed APK file (part 2 of 2)

How to distribute directly to the user

Once you have a signed APK file for an app, you can distribute it directly to your users via a web site or email as described in figure 17-5. Then, your users can install the app without using an established Android marketplace. Since the user doesn't download the app from a marketplace, this is sometimes referred to as *side loading*.

If you are a developer, you need to delete any old versions of the app from your Android device that use a debug key. Otherwise, when you try to install the new version of the app that uses the release key, the installation will fail and Android will display a message that indicates that the installation failed due to different application signatures.

In addition, you may need to select the Unknown Sources item on your Android device to allow apps from sources other than Google Play. Many devices have this item selected by default. However, if this item isn't selected by default, you need to select it.

Once you have removed any debug versions of the app and selected the Unknown Sources item, you can install the app by copying the APK file to the device and clicking on it. Two of the easiest ways to copy the APK file to a device are to download it from a web site or to receive it as an email attachment.

As you review the procedure shown in this figure, note that step 1 only applies to developers who have installed a version of the app that uses a debug key. In addition, on many devices, the Unknown Sources item described in step 2 is selected by default. As a result, most users only need to do steps 3 and 4.

How to distribute via a web site

To distribute via a web site, you can create a mobile-friendly web page that includes a link to download the APK file for the app. Then, a user can download the app by using the browser on the Android device to view that web page and download the APK file.

How distribute via email

To distribute via email, you can attach the APK file to an email and send it to your users. Then, a user can click on the attachment to download the APK file. Although this works seamlessly when your users have a new Android device and a Gmail account, it might not work at all for older Android devices or other types of email accounts. In that case, your users need to find another way to copy the attachment to their device. One easy way to is to download the file from the web as described above.

How to install an app from an APK file

1. If you are a developer, delete any old versions of the app from your Android device that use a debug key.

2. Make sure your Android device allows apps from sources other than Google Play.
   - For Android 4.0 and later, start the Settings app and select the Security→Unknown Sources item.
   - Prior to Android 4.0, start the Settings app and select the Applications→Unknown Sources item.

3. Copy the APK file to your Android device.
   - If the APK file is available from a web page, use the browser on your Android device to view that web page and download the APK file.
   - If the APK file has been sent to you as an email attachment, use your Android device to open the email and click on the attached APK file. This works seamlessly with newer Android devices and Gmail. However, this may not work as well with older devices or other types of email.

4. Click the APK file to install the app.

Description

- Once you have a signed APK file, you can distribute it directly to the user without using an Android marketplace.

- Installing an app from a source other than an established Android marketplace is sometimes referred to as *side loading*.

Figure 17-5   How to distribute directly to the user

How to publish on Google Play

Once you have a signed APK file for an app, you're just a few steps from making it available on the Google Play store. To do that, you need to create some online accounts. Then, you can use the Developer Console for Google Play to upload your app and manage its listing.

How to set up a publisher account

Before you can upload an APK file to Google Play, you must set up a publisher account as shown in figure 17-6. To start, you can sign in with the Google account that you want to use for publishing apps. A Google account is the same kind of account that you use to work with any Google product such as Gmail or the Google Calendar.

If you represent an organization, you may want to create a Google account for the organization. That way, several developers can use that account to publish the app. In this figure, for example, I am signed in as Joel Murach. However, I want to publish this app for an organization (Mike Murach & Associates). To do that, I need to sign in using the Google account for Mike Murach & Associates.

After you sign in with the correct Google account, you can accept the developer agreement, pay the registration fee ($25) for that account, and complete the details for that account. This creates a Google Play account that you can use to publish free apps. However, if you want to create paid apps or use in-app billing, you also need to set up a merchant account. To do that, you can follow one of the links that's displayed after you finish creating the Google Play account.

The Google Play signup page

`https://play.google.com/apps/publish/signup`

The Accept Developer Agreement page

Procedure

1. Go to the web site shown above.
2. Sign in with your Google account.
3. Accept the developer agreement.
4. Pay the registration fee ($25). To do that, you need to create a Google Wallet account if you don't already have one.
5. Complete your account details.
6. If you want to create paid apps or use in-app billing, follow the links to set up a merchant account.

Description

- Before you can upload an APK file to the Google Play store, you must set up a publisher account.

Figure 17-6   How to set up a publisher account

How to use the Developer Console to publish an app

When you finish creating a Google Play account, there is a link for publishing an app on Google Play. You can use this link to navigate to a Developer Console like the one shown in figure 17-7. Or, you can navigate to the Developer Console by entering the URL shown at the top of this figure. Either way, once you display the Developer Console, you can use it to upload your app to Google Play and manage its listing.

To upload your app, you can click the APK link. Then, you can use the three tabs to upload the APK for alpha testing, beta testing, or production. In this figure, the Tip Calculator app has been uploaded to the Production tab. As a result, this app will be released to the general public as soon as it is published.

The Developer Console won't let you publish your app until Google processes your $25 registration fee, and that can take up to 48 hours. Before then, the app has a status of "Draft". This lets you enter the listing and pricing details for the app. However, the app isn't published until you select the "Publish this app" option from the drop-down list to the right of the app's name. After you do that, the status of the app changes to "Published".

To modify the listing for the app, you can click the Store Listing link. Then, you can specify the text and images for the app. This can include text and images like the ones shown in the first two figures in this chapter. At a minimum, the listing must include the name of the app, a brief description of the app, two screenshots of the app, and an image for the app's logo. To create the required text, you may want to work with someone who specializes in marketing. Similarly, to create the required images, you may want to work with someone who specializes in graphic design.

To modify the pricing and distribution for the app, you can click the Pricing and Distribution link. Then, you can specify whether the app is a free or paid app. In addition, you can specify the countries where you want to distribute the app. Of course, you typically want to make sure your app provides for the primary language of each country.

If the app is a paid app, you need to specify the price for each country. This is necessary because every country uses a different currency. Fortunately, if you specify a default price in US dollars, the Developer Console can generate prices for other countries based on current exchange rates.

The Developer Console page

```
https://play.google.com/apps/publish
```

The Developer Console page for the Tip Calculator app

Procedure

1. Log in to the Developer Console.
2. Upload the APK file.
3. Edit the store listing. This includes specifying a title, a description, screenshots, and so on.
4. Edit the pricing and distribution (optional).
5. Edit the in-app products (optional).
6. Select the "Publish this app" item from the combo box to the right of the app's name.

Description

- You can use the Developer Console to publish your app and to manage its listing on Google Play.
- At a minimum, the listing for an app must include the name of the app, a brief description of the app, two screenshots of the app, and an image for the app's logo.
- It may take 48 hours for your $25 registration fee to be processed. The Developer Console won't allow you to publish an app until that fee is processed.

Figure 17-7   How to use the Developer Console to publish an app

How to view the listing for an app

Once you publish an app on Google Play, you can view the listing for the app by going to Google Play and searching for the app. Figure 17-8, for example, shows the listing for the Tip Calculator app. I displayed this listing by using a browser to navigate to Google Play, searching for "murach", and clicking on the link for this app.

If you use the Developer Console to edit the listing for your app, the changes should appear when you reload the Google Play page for your app. For example, if I used the Developer Console shown earlier in this chapter to edit the text or images for this app, those changes should appear in the this listing.

The Google Play listing

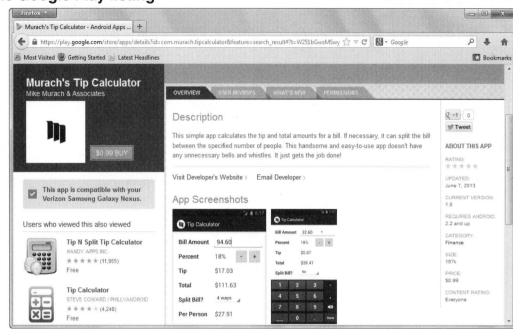

Description

- Once you publish an app on Google Play, you can view the listing for the app by going to Google Play and searching for the app.

Figure 17-8   How to view the listing for an app

Perspective

The skills presented in this chapter should be enough for you to publish a free or paid app on Google Play. However, if you want to add in-app billing or advertising to your app, you'll need to do more research to learn how to do that. Of course, the skills presented in this chapter and throughout this book should provide the foundation you need to learn how to accomplish these tasks.

Terms

monetization	APK file (Android package file)
application marketplace	signed app
paid app	digital certificate
in-app billing	debug key
in-app products	release key
subscriptions	keystore
mobile advertising network	side loading
End User License	
Agreement (EULA)	

Summary

- You can distribute an app through an *application marketplace* such as Google Play, or you can distribute it directly to users via a web site or email.

- Google Play, formerly known as the Android Market, is a digital marketplace for Android apps that's developed and maintained by Google.

- Getting paid for an app is referred to as *monetizing* an app.

- *In-app billing* provides for *in-app products* that have a one-time billing or *subscriptions* that have a recurring, automated billing.

- To run an app on a device or emulator, the app must be stored in an *APK file* (*Android package file*) that's *signed* with a *digital certificate*.

- During development, the IDE automatically generates a *debug key* and uses it to sign the APK file. Before distributing an app, you must generate a *release key* and use it to sign the APK file.

- A *keystore* is a file that contains one or more digital certificates. This file is required to update your app. As a result, it's important to store this file in a secure location that you can remember.

- Installing an app from a source other than an established Android marketplace is sometimes referred to as *side loading*.

Exercise 17-1   Install the Tip Calculator app on your device

In this exercise, you'll install the release build of the Tip Calculator app on an Android device.

1. Select the Android device that you want to use.

2. Uninstall any old versions of the Tip Calculator app that use the debug key.

3. Make sure the Android device allows apps from sources other than Google Play. To do that, you can start the Settings app and select the Unknown Sources item.

4. Copy the APK file that's in the ex_starts\ch17_APK directory to the device. One easy way to do this is to attach it to an email and send it to a Gmail account. Then, you can open Gmail on the device, open the email, and click on the attachment.

5. Click the APK file to install the Tip Calculator app.

6. Start the Tip Calculator app and make sure it's the app presented in this chapter.

Exercise 17-2   Publish an app on Google Play

In this exercise, you'll publish your first app on Google Play.

1. Design and develop an app that you want to publish on Google Play.

2. Test and debug that app on all devices and emulators in your development environment.

3. Prepare the app for release.

4. Create the signed APK file for the app.

5. Create a Google Play publisher account. To do that, you must pay the registration fee ($25).

6. Use the Developer Console to publish your app. To do that, you must upload the APK file and provide the required text and images for the app.

7. Use Google Play to install the app on one or more devices.

8. Test the installed app on one or more devices and make sure it works the way you want.

9. Check Google Play for feedback and continue to improve your app.

18

How to work
with locations and maps

Most mobile devices include hardware that can provide the location of the device to an Android app. As a result, you can use the location of a device to customize your app so it's appropriate for the user's current location.

Often apps that use a device's location also use maps. For example, they might display a map and zoom in on the current location. To do that, the app can use any Android API for mapping. This chapter shows how to use the newest version of the most popular Android API for mapping, Google Maps version 2.

An introduction to locations and maps

This chapter begins by presenting the user interface for an app that tracks a device's location and displays it on a map. Then, it describes two competing Android APIs for working with maps, Google Maps and MapQuest. Finally, this topic describes two versions of the Android API for Google Maps.

Although it's possible to use an emulator to test locations and maps, it's easier to use an actual device. As a result, this chapter doesn't show how to use an emulator to test locations. Instead, it assumes that you have an actual device that you can use for testing. If you need to use an emulator for testing, you can search the Internet for information to learn how to send test location data to an emulator.

The user interface for the Run Tracker app

The skills presented in this chapter are necessary to create the Run Tracker app shown in figure 18-1. When the Run Tracker app starts, the Stopwatch activity displays a stopwatch that you can use to start, stop, and reset a run. Typically, you begin by clicking the Start button to start a run. Then, the Start button becomes a Stop button.

If you want to pause the run, you can click on the Stop button. Then, the Stop button becomes a Start button that you can use to continue the run. However, if you want to delete the current run, you can click the Reset button to do that. Then, you can click the Start button to start a new run.

At any time, you can view the map for the run by clicking the View Map button. This should display a red marker for the user's current location and a black line to indicate the route of the run. In addition, it should display zoom controls that you can use to zoom in or out. When you're ready to switch back to the Stopwatch activity, you can click on the View Stopwatch button. This button is transparent by default, which is usually what you want since it allows you to view the map and compass behind it.

When you start the stopwatch, the app starts a service that uses *GPS (Global Positioning System)* to get the location of the device. As a result, Android displays the GPS icon and a stopwatch icon in the notification area. In this figure, the user has started a run. As a result, it shows both icons in the notification area above both activities.

To remove these notifications, you can stop or reset the stopwatch. This stops the service that uses GPS. As a result, Android removes the GPS icon and the stopwatch icon from the notification area.

Although I designed this app for running or jogging, you can use it for walking, biking, or even driving. It should work for any activity where you are moving and your device can access GPS. However, GPS does not usually work indoors, so you typically need to be outside.

The Stopwatch activity

The Run Map activity

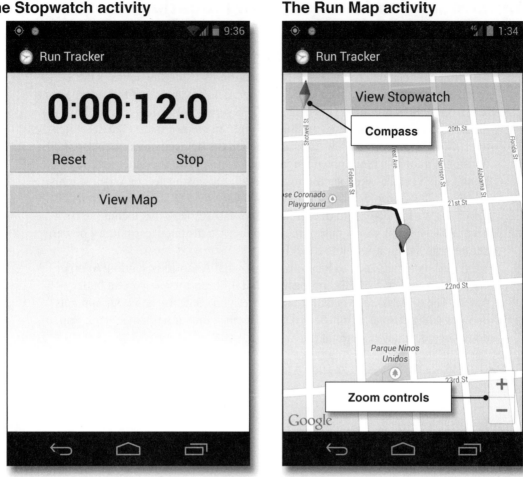

Description

- When the app starts, it displays the Stopwatch activity.
- To start or continue a run, the user can click on the Start button.
- To pause a run, the user can click on the Stop button.
- To delete a run, the user can click on the Reset button.
- To display the Run Map activity, the user can click on the View Map button.
- The Run Map activity displays a red marker for the user's current location and uses a black line to indicate the path of the run.
- When the user starts the stopwatch, the app displays a stopwatch notification icon and starts a service that uses *GPS (Global Positioning System)* to track the user's location. This causes Android to display a notification icon for GPS.
- To remove these notifications, the user can stop or reset the timer.
- To display the Stopwatch activity, the user can click on the View Stopwatch button.

Figure 18-1   The user interface for the Task List app

An introduction to determining location

From an Android developer's point of view, there are three ways to determine the location of a device. Figure 18-2 describes their pros and cons.

First, you can determine a device's location by using the device's GPS receiver to check signals from GPS satellites. The advantage of this approach is that it's the most accurate. However, for GPS to work, the device must have a GPS receiver and an unobstructed line of sight to at least three GPS satellites. As a result, GPS doesn't work well or at all when the device is indoors. Also, GPS consumes a lot of battery power, significantly more than the other techniques. Finally, GPS returns the location more slowly than using cell or Wi-Fi signals.

Second, you can determine a device's location by using signals from cell towers or Wi-Fi hotspots to determine location. Although this technique isn't as accurate as GPS, it works outdoors and indoors, consumes less battery power, and returns the location more quickly.

Third, you can listen to location updates that Android is sending to other apps. The advantage of this approach is that it doesn't use any extra battery power. The disadvantage is that it doesn't let you control when your app gets updates. For example, if other apps don't request location updates, your app doesn't receive any updates at all.

GPS

- Uses signals from GPS (Global Positioning System) satellites to determine the location of the device

Pros

- Most accurate

Cons

- Only works outdoors, not indoors
- Consumes more battery power
- Returns the location slowly

Network

- Uses signals from the cell network or Wi-Fi hotspots to determine location

Pros

- Works outdoors and indoors
- Consumes less battery power
- Returns the location more quickly

Cons

- Less accurate

Passive

- Listens to location updates that are being sent to other apps

Pros

- Doesn't use any extra of battery power

Cons

- Doesn't let you control when your app gets updates

Description

- You can use a device's GPS receiver or its network signals (cell or Wi-Fi) to find a device's current location.
- You can also find a device's current location by listening for location updates that are being sent to other apps.

Figure 18-2   An introduction to determining location

An introduction to maps

You've probably used more than one Google Map by now on a mobile device. If so, you know how easy one is to use. As you might expect, Google Maps is the most popular Android API for mapping. However, there are other Android APIs for mapping. One of the most prominent is MapQuest, one of the oldest mapping services. Figure 18-3 describes some of the pros and cons of these two APIs for mapping.

Google Maps is a complete mapping solution that provides (1) the Android API, (2) the base maps, and (3) the servers that serve the map tiles to the app. On the good side, this makes Google Maps easy for developers to use. On the bad side, this makes it more difficult to use one component of Google Maps without using another component.

Google Maps is available for free for most apps. At the time of this writing, for example, Google Maps allows 25,000 free map requests per day before it starts charging a small fee.

Google Maps also provides many features. It provides for *geocoding*, which is the process of converting latitude/longitude coordinates to street addresses and back. It provides for *routing*, which is the process of providing directions. And it provides for other features such as street view, traffic layers, and 3D buildings.

However, Google Maps isn't free for its heaviest users. As a result, if your app is going to generate more than 25,000 map requests per day, you might want to investigate alternatives to Google Maps. In addition, Google Maps isn't open-source. As a result, the developer doesn't have much control over the API or its pricing.

MapQuest works much like Google Maps, but it has recently embraced open-source mapping, allowing developers to choose between its proprietary maps or open maps that use data from the OpenStreetMap project. The goal of the OpenStreetMap project is to create a free map of the world that's available under the Open Database License.

Although MapQuest doesn't have as many features as Google Maps, the open version of its Android API is free for all users. As a result, if your app is going to generate more than 25,000 map requests per day, you may want to cut costs by using MapQuest. Similarly, if you want to have more control over the map API, or if you are an open-source enthusiast, you may want to try using MapQuest.

Google Maps

Web site
`https://developers.google.com/maps/`

Description
- Google Maps includes (1) the Android API, (2) the base maps, and (3) the servers to serve map tiles.

Pros
- Popular and familiar
- Free for most users
- Many features
  - Geocoding
  - Routing
  - Street view
  - Traffic
  - 3D Buildings

Cons
- Not free for the heaviest users
- Lack of control

MapQuest

Web site
`http://developer.mapquest.com/`

Description
- MapQuest works much like Google Maps, but it has recently embraced open-source mapping, allowing developers to choose between its proprietary maps or open maps that use data from the OpenStreetMap project. The goal of the OpenStreetMap project is to create a free map of the world that's available under the Open Database License.

Pros
- Open version is free for all users

Cons
- Not as many features

Description
- Google Maps is the most popular Android API for working with maps. However, there are several alternatives to Google Maps. One alternative is MapQuest.
- *Geocoding* is the process of converting latitude/longitude coordinates to street addresses and back.
- *Routing* is the process of using map data to provide directions.

Figure 18-3   An introduction to maps

An introduction to the Google Maps Android API

Figure 18-4 describes two versions of the Google Maps Android API. Version 1 (v1) was widely used prior to March 2013 because it makes it easy to add powerful mapping capabilities to an app. As a result, many existing apps use this version of the API.

However, the Android documentation recommends version 2 (v2) for all new development. In fact, Google Maps no longer accepts requests for an API key for version 1. As a result, if you don't already have an API key for version 1, you must use version 2. In addition, as you might expect, version 2 has many new features that improve upon the features available in version 1. For example, version 2 uses fragments to provide more flexibility for working with different screen sizes, displays faster, uses less bandwidth, and supports 3D. For these reasons, this chapter only covers version 2 of the API.

If you need to work on an app that uses version 1, you shouldn't have much trouble finding tutorials and documentation for it. There are plenty of books and online tutorials available for version 1, and the official online documentation for version 1 is still available too.

Google Maps Android API versions

Version 1 (v1)

- Has been officially deprecated as of December 3rd, 2012.
- Requests for API keys are no longer accepted as of March 18th, 2013.
- New features are no longer being added.
- Continues to work for existing apps.
- Uses the MapActivity class to encapsulate maps. This doesn't provide much flexibility for displaying maps on both small and large screens.
- Makes it easy to add powerful mapping capabilities to your application.
- Provides built-in downloading, rendering, and caching of map tiles.
- Provides a variety of display options and controls.
- Uses a MapView widget to display a map with data obtained from the Google Maps service. This widget handles key and touch events to pan and zoom the map automatically, including handling network requests for additional maps tiles.
- Allows several types of overlays on the map.
- Is not part of the standard Android library.
- Is a part of the Google APIs add-on.

Version 2 (v2)

- Is encouraged for all new app development.
- Is distributed as part of the Google Play services SDK.
- Uses the MapFragment class to encapsulate maps. This allows you more flexibility for displaying maps on both small and large screens.
- Uses vector tiles. This allows maps to display faster and use less bandwidth.
- Uses improved caching. This typically allows the map to display without showing empty areas.
- Supports 3D. This allows you to show the map with perspective by moving the user's viewpoint.

Description

- Version 1 of the Google Maps Android API was widely used prior to March 2013. Many existing apps use this version of the API.
- Version 2 of the Google Maps Android API is recommended for all new development as of March 2013.

Figure 18-4   An introduction to the Google Maps Android API

How to configure
the Google Maps Android API v2

Now that you understand some of the options for working with maps, you're ready to learn how to configure version 2 of the Android API for Google Maps. To do that, you need to add the Google Play services library to your project, get an API key for your app, register that key, and set the required permissions for the app.

How to add the Google Play services library to a project

Figure 18-5 shows how to add the Google Play services library to your project. To start, you use the Android SDK Manager dialog box to download the Google Play services library. Then, copy the directory for the library into a subdirectory of your workspace. Next, import the library into your workspace. Finally, add a reference to the library to your project.

When you're done, the Properties dialog box for your project should look something like the one shown in this figure. Here, the project includes a reference to the Google Play services library.

The Properties dialog box

Procedure

1. Start Eclipse as an administrator. To do that, you can right-click on the exe file for Eclipse and select the "Run as administrator" command.

2. In the toolbar, click the button for the Android SDK Manager. This should start the Android SDK Manager.

3. Select the following item:
   - Extras→Google Play services

4. Click the Install button and respond to the resulting dialog boxes.

5. Find the downloaded google-play-services_lib directory and copy it to a libs subdirectory of your workspace directory. If necessary, create the libs subdirectory.

6. Switch back to Eclipse. Then, select the File→Import→Android→Existing Android Code into Workspace item. Then, select the library project named google-play-services_lib.

7. Right-click your project and select the Properties item to display the Properties window. On the left side of the Properties dialog box, select the Android category. In the Library group, click the Add button. Then, use the resulting dialog box to select the library named google-play-services_lib.

The location of the google-play-services_lib directory
```
\android-sdk\extras\google\google_play_services\libproject
```

Description

- To add the Google Play services library to your project, you can download it, import it into your workspace, and add a reference to it from your project.

Figure 18-5   How to add the Google Play services library to a project

How to get the SHA-1 fingerprint

Version 2 of the Android API for Google Maps uses a new system of managing the keys for the Maps API. Existing keys from version 1 don't work with version 2. As a result, if you want to use version 2, you need to get a new key for version 2.

The first step in getting an API key for version 2 is to get the *SHA-1 fingerprint* for your app. This fingerprint is a text string that Google Maps can use to identify your app.

To get a SHA-1 fingerprint, you can follow the procedure shown in figure 18-6. This procedure gets the SHA-1 fingerprint for the debug key that's used during development. Later, when you're ready to release the app, you can modify this procedure to get the SHA-1 fingerprint for the release key.

The main difference between the debug key and the release key is that the release key is typically stored in a directory that's specified by you when you create the key. As a result, to get the SHA-1 fingerprint for a release key, you need to use a different directory and filename in step 3 of the procedure shown in this figure. For more information about debug and release keys, please see chapter 17.

Procedure

Windows

1. Start the Command Prompt window.

2. Change the current directory to Java's bin directory:
   ```
   cd \prgram files\java\jdk1.7.0\bin
   ```

3. Use the keytool command to display the SHA1 fingerprint:
   ```
   keytool -list -v -keystore "C:\Users\Joel\.android\debug.keystore"
   -alias androiddebugkey -storepass android -keypass android
   ```

4. Copy the SHA1 fingerprint to the clipboard. To do that, click the upper left corner of the Command Prompt window, select the Edit→Mark item, select the SHA1 fingerprint, and press Enter.

Mac or Linux

1. Start the Terminal window.

2. Change the current directory to Java's bin directory:
   ```
   cd /Applications/Java/jdk1.7.0/bin
   ```

3. Use the keytool command to display the SHA1 fingerprint:
   ```
   keytool -list -v -keystore ~/.android/debug.keystore
   -alias androiddebugkey -storepass android -keypass android
   ```

4. Copy the SHA1 fingerprint to the clipboard.

Sample output data
```
Certificate fingerprints:
        MD5:  B5:34:D4:70:F3:90:FD:ED:0C:01:A0:E3:85:26:65:96
        SHA1: 2E:D3:20:22:D7:3F:27:28:CD:53:4A:7B:77:7B:48:18:99:FD:AB:ED
        SHA256: A0:39:F3:81:E9:FC:6C:88:25:2F:91:C5:2B:BF:67:C6:4F:B4:06:48:EB:
52:A9:CB:91:27:72:4B:25:07:FB:9A
        Signature algorithm name: SHA256withRSA
        Version: 3
```

The SHA-1 fingerprint
```
2E:D3:20:22:D7:3F:27:28:CD:53:4A:7B:77:7B:48:18:99:FD:AB:ED
```

Description

- The Google Maps Android API v2 uses a new system of managing keys. As a result, keys from v1 don't work with v2.

- A *SHA-1 fingerprint* is a text string generated from the SHA-1 hashing algorithm that Google Maps can use to identify your app.

Figure 18-6   How to get the SHA-1 fingerprint

How to get a Maps API key

Once you have the SHA-1 fingerprint for your app, you can use it to get a Maps API key for version 2 of Google Maps. You need this key to access the Google Maps servers with the Android API. Fortunately, this key is free, works with any app that calls the Maps API, and supports an unlimited number of users.

To get a Maps API key, you can follow the procedure shown in figure 18-7. Part 1 shows how to use the Google APIs Console to create an API project. Then, it shows how to activate the Google Maps Android API v2 service for that project. In this figure, the mouse pointer is positioned and ready to activate the Android API v2 service. Here, the other Google Maps services are for other APIs such as the API for web apps.

The URL for the Google APIs Console

`https://code.google.com/apis/console`

The Google APIs Console with a list of services

How to create an API project

1. Start a web browser and navigate to the Google APIs Console.

2. Create a project that you use to track your usage of the Google Maps Android API. To do that, you typically click the Create Project link to create a new project. By default, this project is named API Project.

3. If necessary, select the Services link from the left navigation bar to display a list of services.

4. Scroll down to the Google Maps Android API v2 item and turn it on.

Figure 18-7   How to get the Maps API key (part 1 of 2)

Part 2 shows how to get the Maps API key. To do that, you paste the SHA-1 fingerprint for your app into the appropriate text box. Then, you type a semicolon at the end of the fingerprint, and type (or copy) the package name for your app. In this figure, I allow two apps to use the same API key. The first is the Run Tracker app presented in this chapter. The second is a Location Viewer app that I used to test some of the code in this chapter. In most cases, you don't want to use the same API key for multiple apps. However, it is possible to do that if you want.

In this figure, I blacked out the second half of the API key, and I blacked out the second half of both SHA-1 fingerpints. I did this because these values should be unique for each app. As a result, you need to use your own API key and SHA-1 fingerprint for your apps.

The Google APIs Console with the API key displayed

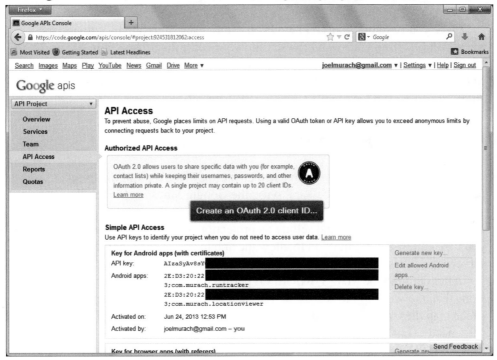

How to get the Maps API key

5. Select the API Access link from the left navigation bar.

6. Click the Create New Android Key button.

7. Enter the SHA-1 fingerprint, then a semicolon, then the package name for your app. It should look something like this:

    ```
    BB:0D:AC:74:D3:21:E1:43:67:71:9B:62:91:AF:A1:66:6E:44:5D:75;com.murach.run
    tracker
    ```

8. Copy the 40-character API key to the clipboard so you can paste it into the AndroidManifest.xml file.

Description

-   The Maps API key is necessary to access the Google Maps servers with the Maps API.

-   The Maps API key is free, works with any app that calls the Maps API, and supports an unlimited number of users.

-   You can obtain a Maps API key from the Google APIs Console by providing your app's SHA-1 fingerprint and package name, separated by a semicolon.

Figure 18-7   How to get the Maps API key (part 2 of 2)

How to register the Maps API key

Once you have the Maps API key, you can register your app by copying the Maps API key from your web browser into your app's manifest file. In figure 18-8, for example, the meta-data element contains the name and value for the Maps API key. This key confirms that you've registered with the Google Maps Android API v2 service via the Google APIs Console.

How to set permissions and features

Figure 18-8 also shows how to use the AndroidManifest.xml file to set permissions and features necessary to use the Google Maps Android API v2 service. To start, you can secure your app by defining a custom permission named MAPS_RECEIVE and setting its protection level to "signature". Then, you can request that permission to allow the app to access the Google Maps servers.

After defining the MAPS_RECEIVE permission, you must request permissions that allow the app to access the Internet, check the Internet connection, cache map tiles, and access Google web-based services. In addition, you typically need to enable the ACCESS_COARSE_LOCATION or ACCESS_FINE_LOCATION permission to allow your app to access the device's current location. Part 2 of this figure describes these permissions in more detail.

Finally, Google Maps Android API version 2 requires OpenGL ES version 2. As a result, you must add a uses-feature element to notify external services of this requirement. This prevents the Google Play store from displaying your app for devices that don't support OpenGL ES version 2.

Part of the AndroidManifest.xml file for the Run Tracker app

```
...
<!-- set up MAPS_RECEIVE permission -->
<permission
    android:name="com.murach.runtracker.permission.MAPS_RECEIVE"
    android:protectionLevel="signature" />
<uses-permission
    android:name="com.murach.runtracker.permission.MAPS_RECEIVE" />

<!-- set other permissions -->
<uses-permission
    android:name="android.permission.INTERNET" />
<uses-permission
    android:name="android.permission.ACCESS_NETWORK_STATE" />
<uses-permission
    android:name="android.permission.WRITE_EXTERNAL_STORAGE" />
<uses-permission
    android:name=
        "com.google.android.providers.gsf.permission.READ_GSERVICES" />
<uses-permission
    android:name="android.permission.ACCESS_FINE_LOCATION" />

<!-- Maps API version 2 requires OpenGL ES version 2 -->
<uses-feature
    android:glEsVersion="0x00020000"
    android:required="true" />

<application
    android:allowBackup="true"
    android:icon="@drawable/ic_launcher"
    android:label="@string/app_name"
    android:theme="@style/AppTheme" >

    <meta-data
        android:name="com.google.android.maps.v2.API_KEY"
        android:value="AIzaSyA4xsQ8kg8Bl8b1pWEoCYnIYm18wUIdEqU" />
...
```

Description

- Once you have the Maps API key, you can register your app by copying the Maps API key from your web browser into your app's manifest file. This key confirms that you've registered with the Google Maps service via the Google APIs Console.

- To secure your app, you can define a custom MAPS_RECEIVE permission for your app and set its protection level.

- When you use Google Maps, your app must define permissions that allow the app to use the Internet and access the Google Maps servers.

- Google Maps Android API version 2 requires OpenGL ES version 2. As a result, when you use Google Maps, you must add a uses-feature element to notify external services of this requirement. This prevents the Google Play store from displaying your app for devices that don't support OpenGL ES version 2.

Figure 18-8   How to register the Maps API and set permissions (part 1 of 2)

Part 2 summarizes the elements and permissions used in part 1. For the most part, these permissions are boilerplate code that you can copy from one app that uses version 2 of the Google Maps Android API to another. However, if you want to learn more about how these elements and permissions work, you can read more about them.

In the permissions, the ACCESS_COARSE_LOCATION permission isn't used in part 1. That's because the Run Tracker app presented in this figure only uses GPS to determine a device's location. As a result, this app only needs the ACCESS_FINE_LOCATION permission. However, if you want to use Wi-Fi or mobile cell data (or both) to determine a device's location, you need to add the ACCESS_COARSE_LOCATION permission to your app.

The meta-data element

Attribute	Description
name	The type of API key.
value	The value of the API key.

The permission element

Attribute	Description
name	The name of the permission that corresponds with the uses-permission element.
protectionLevel	The level of protection for the permission. A value of "signature" allows the system to grant this permission without notifying the user only if the requesting app is signed with the same certificate as the app that declared the permission.

A summary of permissions

Permission	Description
MAPS_RECEIVE	Allows the API to securely contact your app. This prevents other apps from impersonating the API.
INTERNET	Allows the API to download map tiles from Google Maps servers.
ACCESS_NETWORK_STATE	Allows the API to check the connection status in order to determine whether map data can be downloaded.
WRITE_EXTERNAL_STORAGE	Allows the API to cache data such as map tiles in the device's external storage area.
READ_GSERVICES	Allows the API to access Google web-based services.
ACCESS_FINE_LOCATION	Allows the API to use GPS to determine the device's location.
ACCESS_COARSE_LOCATION	Allows the API to use cell tower or Wi-Fi signals to determine the device's location.

The uses-feature element

Attribute	Description
glEsVersion	The OpenGL ES version required by the application. To specify version 2.0, specify a value "0x00020000".
required	Specifies whether the application requires the feature.

Description

- The meta-data element that stores the Maps API key must be declared within the application element.
- If your app needs to access the device's current location, you must enable the ACCESS_COARSE_LOCATION or ACCESS_FINE_LOCATION permissions.

Figure 18-8   How to register the Maps API and set permissions (part 2 of 2)

How to work with locations

Now that you have configured version 2 of the Android API for Google Maps, you're ready to learn how to use it to work with locations. In particular, you're ready to learn how to get and track the device's location. To do that, you begin by connecting to Google Play services.

How to connect to Google Play services

Figure 18-9 shows how to connect to Google Play services. To start, you declare an activity that implements two interfaces: ConnectionCallbacks and OnConnectionFailedListener. In this figure, the methods for these interfaces are shown in part 2. At this point, these methods don't contain any code. However, you'll learn how to add code to these methods in the next few figures.

After implementing these interfaces, you can declare a LocationClient object as an instance variable. Then, in the onCreate method, you can create the LocationClient object. To do that, you can use the *this* keyword to specify the current activity as the object that has the required connection callbacks and is also the failed connection listener.

After you create the LocationClient object, you typically call its connect method from within the onStart method. That way, when the activity is started, it starts an asynchronous thread that opens a connection to Google Play services in the background. When the connection is opened, this calls the onConnected method that's shown in part 2.

When you aren't using a connection, you should close it. To do that, you can call the disconnect method from within the onStop method. That way, when the activity is stopped, it closes the connection to Google Play services. This calls the onDisconnected method that's shown in part 2.

The LocationViewerActivity class   **Page 1**

```
package com.murach.locationviewer;

import com.google.android.gms.common.ConnectionResult;
import com.google.android.gms.common.GooglePlayServicesClient;
import com.google.android.gms.location.LocationClient;

import android.os.Bundle;
import android.app.Activity;
import android.widget.TextView;

public class LocationViewerActivity extends Activity
implements GooglePlayServicesClient.ConnectionCallbacks,
        GooglePlayServicesClient.OnConnectionFailedListener {

    private LocationClient locationClient;
    private TextView coordinatesTextView;

    @Override
    protected void onCreate(Bundle savedInstanceState) {
        super.onCreate(savedInstanceState);
        setContentView(R.layout.activity_location_viewer);

        coordinatesTextView = (TextView)
                findViewById(R.id.coordinatesTextView);

        locationClient = new LocationClient(this, this, this);
    }

    @Override
    protected void onStart() {
        super.onStart();
        locationClient.connect();
    }

    @Override
    protected void onStop() {
        locationClient.disconnect();
        super.onStop();
    }
```

Constructors and methods of the LocationClient class

Constructor/method	Description
LocationClient(context, connectionCallbacks, connectionFailedListener)	Creates a LocationClient object with the specified context, connection callbacks, and failed connection listener.
connect()	Starts an asynchronous thread that opens a connection to Google Play services in the background.
disconnect()	Closes the connection to Google Play services.

Figure 18-9   How to connect to Google Play services (part 1 of 2)

Part 2 begins by showing the onConnected and onDisconnected methods of the ConnectionCallbacks interface. These methods are called when a connection to Google Play services is opened or closed.

In addition, part 2 shows the onConnectionFailed method of the OnConnectionFailedListener interface. This method is called when a connection to Google Play services fails. That doesn't usually happen, but it can happen when the network isn't available. It can happen when the client attempts to connect to the service but the user isn't signed in or is using an invalid account name. And it can happen when Google Play services is missing on a device or is out of date.

The LocationViewerActivity class   Page 2

```
//*************************************************************
// Implement ConnectionCallbacks interface
//*************************************************************
@Override
public void onConnected(Bundle dataBundle) {
    // Put code to run after connecting here
}

@Override
public void onDisconnected() {
    // Put code to run before disconnecting here
}

//*************************************************************
// Implement OnConnectionFailedListener
//*************************************************************
@Override
public void onConnectionFailed(ConnectionResult connectionResult) {
    // Put code to run if connection fails here
}
}
```

Methods of the ConnectionCallbacks interface

Method	Description
onConnected(dataBundle)	Called when the connection is opened.
onDisconnected()	Called when the connection is closed.

Methods of the OnConnectionFailedListener interface

Method	Description
onConnectionFailed()	Called when the connection fails.

Description

- You can use a LocationClient object to open and close a connection to Google Play services.

Figure 18-9   How to connect to Google Play services (part 2 of 2)

How to get the current location

Figure 18-10 shows code that you can add to the onConnected method shown in the previous figure to get the current location. To start, you can call the getLastLocation method from the LocationClient object to get a Location object. This Location object contains detailed information about the best and most recent location of a device.

After you get a Location object, you should check to make sure that it is not null. If that's true, you can call the getLatitude and getLongitude methods of the Location object to get the degree of latitude and longitude for the current location. In this figure, for example, the code gets the latitude and longitude for the current location and displays them on a TextView widget. Later in this chapter, you'll learn how to use these coordinates to display the device's location on a map.

If necessary, you can use other methods of the Location object to get more information about the location. For example, you can use the getTime method to get the time of the location fix in milliseconds. This might be useful for any app that's keeping a historical record of a device's location.

Similarly, you can use the getAccuracy method to determine the accuracy of the location. This might be useful if you want to handle the location differently depending on its estimated accuracy.

Code that gets the current location

```
@Override
public void onConnected(Bundle dataBundle) {
    Location location = locationClient.getLastLocation();
    if (location != null){
        coordinatesTextView.setText(
                location.getLatitude() + "|" + location.getLongitude());
    }
}
```

Another method of the LocationClient class

Method	Description
getLastLocation()	Returns the best and most recent location currently available.

Four methods of the Location class

Method	Description
getLatitude()	Gets the degree of latitude.
getLongitude()	Gets the degree of longitude.
getTime()	Gets the time of this location fix in milliseconds.
getAccuracy()	Gets the estimated accuracy of this location in meters. If you draw a circle centered at this location's latitude and longitude with a radius equal to the accuracy, there is a 68% probability that the true location is inside the circle. If this location does not have an accuracy, this method returns 0.0.

Description

- You can use the LocationClient object to get a Location object that contains detailed information about the best and most recent location of a device.

Figure 18-10   How to get the current location

How to handle a failed connection

Figure 18-11 shows code that you can add to the onConnectionFailed method shown earlier in this chapter to handle a failed connection. But first, you should define a constant for an integer code that uniquely identifies your request for a resolution. In this figure, the code sets the constant to a value of 9000, but you can use any integer value that doesn't conflict with other request codes.

After defining this constant, you can add the code to the onConnectionFailed method like the code shown in this figure. To start, this code calls the hasResolution method from the ConnectionResult parameter to check whether Google Play services knows of a possible solution to the problem that caused the failed connection.

If there is a possible solution, you can call the startResolutionForResult method to start any intents that require user interaction. When you call this method, you pass it the current activity and the constant that identifies your request.

If there is no solution, you can display an error to the user to indicate that the connection has failed. In this figure, for example, the code uses the AlertDialog.Builder class to display an error message that includes the code for the error that caused the connection to fail. To do that, this statement calls the getErrorCode method from the ConnectionResult parameter.

In some cases, you may want to do additional processing when the user returns from the activity that attempts to resolve the problem. To do that, you can add an onActivityResult method like the one shown in this figure. This method is a lifecycle method of the Activity class that's executed when the user returns from an activity that returns a result. In this figure, the onActivityResult method begins by checking if the request code parameter matches the constant for the request code. If so, it can do additional processing. If necessary, this processing can include checking the result code against the constants of the ConnectionResult class to determine the cause of the failed connection.

Code that handles a failed connection

The constant for the request code
```
private final static int CONNECTION_FAILURE_RESOLUTION_REQUEST = 9000;
```

The onConnectionFailed method
```
@Override
public void onConnectionFailed(ConnectionResult connectionResult) {
    // if Google Play services has resolution, display activity
    if (connectionResult.hasResolution()) {
        try {
            connectionResult.startResolutionForResult(this,
                CONNECTION_FAILURE_RESOLUTION_REQUEST);
        }
        catch (IntentSender.SendIntentException e) {
            e.printStackTrace();
        }
    }
    else {
        new AlertDialog.Builder(this)
            .setMessage("Connection failed. Error code: "
                        + connectionResult.getErrorCode())
            .show();
    }
}
```

The onActivityResult method
```
@Override
protected void onActivityResult(int requestCode, int resultCode,
        Intent data) {
    super.onActivityResult(requestCode, resultCode, data);

    if (requestCode == CONNECTION_FAILURE_RESOLUTION_REQUEST) {
        // perform additional processing here
    }
}
```

Methods of the ConnectionResult class

Method	Description
hasResolution()	Returns true if Google Play services can possibly resolve the error.
startResolutionForResult(activity, code)	Attempts to resolve the error by starting any intents requiring user interaction with the specified request code.
getErrorCode()	Gets the code for the error that caused the connection to fail.

Description

- In the onConnectionFailed method, you can use the ConnectionResult parameter to allow Google Play services to attempt to resolve the error that caused the connection to fail.

Figure 18-11   How to handle a failed connection

How to get location updates

Figure 18-12 shows how to get updates of a device's location at a specified interval of time. To start, you implement the LocationListener interface. This interface defines a single method, the onLocationChanged method.

After you implement the interface, you define the constants and instance variables for the class. Here, the UPDATE_INTERVAL constant sets the milliseconds for the requested update interval to 5000, which is 5 seconds. Then, the FASTEST_UPDATE_INTERVAL sets the milliseconds for the fastest rate at which the app can update the location to 2000, which is 2 seconds. Next, this code declares a variable for a LocationRequest object.

In the onCreate method, you can create the LocationRequest object. Then, you can use one of the constants described in part 2 of this figure to set its priority. In this figure, the code sets the priority to high accuracy. This uses GPS, which results in the highest accuracy, but also uses the most battery. In addition, you can set the update interval and the fastest update interval. In this figure, the code sets these intervals to the constants defined in step 2.

In the onConnected method, you can turn on location updates. To do that, you call the requestLocationUpdates method from the LocationClient object and pass it the LocationRequest object and the LocationListener object. In this figure, the current activity is the LocationListener object.

In the onDisconnected method, you can turn off location updates. To do that, you call the removeLocationUpdates method from the LocationClient object and pass it the LocationListener object. But first, you should use the isConnected method of the LocationClient object to make sure the connection is still available.

In the onLocationChanged method, you can use the Location parameter to process the current location. In this figure, the code just displays the latitude and longitude for the location on a TextView widget. Later in this chapter, you'll learn how to update the device's location on a map.

Step 1: Declare the LocationListener interface

```
public class LocationViewerActivity extends Activity
implements GooglePlayServicesClient.ConnectionCallbacks,
        GooglePlayServicesClient.OnConnectionFailedListener,
        LocationListener {
```

Step 2: Declare the constants and instance variables

```
public static final int UPDATE_INTERVAL = 5000;        // 5 seconds
public static final int FASTEST_UPDATE_INTERVAL = 2000; // 2 seconds
private LocationRequest locationRequest;
```

Step 3: Create the location request in the onCreate method

```
locationRequest = LocationRequest.create()
        .setPriority(LocationRequest.PRIORITY_HIGH_ACCURACY)
        .setInterval(UPDATE_INTERVAL)
        .setFastestInterval(FASTEST_UPDATE_INTERVAL);
```

Step 4: Request location updates in the onConnected method

```
locationClient.requestLocationUpdates(locationRequest, this);
```

Step 5: Remove location updates in the onDisconnected method

```
if (locationClient.isConnected()) {
    locationClient.removeLocationUpdates(this);
}
```

Step 6: Implement the LocationListener interface

```
@Override
public void onLocationChanged(Location location) {
    coordinatesTextView.setText(
            location.getLatitude() + "|" + location.getLongitude());
}
```

Description

- You can use a LocationRequest object to configure a request for location updates from Google Play services.

- You can use a LocationListener object to listen for location updates from Google Play services.

- You can use a LocationClient object to turn location requests on and off.

Figure 18-12   How to get location updates (part 1 of 2)

Part 2 begins by showing some of the methods and constants that you can use to create and configure the LocationRequest object. In addition, part 2 shows some more methods of the LocationClient class. If you want to learn more about how these methods and constants work, you can read more about them.

The setPriority method sets the accuracy parameter to one of the constants shown in this figure. If your main priority is accuracy, you should use the PRIORITY_HIGH_ACCURACY constant. This constant uses GPS to request the most accurate locations available. However, this also uses the most battery of all options. If your app only needs accuracy of about 100 meters, use the PRIORITY_BALANCED_POWER_ACCURACY constant instead as it consumes less battery.

Another option is to use the PRIORITY_NO_POWER constant. This requests the best accuracy possible with zero additional battery consumption. This only returns locations when another app is receiving location updates. In that case, your app listens to those locations.

The setInterval method sets the interval that your app prefers to receive location updates. If no other apps are receiving updates, your app receives updates at this rate.

However, if other apps are receiving updates at a faster rate, your app can receive updates at that faster rate too with no additional use of battery. Unfortunately, a faster rate may cause problems with your app. For example, it may cause UI (user interface) flicker or data overflow. That's why you need to use the setFastestInterval method to set the interval at which your app can handle location updates.

Methods of the LocationRequest class

Method	Description
`create()`	Creates and returns a LocationRequest object.
`setPriority(accuracy)`	Sets the accuracy parameter to one of the constants shown below.
`setInterval(millis)`	Sets the rate in milliseconds that your app prefers to receive location updates.
	If no other apps are receiving updates, your app receives updates at this rate.
`setFastestInterval(millis)`	Sets the fastest rate in milliseconds at which your app can handle location updates. This prevents problems with UI flicker or data overflow and helps to save power.
	If other apps have requested a faster rate, you get the benefit of a faster rate with no additional use of battery.

Priority constants from the LocationRequest class

Constant	Description
`PRIORITY_HIGH_ACCURACY`	Requests the most accurate locations available. This uses the most battery of all options and requires the ACCESS_FINE_LOCATION permission.
`PRIORITY_BALANCED_POWER_ACCURACY`	Requests accuracy of about 100 meter accuracy. This consumes less battery and requires the ACCESS_COARSE_LOCATION permission.
`PRIORITY_NO_POWER`	Requests the best accuracy possible with zero additional battery consumption. This only returns locations when another app is receiving location updates. In that case, your app listens to those locations.

More methods of the LocationClient class

Method	Description
`isConnected()`	Returns a boolean value that indicates whether the location client is connected.
`requestLocationUpdates(request, listener)`	Requests location updates using the specified LocationRequest and LocationListener objects.
`removeLocationUpdates(listener)`	Removes location updates from the specified LocationListener object.

Figure 18-12   How to get location updates (part 2 of 2)

How to make sure GPS is enabled

If your app relies on GPS to get the location, you typically want to make sure it's enabled on the user's device before you start your app. To do that, you can use the code shown in figure 18-13.

The code presented in this figure is typically coded within the onCreate, onStart, or onResume method of an activity. That way, it's executed when the activity is first created or anytime its displayed. Here, the first statement uses the getSystemService method to get a LocationManager object. Then, an if statement uses the LocationManager object to check whether GPS is enabled. If not, this code displays a toast that asks the user to enable GPS. Then, it creates an intent for the activity of the Settings app that allows the user to enable and disable GPS. Finally, this code calls the startActivity method and passes it that intent.

Code that makes sure GPS is enabled

```
LocationManager locationManager =
        (LocationManager) getSystemService(LOCATION_SERVICE);

if (!locationManager.isProviderEnabled(LocationManager.GPS_PROVIDER)){
    Toast.makeText(this, "Please enable GPS!",
            Toast.LENGTH_LONG).show();
    Intent intent = new Intent(
            Settings.ACTION_LOCATION_SOURCE_SETTINGS);
    startActivity(intent);
}
```

Methods of the Activity class

Method	Description
getSystemService(serviceType)	Gets the specified system service. You can use the LOCATION_SERVICE constant to get a LocationManager object.

Constants and methods of the LocationManager class

Constant/Method	Description
GPS_PROVIDER	The provider that uses GPS to determine location.
NETWORK_PROVIDER	The provider that uses Wi-Fi or cell networks to determine location.
isProviderEnabled(providerType)	You can use this to check if a type of location provider is available.

Description

- You can use the LocationManager object to check if GPS is enabled. If it isn't, you can display an error message, start the Settings app, and display the activity that allows the user to enable GPS.

Figure 18-13   How to make sure GPS is enabled

How to work with Google Maps

Now that you know how to track a device's location, you're ready to learn how to use that information with a Google Map. To do that, the first step is to add a Google Map widget to your user interface.

How to add a map fragment to a layout

The easiest way to add a Google Map widget to your user interface is to add a fragment element to a layout as shown in figure 18-14. To do that, you can add the XML code for the fragment to the layout. This fragment should include a class attribute that specifies the SupportMapFragment class. For this to work, the project must include the Android support library.

If you switch to the Graphical Layout editor, Eclipse doesn't display the map. However, it displays a grayed out area where the map will be displayed when you run the app.

It's also possible to use a MapFragment class instead of the SupportMapFragment class. However, the MapFragment class requires Android 3.1 (API 12) or higher. As a result, if your app only targets Android 3.1 or higher, you can use the MapFragment class. In that case, you don't need to include the Android support library.

A layout that includes a map fragment

The XML for map fragment

```
<fragment
    android:id="@+id/map"
    android:layout_width="match_parent"
    android:layout_height="match_parent"
    class="com.google.android.gms.maps.SupportMapFragment" />
```

Description

- To add a map to a layout, add a fragment element for the SupportMapFragment class. For this to work, your project must include the Android support library.

- If your app only targets Android 3.1 (API 12) or higher, you can use the MapFragment class instead of the SupportMapFragment class and you don't need to include the Android support library.

Figure 18-14   How to add a map fragment to a layout

How to display a map

Figure 18-15 shows how to display a map. First, you create an activity that extends the FragmentActivity class. In this figure, for example, the activity named LocationViewerActivity extends the FragmentActivity class.

Second, you declare an instance variable for the GoogleMap object. This is the primary object for working with a Google Map. In this figure, for example, the code declares an instsance variable for a GoogleMap object named map.

Third, you add code to the onStart method of the activity to get the GoogleMap object. That way, when the activity is displayed, the code gets the map and displays it. In this figure, the code begins by checking whether the map is null. If so, it gets a FragmentManager object, and uses it to get the SupportMapFragment object for the map. Then, it calls the getMap method from that object to get the GoogleMap object.

Once you have a GoogleMap object, you can call its methods to change its type or its UI (user interface) settings. In this figure, for example, the code checks to make sure a the GoogleMap object exists. Then, it calls the getUiSettings method from the GoogleMap object to return a UiSettings object. From that object, this code calls the setZoomControlsEnabled method to enable the zoom controls. As a result, the map should include zoom controls (plus and minus buttons) like the ones shown on the Run Activity in figure 18-1.

In this figure, the code uses the default map type (the normal type). This displays a map like the one shown on the Run Map activity in figure 18-1. However, if you want to use a different map type, you can call the setMapType method from the GoogleMap object to change the map type to one of the constants shown in this figure. For example, you can change the map type to a satellite map, a hybrid map, or a terrain map.

Similarly, the code in this figure uses most of the default UI settings for the map. The only UI setting change this code makes is to enable the zoom controls. However, if you want, you can call methods from the GoogleMap object to customize most UI settings of the map so they're appropriate for your app. For example, you can use the methods of the UiSettings object shown in this figure to disable the compass or zoom gestures.

At this point, the code should display a map of the world with zoom controls. In the next figure, you'll learn how to zoom in on a particular location.

Step 1: Extend the FragmentActivity class

```
public class LocationViewerActivity extends FragmentActivity
implements GooglePlayServicesClient.ConnectionCallbacks,
          GooglePlayServicesClient.OnConnectionFailedListener,
          LocationListener {
```

Step 2: Declare the instance variable for the GoogleMap object

```
private GoogleMap map;
```

Step 3: Get the GoogleMap object in the onStart method

```
// if GoogleMap object is not already available, get it
if (map == null) {
    FragmentManager manager = getSupportFragmentManager();
    SupportMapFragment fragment = (SupportMapFragment)
        manager.findFragmentById(R.id.map);
    map = fragment.getMap();
}

// if GoogleMap object is available, configure it
if (map != null) {
    map.getUiSettings().setZoomControlsEnabled(true);
}
```

A method of the SupportMapFragment class

Method	Description
getMap()	Gets a GoogleMap object.

Some constants and methods of the GoogleMap class

Constant/Method	Description
MAP_TYPE_NORMAL	The default map.
MAP_TYPE_HYBRID	A satellite map with a transparent layer of major streets.
MAP_TYPE_SATELLITE	A satellite map with no labels.
MAP_TYPE_TERRAIN	A terrain map.
setMapType(mapType)	Sets the type of map tiles that should be displayed.
getUiSettings()	Gets a UiSettings object for the user interface settings for the map.

Some methods of the UiSettings class

Method	Description
setCompassEnabled(bool)	Enables or disables the compass.
setZoomControlsEnabled(bool)	Enables or disables the zoom controls.
setZoomGesturesEnabled(bool)	Enables or disables zoom gestures.

Figure 18-15   How to display a map

How to zoom in on a location

Figure 18-16 shows how to use a GoogleMap object to zoom in on a location. In this figure, the code begins by checking to make sure the GoogleMap object exists before calling any methods from that object. This is a good practice as it prevents a NullPointerException from being thrown.

If the GoogleMap object exists, this code calls the animateCamera method from it to zoom in on the specified target. To help visualize how this works, you can think of a virtual camera pointing at the specified target from the specified position. Here, the target is the latitude and longitude that are stored in the Location object named location. The zoom level is 16.5, which zooms in so the map displays a few city blocks. The bearing is 0, which means the camera is pointing due north. And the tilt is 25, which means that the camera is tilted 25 degrees towards the bearing (as opposed to the default setting of 0, which points straight down). This displays a map like the one shown in the Run Map activity in figure 18-1.

When you use the animateCamera method to zoom in on the current location, the API uses an animation to zoom in as if the camera is flying. If you don't like that, you can use the moveCamera method. It works the same as the animateCamera method, but it just displays the map from the camera position without using any animations to get there.

You may need to experiment a bit to find the best zoom level, bearing, and tilt for your app. The maximum zoom level is limited by the underlying map, and varies for the different locations around the world. For example, you can zoom in more closely on populated areas and can't zoom in as closely on other areas such as the north pole. Also, the maximum zoom level varies depending on the map type. If you're curious, you can use the getMaxZoomLevel to return the maximum zoom level for the current camera position and map type.

The maximum tilt value depends on the zoom level. However, it's never more than 67.5. Fortunately, if you specify a value that's greater than the maximum, it doesn't cause an error. Instead, the API just uses the maximum value.

How to zoom in on a map

```
if (map != null) {
    map.animateCamera(
        CameraUpdateFactory.newCameraPosition(
            new CameraPosition.Builder()
                .target(new LatLng(location.getLatitude(),
                                   location.getLongitude()))
                .zoom(16.5f)
                .bearing(0)
                .tilt(25)
                .build()));
}
```

More methods of the GoogleMap class

Method	Description
animateCamera(cameraPosition)	Animates a move to the specified latitude and longitude, zooms in to the specified amount, and sets the bearing and the tilt of the camera.
moveCamera(cameraPosition)	Works like the animateCamera method, but doesn't animate the move.
getMaxZoomLevel()	Returns the maximum zoom level for the current camera position and map type.

Some methods of the CameraPosition.Builder class

Method	Description
target(latLng)	Sets the location to the specified LatLng object.
zoom(level)	Sets the zoom level to the specified float value. At level 0, a map that's 256dp wide can display the whole world. At level 16.5, a map that's 256dp wide can only display a few blocks.
bearing(degrees)	Sets the direction that the camera is pointing, in degrees clockwise from north where 0 is north and 180 is south.
tilt(degrees)	Sets the angle of the camera in degrees from directly facing the Earth. The minimum degrees value is 0 (no tilt), and the maximum degrees value varies depending on the zoom level but is never more than 67.5.
build()	Returns the CameraPosition object.

Constructor of the LatLng class

Constructor	Description
LatLng(latitude, longitude)	Creates a LatLng object with the specified latitude and longitude.

Description

- You can use a GoogleMap object to zoom in on a location.

Figure 18-16   How to zoom in on a location

How to add markers

Figure 18-17 shows how to add one or more markers to a map. To do that, you can use the addMarker method of the GoogleMap class. If necessary, you can use the clear method of the GoogleMap class to clear any existing markers.

In this figure, the code calls the clear method to clear any old markers from the map. Then, it uses the addMarker method to add a marker for the device's current location. To do that, this code creates a new MarkerOptions object, sets its position to the latitude and longitude that are stored in the Location object named location, and sets its title to a value of "You are here". Then, this code passes the newly created MarkerOptions object to the addMarker method. This displays a default red marker icon like the one shown in the Run Map activity in figure 18-1.

If you don't want to use the default marker icon, you can use the other methods of the MarkerOptions class to customize it. For example, you can use the icon method to change the default icon. To do that, you need to add an appropriate icon to the project's resources. Then, you can use the BitmapDescriptorFactory to get a BitmapDescriptor object from that resource. In this figure, the second code example gets a BitmapDescriptor object from a resource named ic_runner.png that's stored in the project's res\drawable directory.

How to add a marker for the current location

```
if (map != null) {
    map.clear();   // clear any old markers
    map.addMarker(
        new MarkerOptions()
            .position(new LatLng(location.getLatitude(),
                                 location.getLongitude()))
            .title("You are here"));
}
```

An example that uses the icon method to set a new icon for a marker

```
.icon(BitmapDescriptorFactory.fromResource(R.drawable.ic_runner))
```

More methods of the GoogleMap class

Method	Description
clear()	Clears all markers from the map.
addMarker(marker)	Adds a marker to the map at the specified latitude and longitude.

Some methods of the MarkerOptions class

Method	Description
position(latLng)	Sets the location for the marker.
title(title)	Sets the title for the marker.
icon(bitmapDescriptor)	Sets the icon for the marker. If you don't set an icon, Google Maps uses the default marker icon.

Description

- You can use a GoogleMap object to add one or more markers to a map.

Figure 18-17   How to add markers

How to add lines

Figure 18-18 shows how to display a series of connected lines on a map. To do that, you can call the addPolyline method from the GoogleMap object and pass it a PolylineOptions object that contains a series of points. Then the API draws a line that connects these points in the order in which they were added.

In this figure, the code begins by creating a new PolylineOptions object. Then, an if statement checks whether an ArrayList<Location> object named locationsList contains any locations. If so, this code loops through this list, converts each Location object to a LatLng object, and adds each LatLng object to the PolylineOptions object. After the loop, this code uses the addPolyline method to add this polyline to the map. This displays a crooked black line that connects a series of points like the line shown in the Run Map activity in figure 18-1.

If you don't want to use the default settings for the line, you can use the other methods of the PolylineOptions class to customize it. For example, you can use the width method to make the line thicker or thinner. In this figure, the second code example sets the width of a line to 10 pixels. Similarly, you can use the color method to change the color of the line. In this figure, the third example sets the color of the line to red. To do that, it uses the RED constant that's available from the Color class of the android.graphics package.

How to add lines to a map

```
if (map != null) {
    PolylineOptions polyline = new PolylineOptions();
    if (locationList.size() > 0) {
        for (Location l : locationList) {
            LatLng point = new LatLng(
                l.getLatitude(), l.getLongitude());
            polyline.add(point);
        }
    }
    map.addPolyline(polyline);
}
```

An example that sets the width of a line

```
polyline.width(10);
```

An example that sets the color of a line

```
polyline.color(Color.RED);
```

More methods of the GoogleMap class

Method	Description
addPolyline(polyline)	Adds a line that connects the points in the specified PolylineOptions object.

Constructor/methods of the PolylineOptions class

Constructor/Method	Description
PolylineOptions()	Creates a PolylineOptions object.
add(latLng)	Adds the point specified by the LatLng object.
width(pixels)	Specifies the width of the line to the specified number of pixels.
color(argbColor)	Sets the color of the polyline as a 32-bit ARGB color. The default color is black.

Description

- You can use a GoogleMap object to display a series of connected lines on a map.

Figure 18-18   How to add lines to a map

The Run Tracker app

At this point, you have all the skills you need to understand the code for the Run Tracker app. Since the Stopwatch activity for this app doesn't contain much code that works with locations or maps, this chapter doesn't present this activity. Instead, this chapter presents the layout and class for the Run Map activity. In addition, it presents the class for the Run Tracker service. These files contain most of the code that works with locations and maps. Of course, if you want to see the complete code, you can open the project for this app and view its code.

The activity_run_map layout

Figure 18-19 shows the XML for the activity_run_map layout. This layout displays the View Stopwatch button over a map. To do that, this XML uses a frame layout. Within the frame layout, the fragment element adds the map. Then, the Button element adds a button across the top of the map.

The activity_run_map layout

```xml
<?xml version="1.0" encoding="utf-8"?>
<FrameLayout xmlns:android="http://schemas.android.com/apk/res/android"
    android:layout_width="match_parent"
    android:layout_height="match_parent" >

    <fragment
        android:id="@+id/map"
        android:layout_width="match_parent"
        android:layout_height="match_parent"
        class="com.google.android.gms.maps.SupportMapFragment" />

    <Button
        android:id="@+id/viewStopwatchButton"
        android:layout_width="match_parent"
        android:layout_height="wrap_content"
        android:layout_marginTop="10dp"
        android:gravity="center"
        android:text="@string/stopwatch"
        android:textSize="20sp" />

</FrameLayout>
```

Figure 18-19   The activity_run_map layout

The RunMapActivity class

Figure 18-20 shows the code for the RunMapActivity class. To start, this class imports some classes from the Java API. Then, it imports some classes from the standard Android API. Finally, this class imports classes from the Google Maps Android API v2. By now, you should be familiar with all of these classes.

The RunMapActivity class extends the FragmentActivity class and implements three interfaces. The first interface is necessary to execute code when the user clicks on the View Stopwatch button. The next two interfaces were described earlier in this chapter and are necessary to connect to Google Play services and get the current location.

After the declaration for the the RunMapActivity class, this code defines some constants and instance variables that are needed by the class. This includes a constant named INTERVAL_REFRESH that specifies that the activity should refresh itself every 10 seconds. In addition, it includes a RunTrackerDB object named db that can be used to read and write data from a database. Although the RunTrackerDB class isn't shown in this chapter, it works similarly to the other database classes presented in this book.

The onCreate method displays the layout for the activity. Then, it initializes the View Stopwatch button, its intent, and the RunTrackerDB object. Next, it checks whether GPS is enabled. If not, it displays a toast asking the user to enable GPS and an activity within the Settings app that allows the user to enable GPS. Finally, this code initializes a LocationClient object.

The onStart method begins by getting the GoogleMap object if it isn't already available. Then, it enables the zoom controls on this map object. Finally, it attempts to connect to Google Play services. If successful, this executes the onConnected method shown in part 4. In other words, the app should connect to Google Play services when the user displays the activity.

The onStop method disconnects from Google Play services. This executes the onDisconnected method shown in part 4. In other words, the app should disconnect from Google Play services if the user navigates away from this activity.

The updateMap method begins by checking whether the app is connected to Google Play services. If so, it calls another method in this class to set the marker for the current location. Then, it calls another method in this class to display the route for the run.

The setCurrentLocationMarker method gets the current location, zooms in on that location, and adds a marker to identify the location. Since this method uses code that was described earlier in this chapter, you shouldn't have much trouble understanding how it works.

The displayRun method gets a list of locations from the database and adds a line that connects those locations to the map. Again, this method uses code that was described earlier in this chapter.

The RunMapActivity class   **Page 1**

```java
package com.murach.runtracker;

import java.util.List;
import java.util.Timer;
import java.util.TimerTask;

import android.app.AlertDialog;
import android.content.Intent;
import android.content.IntentSender;
import android.location.Location;
import android.location.LocationManager;
import android.os.Bundle;
import android.provider.Settings;
import android.support.v4.app.FragmentActivity;
import android.support.v4.app.FragmentManager;
import android.view.View;
import android.view.View.OnClickListener;
import android.widget.Button;
import android.widget.Toast;

import com.google.android.gms.common.ConnectionResult;
import com.google.android.gms.common.GooglePlayServicesClient;
import com.google.android.gms.location.LocationClient;
import com.google.android.gms.maps.CameraUpdateFactory;
import com.google.android.gms.maps.GoogleMap;
import com.google.android.gms.maps.SupportMapFragment;
import com.google.android.gms.maps.model.CameraPosition;
import com.google.android.gms.maps.model.LatLng;
import com.google.android.gms.maps.model.MarkerOptions;
import com.google.android.gms.maps.model.PolylineOptions;

public class RunMapActivity extends FragmentActivity
implements OnClickListener,
            GooglePlayServicesClient.ConnectionCallbacks,
            GooglePlayServicesClient.OnConnectionFailedListener {

    private final static int CONNECTION_FAILURE_RESOLUTION_REQUEST = 9000;
    private static final int INTERVAL_REFRESH = 10 * 1000;    // 10 seconds

    private GoogleMap map;
    private LocationClient locationClient;
    private List<Location> locationList;

    private RunTrackerDB db;

    private Button stopwatchButton;
    private Intent stopwatchIntent;

    private Timer timer;
```

Figure 18-20   The RunMapActivity class (part 1 of 4)

The RunMapActivity class

```java
@Override
protected void onCreate(Bundle savedInstanceState) {
    super.onCreate(savedInstanceState);
    setContentView(R.layout.activity _ run _ map);

    stopwatchButton = (Button) findViewById(R.id.viewStopwatchButton);
    stopwatchButton.setOnClickListener(this);
    stopwatchIntent = new Intent(getApplicationContext(),
        StopwatchActivity.class).addFlags(Intent.FLAG _ ACTIVITY _ CLEAR _ TOP);

    db = new RunTrackerDB(this);

    // if GPS is not enabled, start GPS settings activity
    LocationManager locationManager =
            (LocationManager) getSystemService(LOCATION _ SERVICE);
    if (!locationManager.isProviderEnabled(LocationManager.GPS _ PROVIDER)){
        Toast.makeText(this, "Please enable GPS!",
                Toast.LENGTH _ LONG).show();
        Intent intent =
                new Intent(Settings.ACTION _ LOCATION _ SOURCE _ SETTINGS);
        startActivity(intent);
    }

    locationClient = new LocationClient(this, this, this);
}

@Override
protected void onStart() {
    super.onStart();

    // if GoogleMap object is not already available, get it
    if (map == null) {
        FragmentManager manager = getSupportFragmentManager();
        SupportMapFragment fragment =
                (SupportMapFragment) manager.findFragmentById(R.id.map);
        map = fragment.getMap();
    }

    // if GoogleMap object is available, configure it
    if (map != null) {
        map.getUiSettings().setZoomControlsEnabled(true);
    }

    locationClient.connect();
}

@Override
protected void onStop() {
    locationClient.disconnect();

    super.onStop();
}
```

Figure 18-20   The RunMapActivity class (part 2 of 4)

The RunMapActivity class   **Page 3**

```
private void updateMap(){
    if (locationClient.isConnected()){
        setCurrentLocationMarker();
    }
    displayRun();
}

private void setCurrentLocationMarker(){
    if (map != null) {
        // get current location
        Location location = locationClient.getLastLocation();

        if (location != null) {
            // zoom in on current location
            map.animateCamera(
                CameraUpdateFactory.newCameraPosition(
                    new CameraPosition.Builder()
                        .target(new LatLng(location.getLatitude(),
                                            location.getLongitude()))
                        .zoom(16.5f)
                        .bearing(0)
                        .tilt(25)
                        .build()));

            // add a marker for the current location
            map.clear();        // clear old marker(s)
            map.addMarker(      // add new marker
                new MarkerOptions()
                    .position(new LatLng(location.getLatitude(),
                                          location.getLongitude()))
                    .title("You are here"));
        }
    }
}

private void displayRun(){
    if (map != null) {
        locationList = db.getLocations();
        PolylineOptions polyline = new PolylineOptions();
        if (locationList.size() > 0) {
            for (Location l : locationList) {
                LatLng point = new LatLng(
                    l.getLatitude(), l.getLongitude());
                polyline.add(point);
            }
        }
        map.addPolyline(polyline);
    }
}
```

Figure 18-20   The RunMapActivity class (part 3 of 4)

The setMapToRefresh method starts a timer that updates the map at the specified refresh interval, which was set to 10 seconds earlier in the class. To do that, this method uses the Timer and TimerTask classes that were described in chapter 10. As a result, you shouldn't have much trouble understanding how this works.

The onConnected method calls the updateMap method to update the map. Then, it calls the setMapToRefresh method so that map will update every 10 seconds.

The onDisconnected method cancels the timer. This stops the thread for the timer when this activity isn't displayed, which is what you want.

The onConnectionFailed method attempts to resolve the problem that caused the connection to fail. If it can't resolve the problem, it displays a dialog box that displays an error message that includes the code for the error that caused the connection to fail.

The onClick method is executed when the user clicks on the View Stopwatch button. The code for this method starts the Stopwatch activity.

The RunMapActivity class   **Page 4**

```java
    private void setMapToRefresh(){
        timer = new Timer();
        TimerTask task = new TimerTask() {
            @Override
            public void run() {
                RunMapActivity.this.runOnUiThread(new Runnable() {
                    @Override
                    public void run() {
                        updateMap();
                    }
                });
            }
        };
        timer.schedule(task, INTERVAL_REFRESH, INTERVAL_REFRESH);
    }

    @Override
    public void onConnected(Bundle dataBundle) {
        updateMap();
        setMapToRefresh();
    }

    @Override
    public void onDisconnected() {
        timer.cancel();
        Toast.makeText(this, "Disconnected", Toast.LENGTH_SHORT).show();
    }

    @Override
    public void onConnectionFailed(ConnectionResult connectionResult) {
        // see if Google Play services can possibly resolve the error
        if (connectionResult.hasResolution()) {
            try {
                // start an Activity that tries to resolve the error
                connectionResult.startResolutionForResult(this,
                        CONNECTION_FAILURE_RESOLUTION_REQUEST);
            }
            catch (IntentSender.SendIntentException e) {
                e.printStackTrace();
            }
        }
        else {
            new AlertDialog.Builder(this)
                .setMessage("Connection failed. Error code: "
                            + connectionResult.getErrorCode())
                .show();
        }
    }

    @Override
    public void onClick(View v) {
        startActivity(stopwatchIntent);
    }
}
```

Figure 18-20   The RunMapActivity class (part 4 of 4)

The RunTrackerService class

Figure 18-21 shows the code for the RunTrackerService class. This class defines a service that continues running as long as the stopwatch is running, even if the user navigates away from the Stopwatch activity. However, the user can stop this service from running by clicking on the Stop or Reset buttons.

To start, this class imports some classes from the Android API. Then, this class imports classes from the Google Maps Android API v2. If you read chapter 10 and this chapter, you should understand how these classes work.

The RunTrackerService class implements three interfaces. All three of these interfaces were described earlier in this chapter. The first interface is necessary to track the device's location, and the next two are necessary to connect to Google Play services and get the current location.

After declaring the RunTrackerService class, this code defines some constants and instance variables that are needed by the class. To start, this code defines the constant that's used to request location updates every 5 seconds. Then, it defines the constant that indicates that the app can handle location updates every 2 seconds if necessary. Next, it defines three instance variables including a RunTrackerDB object named db that can be used to work with a database. Although the RunTrackerDB class isn't shown in this chapter, it works similarly to the other database classes presented in this book. To see the code for this class, you can view the downloadable source code for this app.

The onCreate method initializes the RunTrackerDB object, the LocationClient object, and the LocationRequest object. Here, the LocationRequest object specifies that its priorty is high accuracy. As a result, this app uses GPS to determine the location of the device. In addition, the LocationRequest object uses the constants defined earlier in this class to specify the requested update interval and the fastest allowable update interval.

The RunTrackerService class   Page 1

```
package com.murach.runtracker;

import android.app.Service;
import android.content.Intent;
import android.location.Location;
import android.os.Bundle;
import android.os.IBinder;
import android.widget.Toast;

import com.google.android.gms.common.ConnectionResult;
import com.google.android.gms.common.GooglePlayServicesClient;
import com.google.android.gms.location.LocationClient;
import com.google.android.gms.location.LocationListener;
import com.google.android.gms.location.LocationRequest;

public class RunTrackerService extends Service
implements GooglePlayServicesClient.ConnectionCallbacks,
          GooglePlayServicesClient.OnConnectionFailedListener,
          LocationListener {

    public static final int UPDATE_INTERVAL = 5000;        // 5 seconds
    public static final int FASTEST_UPDATE_INTERVAL = 2000; // 2 seconds

    private LocationClient locationClient;
    private LocationRequest locationRequest;

    private RunTrackerDB db;

    @Override
    public void onCreate() {
        super.onCreate();

        // get database
        db = new RunTrackerDB(getApplicationContext());

        // get location client
        locationClient = new LocationClient(this, this, this);

        // get location request and set it up
        locationRequest = LocationRequest.create()
            .setPriority(LocationRequest.PRIORITY_HIGH_ACCURACY)
            .setInterval(UPDATE_INTERVAL)
            .setFastestInterval(FASTEST_UPDATE_INTERVAL);
    }

    @Override
    public IBinder onBind(Intent intent) {
        return null;
    }
```

Figure 18-21   The RunTrackerService class (part 1 of 2)

The onStartCommand method is executed when the service is started. The first statement in this method attempts to connect to Google Play services. This causes the onConnected method shown later in this class to be executed.

The onDestroy method is executed when the service is stopped. The statements in this method attempt to disconnect from Google Play services. This causes the onDisconnected method shown later in this class to be executed.

The onConnected method is executed when the device connects to Google Play services. This method begins by getting the current location for the user. Then, it writes that location to the database. Next, it requests location updates.

The onDisconnected method is executed when the device disconnects from Google Play services. This method removes its request for location updates, which is what you want when you disconnect.

The onConnectionFailed method is executed if the connection fails. Since it's difficult to display a dialog box from a service, this method displays a toast that indicates that the connection has failed. This works differently than the code for the same method that's in the Run Map activity where it's easy to display a dialog box.

The onLocationChanged method is executed every time Android gets a fix on a new location. The code within this method writes some of the data that's stored in the Location object to the database for this app. That way, the Run Map activity can read this data from the database.

The RunTrackerService class   Page 2

```
    @Override
    public int onStartCommand(Intent intent, int flags, int startId) {
        locationClient.connect();
        return super.onStartCommand(intent, flags, startId);
    }

    @Override
    public void onDestroy() {
        if (locationClient.isConnected()) {
            locationClient.disconnect();
        }
        super.onDestroy();
    }

    //*************************************************************
    // Implement ConnectionCallbacks interface
    //*************************************************************
    @Override
    public void onConnected(Bundle dataBundle) {
        Location location = locationClient.getLastLocation();
        if (location != null){
            db.insertLocation(location);
        }
        locationClient.requestLocationUpdates(locationRequest, this);
    }

    @Override
    public void onDisconnected() {
        if (locationClient.isConnected()) {
            locationClient.removeLocationUpdates(this);
        }
    }

    //*************************************************************
    // Implement OnConnectionFailedListener
    //*************************************************************
    @Override
    public void onConnectionFailed(ConnectionResult connectionResult) {
        Toast.makeText(this, "Connection failed! " +
            "Please check your settings and try again.",
            Toast.LENGTH_SHORT).show();
    }

    //*************************************************************
    // Implement LocationListener
    //*************************************************************
    @Override
    public void onLocationChanged(Location location) {
        if (location != null){
            db.insertLocation(location);
        }
    }
}
```

Figure 18-21   The RunTrackerService class (part 2 of 2)

Perspective

The skills presented in this chapter should be enough to get you started with locations and maps. However, this is a huge topic, and there's plenty more to learn about it. For example, you may want to learn more about the new and exciting features that are available from version 2 of the Android API for Google Maps.

One new feature is geofencing. This feature lets your app specify a boundary around a location. Then, your app can receive notifications when users enter or leave that boundary.

Another new feature is activity recognition. This feature lets you determine the user's current activity. For example, you can determine whether the user is walking, cycling, or riding in a vehicle.

On the other hand, you may need to learn more about less exciting features that have been available for years. For example, you may need to use geocoding to convert latitude and longitude to a street address. You may need to use routing to get directions for your users. Or, you may need to learn how to test an app that uses locations in an emulator. To do that, you can search the Internet for more information.

Terms

GPS (Global Positioning System)	Routing
Geocoding	SHA-1 fingerprint

Summary

- You can use a device's *GPS* (*Global Positioning System*) receiver or its network signals (cell or Wi-Fi) to find a device's current location.

- Google Maps is the most popular Android API for working with maps. One alternative is MapQuest.

- *Geocoding* is the process of converting latitude/longitude coordinates to street addresses and back.

- *Routing* is the process of using map data to provide directions.

- Version 2 of the Google Maps Android API is recommended for all new development as of March 2013.

- To add the Google Play services library to your project, you can download it, import it into your workspace, and add a reference to it from your project.

- A *SHA-1 fingerprint* is a text string that Google Maps can use to identify your app.

- The Maps API key is necessary to access the Google Maps servers with the Maps API.

- You can use a LocationClient object to open and close a connection to Google Play services, and to get a Location object that contains detailed information about the best and most recent location of a device.

- You can use a LocationRequest object to configure a request for location updates from Google Play services.

- You can use a LocationListener object to listen for location updates from Google Play services.

- You can use the LocationManager object to check if GPS is enabled.

- You can use a GoogleMap object to zoom in on a location, add one or more markers to a map, and to display a series of connected lines on a map.

Exercise 18-1   Test and modify the Run Tracker app

In this exercise, you'll test the Run Tracker app on a physical device. To do that, you'll need to get your own Maps API key from the Google APIs Console so you can display a map. Then, you'll modify some of the settings for the map to change its appearance.

This exercise only describes how to test on an actual device, not on an emulator. In addition, this device must have a valid version of Google Play installed, and it must support OpenGL ES version 2.

Test the Run Tracker app on your device

1. Start Eclipse and import the library project named google-play-services_lib that's in the libs subdirectory of the workspace directory. To do that, you can being by selecting the File→Import→Android→Existing Android Code into Workspace item.

2. Import the project named ch18_ex1_RunTracker.

3. Display the Properties dialog box for this project, select the Android item, and make sure the project contains a working reference to the Google Play services library. If it doesn't, add this reference.

4. Connect the device you want to use for testing. If you have already installed the Run Tracker app on this device, uninstall it now.

5. Run the Run Tracker app on this device. Then, click on the View Map button. This should display a blank map, and the LogCat view should display an error message for an authentication error. That's because the app hasn't registered a valid API key for Google Maps.

6. Start a command prompt and use the keytool command to get the SHA-1 fingerprint for this app.

7. Start a web browser and use the Google APIs Console to get an API key for this app.

8. Start Eclipse, open the AndroidManifest.xml file, and paste the API key into the value attribute of the meta-data element.

9. Uninstall the Run Tracker app that's currently on the device.

10. Run the Run Tracker app with the valid API key on device. Then, click on the View Map button. This time, the app should display a map and zoom in on the current location. If it doesn't, check for error messages in the LogCat view and do your best to troubleshoot the problem.

11. Click the View Stopwatch button. Then, click on the Start button to start a run. Go outside and walk for a while. Switch back and forth between the Stopwatch and Run Map activities. The map should display a marker for your current location and track your route (or at least somewhat close to your route).

12. Click the Stop button to stop the run. At this point, the map should still display the route of the run.

13. Click the Reset button to delete the run. At this point, the map should only display your current location, not your route.

Modify the settings for the map

14. Open the RunMapActivity class and review its code.

15. Change the map type from the default type to a hybrid map. Then, run the app. Note how this changes the appearance of the map.

16. Use the MAP_TYPE_NORMAL constant to change the map type back to the default type.

17. Use the moveCamera method to change the position of the camera instead of using the animateCamera method. This should display the current location without using any animation to zoom in on it.

18. Change zoom level to 18 and change the tilt to 10 degrees. This should zoom in even further with less tilt.

19. Modify the code so the marker uses the icon named ic_runner that's in the res\drawable directories instead of using the default icon for the marker. This should display a blue runner.

20. Change the width of the line that marks the route to 10 pixels.

21. Change the color of the line that marks the route to blue.

Exercise 18-2   Refactor the Run Tracker app

In this exercise, you'll refactor the Run Tracker app so its code is more consistent and has less duplication.

Use a location listener instead of a timer

1. Start Eclipse and import the project named ch18_ex2_RunTracker.

2. Open the RunTrackerService class. Note how it implements the LocationListener interface from the Google Play services library (the com.google.android.gms.location package).

3.   Open the RunMapActivity class and modify it so it also implements the LocationListener interface. To do that, you can copy most of the code that sets the update intervals and creates the update request object.

4.   Add code to the onLocationChanged method that updates the marker for the current position as well as the line for the route.

5.   Delete the setMapToRefresh method and the rest of the code that works with the Timer and TimerTask objects.

Use an Application object to reduce code duplication

6.   Add a class named RunTrackerApp to the project. This class should extend the Application class.

7.   Register this class and test it to make sure Android executes its onCreate method when you start the app. If necessary, you can refer to chapter 11 for information about how to work with an Application object.

8.   Move the UPDATE_INTERVAL and FASTEST_UPDATE_INTERVAL constants to the Application class.

9.   Modify the code for the RunMapActivity and RunTrackerService classes so they use the constants from the Application class instead of defining their own constants.

10.   Move the LocationRequest object and the code that creates it to the Application class. Add a public method named getLocationRequest that returns the LocationRequest object.

11.   Modify the code for the RunMapActivity and RunTrackerService classes to they use the getLocationRequest method to get the LocationRequest object.

12.   Move the RunTrackerDB object and the code that creates it to the Application class. Add a public method named getDB that returns the RunTrackerDB object.

13.   Modify the code for the RunMapActivity and RunTrackerService classes so they use the getDB method to get the RunTrackerDB object.

14.   Add a public method named enableGPS to the Application class. This method should accept one Context parameter. Move the code that helps the user enable GPS into this method.

15.   Modify the code for the StopwatchActivity and RunMapActivity classes so they use the enableGPS method instead of duplicating this code.

Appendix A

How to set up your PC for this book

This appendix shows how to install and configure the software that we recommend for developing Android apps on a PC. This appendix also shows how to install the source code for this book.

This appendix is designed for a PC that's running the Windows operating system. For directions on setting up a Mac, please see appendix B.

As you read this appendix, please remember that most web sites and install programs are continually updated. As a result, some of the procedures in this appendix may have changed since this book was published. Nevertheless, these procedures should still be good guides to installing the software. And if there are significant changes to these setup instructions, we will post updates on the book's FAQs page at our web site (www.murach.com).

How to install the source code for this book

Figure A-1 shows how to download and install the source code for this book. This includes the source code for the applications presented in this book, the starting points for the chapter exercises, and the solutions to the exercises.

When you finish this procedure, the book applications as well as the starting points and solutions for the exercises should be in the directories shown in this figure. Then, you can review the applications presented in this book, and you're ready to do the exercises in this book.

The Murach web site

`www.murach.com`

The directories for the book applications and exercise starts

```
C:\murach\android\workspace\book_apps
C:\murach\android\workspace\ex_starts
C:\murach\android\workspace\ex_solutions
```

How to download and install the files for this book

1. Go to www.murach.com, and go to the page for *Murach's Android Programming*.

2. Click the link for "FREE download of the book applications." Then, click the "All book files" link for the self-extracting zip (exe) file. This should download a file named andp_allfiles.exe to your C drive.

3. Use the Windows Explorer to find the exe file on your C drive. Then, double-click this file and respond to the dialog boxes that follow. This installs the files for this book in directories that start with C:\murach\android.

How to use a zip file instead of a self-extracting zip file

- Although we recommend using the self-extracting zip (exe) file to install the down-loadable files as described above, some systems won't allow self-extracting zip files to run. In that case, you can download a regular zip file (andp_allfiles.zip) from our web site. Then, you can extract the files stored in this zip file into the C:\murach directory. If the C:\murach directory doesn't already exist, you will need to create it.

Description

- You can install the source code for this book by downloading it from murach.com.

Figure A-1   How to install the source code for this book

How to install the Java SE JDK

For Android development, you need to have the *Java Development Kit* (*JDK*) for *Java SE* (*Standard Edition*) installed on your computer. If you've already done some Java development, you probably already have the JDK installed on your system. If not, figure A-2 shows how to install the JDK. To start, you download the exe file for the setup program for the most recent version of the JDK from the Java web site. Then, you navigate to the directory that holds the JDK, run the setup file, and respond to the resulting dialog boxes.

Since the Java web site may change after this book is printed, we've kept the procedure shown in this figure somewhat general. As a result, you may have to do some searching to find the current version of the JDK. In general, you can start by searching the Internet for the download for Java SE. Then, you can find the most current version of the JDK for your operating system.

The Android apps in this book only use Java SE 6 features. As a result, you can use JDK 1.6 or later. For example, you can use JDK 1.7, which corresponds with Java SE 7, for this book. This works because Java has a good record of being backwards compatible.

The download page for Java SE

`www.oracle.com/technetwork/java/javase/downloads`

Procedure

1. Go to the download page for Java SE. The easiest way to find this page is to search the Internet for "Java SE download".
2. Click on the Download button for the JDK and follow the instructions for your operating system.
3. Save the download file to your hard disk. This file should be an exe file.
4. Double-click on the exe file to run it and respond to the resulting dialog boxes. When you're prompted for the JDK directory, use the default directory.

The default Java SE 7 directory on most Windows systems

`C:\Program Files\Java\jdk1.7.0`

Description

- For Android development, you need to install *Java SE* (*Standard Edition*). To do that, you can download and install the *JDK* (*Java Development Kit*).
- For more information about installing the JDK, you can refer to the Oracle web site.

Figure A-2   How to install Java

How to use the ADT bundle to install Android and Eclipse

Before you can develop Android apps, you need to install several pieces of software. First, you need to install the *Android SDK* (*Software Development Kit*). Second, you need to install the *Eclipse IDE* (*Integrated Development Environment*). Third, you need to install the *ADT* (*Android Developer Tools*) *plugin*. This plugin allows Eclipse to work with Android projects.

In the early days of Android development, developers had to install all three of these software pieces separately, which was a time-consuming and error-prone process. Fortunately, developers can now use the Android SDK ADT bundle to streamline this process as shown in figure A-3.

In this procedure, you can copy the "adt-bundle" directory to any appropriate directory on your computer. However, you should *not* move any of the files or directories within the "adt-bundle" directory. If you do, the ADT plugin won't be able to locate the SDK.

How to start Eclipse

Unlike most programs, there is no installer for Eclipse. Instead, you start Eclipse by navigating to the eclipse directory. Then, you double-click on the application file for Eclipse (eclipse.exe).

Since this isn't the most convenient way to start a program, you may want to create an easier way to start Eclipse for subsequent sessions. For example, you can pin Eclipse to your task bar by right-clicking on the application file for Eclipse and selecting the Pin to Taskbar item. Or, you can create a shortcut on your desktop by right-clicking on the application file for Eclipse and selecting the Send To→Desktop (create shortcut) item.

When you first start Eclipse, you may notice that it displays a splash screen that says, "Android Developer Tools". In addition, the title bar for the application uses a green icon instead of the traditional blue Eclipse icon and says ADT instead of Eclipse. That's because this appendix shows how to install a version of Eclipse that has been customized to include the Android Developer Tools (ADT).

The download page for the Android ADT bundle

`developer.android.com/sdk`

How to install the Android SDK and Eclipse with the ADT plugin

1. Go to the download page for the Android SDK with the ADT Bundle. The easiest way to find this page is to search the Internet for "Android SDK download".

2. Click on the link to download the SDK with the ADT bundle and follow the instructions on the web site. This should download a zip file with a name something like adt-bundle-windows-x86_64.zip and save it in the directory you specify. This is a large file and will probably take a while to download.

3. Extract the contents of the zip file into an appropriate directory on your hard drive.

4. Navigate to the eclipse directory, double-click the eclipse.exe file, and run it. This should start Eclipse.

5. When Eclipse starts for the first time, it displays the Workspace Launcher dialog box and displays the default directory for storing your Eclipse projects. In most cases, you can click the OK button to accept the default directory.

6. If Eclipse displays a Welcome page, close the tab for that page. This should display the view that you typically use when you work with Eclipse.

7. To make it easy to start Eclipse again, you can pin the eclipse.exe file to your taskbar, pin it to your Start menu, or create a shortcut to it on your desktop.

A possible location for the ADT bundle

`C:\adt-bundle-windows-x86_64-20130219\eclipse`
`\sdk`

Description

- For Android development, you need to install the *Android SDK* (*Software Development Kit*).

- To use Eclipse for Android development, you need to install the *ADT* (*Android Developer Tools*) *plugin*, which is now part of the bundle.

Figure A-3   How to install the Android SDK and Eclipse with the ADT plugin

How to configure Eclipse for this book

When you install Eclipse, you can accept the default directory for the Eclipse *workspace*, which is where Eclipse stores its projects. Then, if necessary, you can switch the workspace to another directory later. When you're working with the source code for this book, for example, you can use the procedure shown in figure A-4 to switch the workspace to the directory that stores the source code for this book.

In addition, when you're working with the source code for this book, you should set a couple of other preferences for Eclipse. First, you should set the JDK compliance to 1.6 as shown in this figure. If you don't, Eclipse will display error messages when you import the projects for this book as described later in this appendix.

Second, you should make sure the Java Proposals item is selected as shown in this figure. This item is selected by default on most systems, but not on all. If this item isn't selected, the code completion feature won't work as described in chapter 2.

How to configure Eclipse for this book

1. Start Eclipse.
2. Select the File→Switch Workspace→Other item from the menu bar. This should display the Workspace Launcher dialog box.
3. Click the Browse button and use the resulting dialog box to select this directory:

   `C:\murach\android\workspace`

   When you click OK, Eclipse should restart and load the new workspace.
4. Select the Window→Preferences item from the menu bar. Then, select the Java→Compiler node. In the JDK Compliance category, select 1.6 as the Compiler compliance level. Click OK button to confirm the change.
5. Select the Window→Preferences item from the menu bar. Then, select the Java→Editor →Content Assist→Advanced node. In the first list box, select the Java Proposals item if it isn't already selected.

Description

- When you use Eclipse with this book, you typically want to change the workspace to the directory shown above, set the JDK compliance to 1.6 for the workspace, and make sure the Content Assist feature includes Java proposals.

Figure A-4   How to configure Eclipse for this book

How to use the Android SDK Manager

The Android SDK ADT bundle includes a starting set of Android tools and platforms. However, we recommend following the procedure presented in figure A-5 to make sure that you have the tools and platforms that you'll need as you work through this book.

When you use the Android SDK Manager to install packages, you may get errors that say:

`(Access is denied)`

This is usually because you did not start Eclipse as an administrator as described in step 1.

When you use the Android SDK Manager, it displays "Installed" in the Status column if the most current version of a tool or platform has been installed. Otherwise, it displays a status of "Not Installed" or "Update Available".

Throughout this book, we have used Android 4.2 for testing new devices and Android 2.2 for testing old devices. As a result, we recommend installing these platforms to get started. We recommend installing the Google USB Driver item in the Extras category because you may need to configure Android devices for testing and debugging. In addition, we recommend installing the Android Support Library item in the Extras category because it's necessary to support some newer features on older devices.

If you prefer to use versions of Android other than 4.2 and 2.2, you can do that too. For example, you may want to use Android 4.3 instead of 4.2. Or, you may want to use Android 2.3 instead of 2.2. However, it's usually easier to follow along with the book if you use Android 4.2 and 2.2. That way, this book will match your system, and you will encounter fewer bugs since we have tested the apps in this book on these versions of Android.

This figure shows how to install the tools and platforms necessary to work with this book. If you want to install more tools or platforms, you can use the Android SDK Manager to do that. Of course, additional tools and platforms require more disk space and take longer to download. As a result, if you want to get started as fast as possible, you can follow the instructions in this figure.

The Android SDK Manager

Procedure

1. Start Eclipse as an administrator. To do that, you can right-click on the exe file for Eclipse and select the "Run as administrator" command.

2. In the toolbar, click the button for the Android SDK Manager. This should start the Android SDK Manager.

3. To view the tools and platforms that are installed, expand or collapse the category nodes.

4. To install more tools or platforms, select the tools or platforms. For this book, we recommend installing all of the defaults, including the following items:
   - Tools (all default items)
   - Android 4.2.2→SDK Platform
   - Android 4.2.2→ARM EABI v7a System Image
   - Android 2.2→SDK Platform
   - Extras→Android Support Library
   - Extras→Google USB driver

5. Click the Install button. This should display another dialog box.

6. Select the Accept License radio button. Then, click the Install button. On some systems, this may take an hour or longer. As a result, you may want to run this installation in the background while you do other things. In addition, you may need to close Eclipse while the Android SDK Manager finishes the download.

Description

- You can use the Android SDK Manager to install Android tools and platforms.

Figure A-5   How to use the Android SDK Manager

How to create an emulator

Figure A-6 shows how to create an *Android Virtual Device* (*AVD*) that you can use to test your apps. Since an AVD emulates a physical device, an AVD can also be called an *emulator*. In general, you want to create an emulator for each platform that you want to test. For this book, you can get started by creating an emulator for Android 4.2 to test new Android features.

When you create an emulator, you must provide a name for the emulator and the target platform. In this figure, I created an emulator named Android_4.2_Phone. This emulator is based upon a device definition for a device that has a 4-inch screen that's 480 by 800 pixels wide. As a result, this screen uses a high number of dots per inch (hdpi). This emulator uses the Android 4.2 (API 17) platform. In addition, it uses an ARM processor, which is the most typical type of processor for smart phones.

When you create an emulator, I recommend deselecting the "Hardware keyboard present" option. That way, you can use an on-screen keyboard, which is known as a *soft keyboard*, to enter text. This is the way that you typically enter text for most modern devices. If you leave the "Hardware keyboard present" option selected, you can use your computer's keyboard to enter text. This may make it easier to enter text, but it doesn't accurately emulate how most devices work.

The emulator shown in this figure is for a generic device that supports the Android 4.2 platform. If you wanted to closely emulate a particular device, you need to modify the underlying device definition for the emulator so it matches the specs for the device that you want to emulate. You can learn more about how that works in chapter 4.

The dialog box for creating an emulator

Procedure

1. Start Eclipse.
2. Click the toolbar button for the Android Virtual Device Manager. This should start the Android Virtual Device Manager and display the Android Virtual Devices tab.
3. Select the New button. This should display the dialog box shown above.
4. Enter a name for the emulator, select the device definition that specifies the screen size and density, and select the target platform.
5. If necessary, select the processor (CPU).
6. Deselect the "Hardware keyboard present" option.
7. Click the OK button.

Description

- To test your Android applications, you can create an *Android Virtual Device* (*AVD*) for each platform that you wish to test. An Android Virtual Device can also be referred to as an *emulator*.

Figure A-6   How to create an emulator

How to configure a device for development

Figure A-7 shows how to configure a physical device so that you can test Android apps on it. This is preferable to testing on an emulator for two reasons. First, apps install and run more quickly on a physical device than on an emulator. Second, testing on a physical device is the only way to truly see how an app works on that device.

To configure a physical device for development, you must connect your device to the computer with a USB cable and turn on the "USB debugging" option on your device. The procedure for doing this varies depending on the version of Android that's running on the phone.

In addition, you must install a driver for the device. Again, the procedure for doing this varies depending on the device. For some devices, you can install the *Google USB Driver*. For other devices, you need to install an *OEM (Original Equipment Manufacturer) driver*. To determine what type of driver you need for your device, you can search the Internet. Or, you can consult the OEM USB Drivers document by going to the link shown at the bottom of this figure.

How to configure a device for development

1. Turn on the "USB debugging" option on your device.
   - For Android 4.0 and later, you can find this option by selecting the Settings→Developer item.
   - Prior to Android 4.0, you can find this option by selecting the Settings→Applications→Development item.
2. Connect your Android device to your computer's USB port.
3. Download a driver for the device.
   - For an Android Developer Phone, a Nexus One, or a Nexus S, you can use the *Google USB Driver*. If you followed the steps in figure A-5, you should have already downloaded this driver.
   - For the Galaxy Nexus, go to the Samsung web site and download the Google USB Driver for that phone, which is listed as model SCH-I515.
   - For most other devices, install an *OEM (Original Equipment Manufacturer) driver* as described below.
4. Open the Windows Explorer, right-click on Computer, and select the Manage command. This should display the Computer Management window.
5. Select the Device Manager category in the left pane, expand the Portable Devices node in the right pane, right-click the device name (such as Galaxy Nexus or Android Phone) and select the Update Driver Software command.
6. Select the Browse option, browse to the directory that contains the file for the USB driver, and click the Next button to install the driver.

How to download the Google USB Driver

- Start the Android SDK Manager as described in figure A-5 and use it to download the Google USB Driver. To do that, expand the Extras category, select the Google USB Driver check box, and click the Install button. By default, this downloads the driver to this directory:

  `C:\Users\YourName\android-sdks\extras\google\usb_driver`

How to install an OEM driver

- Search the Internet or use the OEM USB Drivers document to find an OEM driver for your device. Then, download the driver to your computer.

The link to the OEM USB Drivers document

`developer.android.com/tools/extras/oem-usb.html`

Description

- Since an actual device runs more quickly than an emulator, we recommend using an actual device whenever possible.

Figure A-7   How to configure a device for development

How to verify that your PC is set up correctly

At this point, you have installed all of the software and source code that you need to work with this book, but you haven't tested it to make sure it's working correctly. To do that, you can run the Tip Calculator application for chapter 3 as shown by the procedure in figure A-8.

To begin, you can start Eclipse and import the project for the Tip Calculator app as shown in part 1 of the figure. At this point, the name of the project will appear in the Package Explorer window of Eclipse, but the emulator won't appear on your screen until you've gone through the additional steps in part 2.

Eclipse and an emulator after the Tip Calculator app has been run

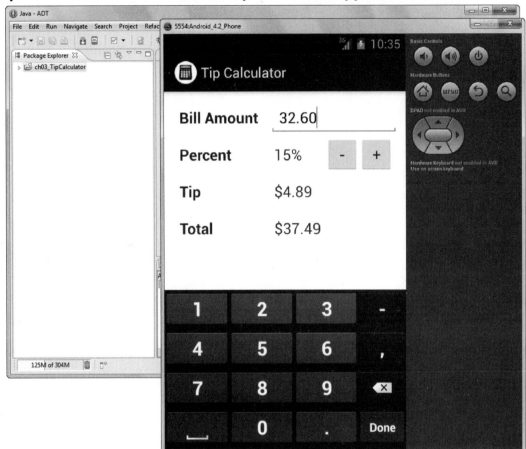

Import a project

1. Start Eclipse.
2. Select the File→Import item from the menu bar.
3. Select the General→Existing Projects into Workspace option from the first dialog box and click the Next button.
4. Click the Browse button and browse to the directory that contains the Tip Calculator project for chapter 3:

    `C:\murach\android\workspace\book_apps\ch03_TipCalculator`

5. Click the OK button.
6. Click the Finish button to import the project. This should display the ch03_TipCalculator project in the Package Explorer window.

Description

- To verify that you have set up your PC correctly for this book, you can run the Tip Calculator app for chapter 3 in an emulator and, optionally, on an actual device.

Figure A-8   How to verify that your PC is set up correctly (part 1 of 2)

After importing the project for the app into Eclipse, you can run the app in an emulator as shown in part 2. Finally, if you have an Android device that you have configured for development, you can run the app on that device. If you are able to successfully accomplish all of these tasks, your PC is set up correctly for this book. Congratulations!

The Android Device Chooser dialog box

Test on an emulator

7. Start the emulator. To do that, click on the Android Virtual Device Manager button in the toolbar, select the emulator named Android_4.2_Phone described in figure A-6, click the Start button and respond to the resulting dialog boxes. Depending on your system, it may take a long time for the emulator to launch, so be patient! After loading, you need to unlock the emulator screen to start using it by dragging the lock icon.

8. Right-click on the ch03_TipCalculator project and select the Run As→Android Application command. This should run the app on the emulator that you started in the previous step.

9. If the Tip Calculator app displays correctly and you can use it to calculate a tip, your system is set up correctly for this book!

Test on a device (optional)

10. Connect the device you configured in figure A-7 to the computer with a USB cable.

11. Right-click on the ch03_TipCalculator project and select the Run As→Run Configurations item. This should display the Android Run Configurations dialog box.

12. Click the Target tab, select the "Always prompt" item, and click the Run button. This should display the Android Device Chooser dialog box.

13. Select the "Choose" option. This should show two running devices: the emulator named Android_4.2_Phone and your device. Then, select your device and click the OK button. This should start the Tip Calculator app on your device.

14. If the Tip Calculator app displays correctly and you can use it to calculate a tip, your device is configured correctly for testing and debugging!

Figure A-8   How to verify that your PC is set up correctly (part 2 of 2)

Appendix B

How to set up your Mac for this book

This appendix shows how to install and configure the software that we recommend for developing Android apps on a Mac. This appendix also shows how to install the source code for this book.

Please note that this appendix is designed for a Mac that's running the OS X operating system. For directions on setting up a PC, see appendix A.

As you read this appendix, please remember that most web sites are continually updated. As a result, some of the procedures in this appendix may have changed since this book was published. Nevertheless, these procedures should still be good guides to installing the software. And if there are significant changes to these setup instructions, we will post updates on our web site (www.murach.com).

How to install the source code for this book

Figure B-1 shows how to download and install the source code for this book. This includes the source code for the applications presented in this book, the starting points for the exercises, and the solutions to the exercises.

When you finish this procedure, the book applications as well as the starting points and solutions for the exercises should be in the directories shown in this figure. Then, you can review the applications presented in this book, and you're ready to do the exercises in this book.

As you read this book, you'll notice that it often instructs you to right-click, which is a common technique on PCs. On a Mac, right-clicking is not enabled by default. Instead, you can use the Ctrl-click instead of the right-click. Or, if you prefer, you can enable right-clicking by editing the system preferences for your mouse. Then, you can follow the instructions in this book more closely.

The Murach web site

<u>www.murach.com</u>

The directories for the book applications and exercise starts

```
/murach/android/workspace/book_apps
/murach/android/workspace/ex_starts
/murach/android/workspace/ex_solutions
```

How to download and install the files for this book

1. Go to the Murach web site.

2. Find the page for *Murach's Android Programming*.

3. Click the link for "FREE download of the book applications." Then, click the "All book files" link for the regular zip file. This will download a zip file named andp_allfiles.zip onto your hard drive.

4. Use Finder to browse to this file and double-click on it to unzip it. This creates the android directory and its subdirectories.

5. Use Finder to create the murach directory directly on the Mac hard drive.

6. Use Finder to move the android directory into the murach directory.

A note about right-clicking

• This book often instructs you to right-click, because that's common on PCs. On a Mac, right-clicking is not enabled by default. Instead, you can use the Ctrl-click instead of the right-click. Or, if you prefer, you can enable right-clicking by editing the system preferences for your mouse.

Figure B-1   How to install the source code for this book

How to install the Java SE JDK

By default, the *Java Development Kit* (*JDK*) for *Java SE* (*Standard Edition*) is installed on Mac OS X. Figure B-2 shows how to use the Software Update feature to make sure you have the most current version that's available for Mac OS X. Then, you can use the Finder to check what version of the JDK is currently installed on your system.

All of the examples in this book have been tested against version 6 of the JDK. Since Java has a good track record of being backwards compatible, all of the examples in this book should work equally well with later versions of the JDK.

If the JavaVirtualMachines directory doesn't exist on your computer, the JDK isn't installed on your computer. In that case, you can wait until you attempt to start Eclipse as shown later in this appendix. When you do that, Eclipse will prompt you to install the JDK. If you accept this prompt, your system will install the JDK for you automatically.

The Apple Developer web site with a Java search

`developer.apple.com/java`

How to install and check the JDK

1. Select the Software Update command from the Apple menu and respond to the resulting dialog boxes.

2. You can check which versions of the JDK are installed on your system by using the Finder to view the contents of this directory:

    `/System/Library/Java/JavaVirtualMachines`

Description

*   By default, Mac OS X comes with a preinstalled version of the *JDK* (*Java Development Kit*) for *Java SE* (*Standard Edition*). To get the most current version, you can use the Software Update command.

*   For more information about installing the JDK, you can refer to the Apple Developer web site.

Figure B-2   How to install Java

How to use the ADT bundle to install Android and Eclipse

Before you can develop Android apps, you need to install several pieces of software. First, you need to install the *Android SDK* (*Software Development Kit*). Second, you need to install the *Eclipse IDE* (*Integrated Development Environment*). Third, you need to install the *ADT* (*Android Developer Tools*) *plugin*. This plugin allows Eclipse to work with Android projects.

In the early days of Android development, developers had to install all three of these software pieces separately, which was a time-consuming and error-prone process. Fortunately, developers can now use the Android SDK ADT bundle to streamline this process as shown in figure B-3.

In this procedure, you can copy the "adt-bundle" directory to any appropriate directory on your computer. However, you should *not* move any of the files or directories within the "adt-bundle" directory. If you do, the ADT plugin won't be able to locate the SDK.

How to start Eclipse

Unlike most programs, there is no installer for Eclipse. Instead, you start Eclipse by navigating to the eclipse directory. Then, you double-click on the application file for Eclipse.

Since this isn't the most convenient way to start a program, you may want to create an easier way to start Eclipse for subsequent sessions. For example, you can dock Eclipse by right-clicking on the application file for Eclipse and selecting the Options→Keep in Dock item. Or, you can create an alias for your desktop, by right-clicking on the application file for Eclipse and selecting the Make Alias item. Then, drag the alias to the desktop.

When you first start Eclipse, you may notice that it displays a splash screen that says, "Android Developer Tools". In addition, the title bar for the application uses a green icon instead of the traditional blue Eclipse icon and says ADT instead of Eclipse. That's because this appendix shows how to install a version of Eclipse that has been customized to include the Android Developer Tools (ADT).

The download page for the Android SDK

`developer.android.com/sdk`

How to install the Android SDK and Eclipse with the ADT plugin

1. Go to the download page for the Android SDK. The easiest way to find this page is to search the Internet for "Android SDK download".

2. Click on the link to download the SDK with ADT bundle for Mac and follow the instructions on the website. This should download a zip file with a name something like adt-bundle-mac-x86_64.zip.

3. Use Finder to browse to this file and double-click on it to unzip it. This creates the android-sdk-macosx directory and its subdirectories. Then, copy this directory into an appropriate directory, such as the Applications directory.

4. Navigate to the eclipse directory and double-click the Eclipse application file. This should start Eclipse.

5. When Eclipse starts for the first time, it displays the Workspace Launcher dialog box and displays the default directory for storing your Eclipse projects. In most cases, you can click the OK button to accept the default directory.

6. If Eclipse displays a Welcome page, close the tab for that page. This should display the view that you typically use when you work with Eclipse.

7. To make it easy to start Eclipse again, you can choose to keep an icon for Eclipse in your dock, or create an alias to keep on your desktop.

A typical directory for the Android SDK

`/Applications/adt-bundle-mac-x86_64-20130729/eclipse`
                                                                        `/sdk`

Description

- For Android development, you need to install the *Android SDK* (*Software Development Kit*).

- For more information about installing the Android SDK, you can refer to the Android web site here:

  `developer.android.com/sdk/installing`

Figure B-3   How to install the Android SDK and Eclipse with the ADT plugin

How to configure Eclipse for this book

When you install Eclipse, you can accept the default directory for the Eclipse *workspace*, which is where Eclipse stores its projects. Then, if necessary, you can switch the workspace to another directory later. When you're working with the source code for this book, for example, you can use the procedure shown in figure B-4 to switch the workspace to the directory that stores the source code for this book.

How to configure Eclipse for this book

1. Start Eclipse.
2. Select the File→Switch Workspace→Other item from the menu system. This should display the Workspace Launcher dialog box.
3. Click the Browse button and use the resulting dialog box to select this directory:

   `/murach/android/workspace`

   When you do, Eclipse should restart and load the new workspace.

Description

- When you use Eclipse with this book, you typically want to change the workspace to the directory shown above.

Figure B-4   How to configure Eclipse for this book

How to use the Android SDK Manager

The Android SDK ADT bundle includes a starting set of Android tools and platforms. However, we recommend following the procedure presented in figure B-5 to make sure that you have the tools and platforms that you'll need as you work through this book.

When you use the Android SDK Manager, it displays "Installed" in the Status column if the most current version of a tool or platform has been installed. Otherwise, it displays a status of "Not Installed" or "Update Available".

Throughout this book, we have used Android 4.2 for testing new devices and Android 2.2 for testing old devices. As a result, we recommend installing these platforms to get started. We recommend installing the Android Support Library item in the Extras category because it's necessary to support some newer features on older devices.

If you prefer to use versions of Android other than 4.2 and 2.2, you can do that too. For example, you may want to use Android 4.3 instead of 4.2. Or, you may want to use Android 2.3 instead of 2.2. However, it's usually easier to follow along with the book if you use Android 4.2 and 2.2. That way, this book will match your system, and you will encounter fewer bugs since we have tested the apps in this book on these versions of Android.

This figure shows how to install the tools and platforms necessary to work with this book. If you want to install more tools or platforms, you can use the Android SDK Manager to do that. Of course, additional tools and platforms require more disk space and take longer to download. As a result, if you want to get started as fast as possible, you can follow the instructions in this figure.

The Android SDK Manager

Procedure

1. Start Eclipse.
2. In the toolbar, click the button for the Android SDK Manager. This should start the Android SDK Manager.
3. To view the tools and platforms that are installed, expand or collapse nodes.
4. To install more tools or platforms, select the tools or platforms. For this book, we recommend installing all of the defaults including the following items:
   - Tools (all default items)
   - Android 4.2.2→SDK Platform
   - Android 4.2.2→ARM EABI v7a System Image
   - Android 2.2→SDK Platform
   - Extras→Android Support Library
5. Click the Install button. This should display another dialog box.
6. Select the Android SDK License option and the Accept License radio button. Then, select the Install button. On some systems, this may take an hour or longer. As a result, you may want to run this installation in the background while you do other things. In addition, you may need to close Eclipse while the Android SDK Manager finishes the download.

Description

- You can use the Android SDK Manager to install Android tools and platforms.

Figure B-5   How to use the Android SDK Manager

How to create an emulator

Figure B-6 shows how to create an *Android Virtual Device* (*AVD*) that you can use to test your apps. Since an AVD emulates an actual device, an AVD can also be called an *emulator*. In general, you want to create an emulator for each platform that you want to test. For this book, I recommend creating an emulator for Android 4.2 to test new Android features and one for Android 2.2 to test compatibility with older Android devices.

When you create an emulator, you must provide a name for the emulator and the target platform. In this figure, I created an emulator named Android_4.2_Phone. This emulator is based upon a device definition for a device that has a 4-inch screen that's 480 by 800 pixels wide. As a result, this screen uses a high number of dots per inch (hdpi). This emulator uses the Android 4.2 (API 17) platform. In addition, it uses an ARM processor, which is the most typical type of processor for smart phones.

When you create an emulator, I recommend deselecting the "Hardware keyboard present" option. That way, you can use an on-screen keyboard, which is known as a *soft keyboard*, to enter text. This is the way that you typically enter text for most modern devices. If you leave the "Hardware keyboard present" option selected, you can use your computer's keyboard to enter text. This may make it easier to enter text, but it doesn't accurately emulate how most devices work.

The emulator shown in this figure is for a generic device that supports the Android 4.2 platform. If you wanted to closely emulate a particular device, you need to modify the underlying device definition for the emulator so it matches the specs for the device that you want to emulate. You can learn more about how that works in chapter 4.

The dialog box for creating an emulator

Procedure

1. Start Eclipse.
2. Click the toolbar button for the Android Virtual Device Manager. This should start the Android Virtual Device Manager and display the Android Virtual Devices tab.
3. Select the New button. This should display the dialog box shown above.
4. Enter a name for the emulator, select the device definition that specifies the screen size and density, and select the target platform.
5. If necessary, select the processor (CPU).
6. Deselect the "Hardware keyboard present" option.
7. Click the OK button.

Description

- To test your Android applications, you can create an *Android Virtual Device* (*AVD*) for each platform that you wish to test. An Android Virtual Device can also be referred to as an *emulator*.

Figure B-6   How to create an emulator

How to configure a device for development

Figure B-7 shows how to configure a physical device so that you can test Android apps on it. This is preferable to testing on an emulator for two reasons. First, apps install and run more quickly on a physical device than on an emulator. Second, testing on a physical device is the only way to truly see how an app works on that device.

On a Mac, configuring a device for development is easy. All you need to do is to connect your device to the computer with a USB cable and turn on the "USB debugging" option on your device. The procedure for doing this varies depending on the version Android that's running on the phone.

How to configure a device for development

1. Turn on the "USB debugging" option on your device.

    - For Android 4.0 and later, you can find this option by selecting the Settings→Developer item.

    - Prior to Android 4.0, you can find this option by selecting the Settings→Applications→Development item.

2. Connect your device to your computer's USB port.

Description

-   Since a physical device runs more quickly than an emulator, we recommend using a physical device whenever possible.

Figure B-7   How to configure a device for development

How to verify that your Mac is set up correctly

At this point, you have installed all of the software and source code that you need to work with this book, but you haven't tested it to make sure it's working correctly. To do that, you can run the Tip Calculator application for chapter 3 as shown by the procedure in figure B-8.

To begin, you can start Eclipse and import the project for the Tip Calculator app as shown in part 1 of the figure. At this point, Eclipse will display the name of the project in the Package Explorer, but the emulator won't appear on your screen until you've gone through the steps in part 2.

Eclipse and an emulator after the Tip Calculator app has been run

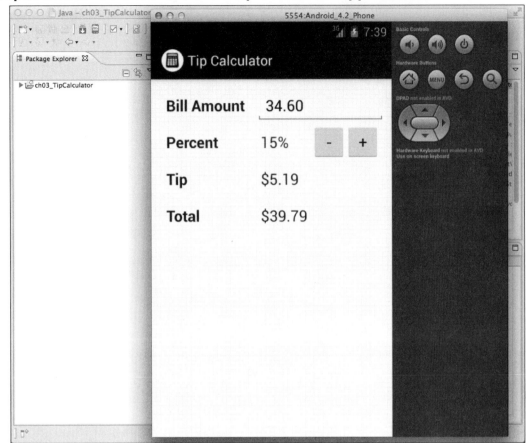

Import a project

1. Start Eclipse.
2. Select the File→Import item from the menu system.
3. Select the General→Existing Projects into Workspace option from the first dialog box and click the Next button.
4. Click the Browse button and browse to the directory that contains the Tip Calculator project for chapter 3:

   `/murach/android/workspace/book_apps/ch03_TipCalculator`
5. Click the Open button.
6. Click the Finish button to import the project. This should display the ch03_TipCalculator project in the Package Explorer window.

Description

- To verify that you have set up your Mac correctly for this book, you can run the Tip Calculator app for chapter 3 in an emulator and, optionally, on an actual device.

Figure B-8   How to verify that your Mac is set up correctly (part 1 of 2)

After importing the project for the app into Eclipse, you can run the app in an emulator as shown in part 2. Finally, if you have an Android device that you have configured for development, you can run the app on that device. If you are able to successfully accomplish all of these tasks, your Mac is set up correctly for this book. Congratulations!

The Android Device Chooser dialog box

Test the emulator

7. Right-click on the Tip Calculator app and select the Run As→Run Configurations command. This should display the Android Run Configurations dialog box.

8. Click on the Target tab, select the "Always prompt" item, and click the Run button. This should display the Android Device Chooser dialog box.

9. Select the "Launch" option, select the emulator named Android_4.22_Phone that you created in figure B-6, and click the OK button. This should start the emulator and launch the Tip Calculator app within it. Depending on your system, this can take 5 minutes or more, so please be patient!

10. If the Tip Calculator app displays correctly, your system is set up correctly for this book!

Test a physical device (optional)

11. Make sure the physical device that you configured in figure B-8 is connected to the computer with a USB cable.

12. Right-click on the Tip Calculator app and select the Run As→Android Application command. This should display the Android Device Chooser dialog box.

13. Use the "Choose" option to select the physical device. This should launch the Tip Calculator app on that device.

14. If the Tip Calculator app displays correctly, and you can use it to calculate a tip, your device is configured correctly!

Figure B-8   How to verify that your Mac is set up correctly (part 2 of 2)

Appendix J

How to use IntelliJ with this book

This book shows how to use the Eclipse IDE (Integrated Development Environment) to develop Android apps. However, some Android developers prefer other IDEs. In particular, the IntelliJ IDEA has been gaining popularity with Android developers in recent years. This appendix shows how to install and configure the IntelliJ IDEA so you can use it with this book.

As you read this appendix, please remember that most web sites and install programs are continually updated. As a result, some of the procedures in this appendix may have changed since this book was published. Nevertheless, these procedures should still be good guides to installing the software. And if there are significant changes to these setup instructions, we will post updates on our web site (www.murach.com).

A comparison of Eclipse, IntelliJ, and Android Studio

There are several popular IDEs for Android development. Of these IDEs, Eclipse is currently the most popular. Eclipse is open-source, has been around since the early days of Android, has received a lot of support from Google developers, and provides many mature features. For these reasons, we recommend using Eclipse. That's why we show how to use Eclipse throughout this book. And that's why we provide the apps and exercises for this book as Eclipse projects.

Unfortunately, Eclipse has an unintuitive user interface that can be confusing for developers who are new to it. For example, Eclipse asks you to import or delete a project instead of asking you to open or close a project. For this reason or for others, you may want to use another IDE.

One IDE that has been gaining in popularity in recent years is IntelliJ IDEA. Its Community Edition is open-source, and many developers consider it to be more intuitive than Eclipse. If you want to use IntelliJ instead of Eclipse, this appendix shows how to install IntelliJ and import the source code for this book. After that, you should be able to use this book with IntelliJ. If you hit any problems, you can search the Internet for solutions.

In May of 2013, IntelliJ announced that it is collaborating with Google to create a new IDE called Android Studio. This IDE is based on the IntelliJ source code but streamlined for Android development. As this book goes to press, Android Studio is only available as an early access preview. However, since it's backed by Google and IntelliJ, it might become very popular after a production version is released. As a result, it's worth keeping an eye on this IDE for the next few years.

How to prepare for using IntelliJ with this book

Before you can use IntelliJ with this book, you need to install several pieces of software including the source code for this book, the *Android SDK* (*Software Development Kit*), and the *ADT* (*Android Developer Tools*) *plugin*. This plugin allows IntelliJ to work with Android projects.

The easiest way to install these pieces of software is to install the ADT Bundle that includes the Eclipse IDE. Please see Appendix A (PC) or B (Mac) for installation instructions.

How install IntelliJ IDEA Community Edition

Figure J-1 shows how to install IntelliJ IDEA. Unlike Eclipse, IntelliJ installs like most other applications. When it finishes installing, you can create a shortcut (PC) or alias (Mac) on your desktop. Since this provides a convenient way to start IntelliJ for subsequent sessions, we recommend doing that.

The download page for the IntelliJ Community Edition

`http://www.jetbrains.com/idea/download/index.html`

How to prepare for using IntelliJ with this book

1. Install the source code for this book as described in appendix A or B.
2. Install the Java SDK as described in appendix A or B.
3. Install the ADT Bundle as described in appendix A or B.

How to install the IntelliJ Community Edition

1. Go to the download page for the IntelliJ Community Edition. An easy way to find this page is to search the Internet for "IntelliJ Community Edition".
2. Click on the appropriate link or button to download the Community Edition and follow the instructions on the web site. This should download an install file with a name something like ideaIC-12.1.4.exe (PC) or ideaIC-12.1.4.dmg (Mac). This is a large file and may take more than 10 minutes to download.
3. Double-click the install file and respond to the resulting dialog boxes. This should install IntelliJ.
4. To make it easy to start IntelliJ again, you can create a shortcut or alias for IntelliJ on your desktop.

Description

- An *IDE* (*Integrated Development Environment*) is a tool for developing software.
- Eclipse is currently the most popular IDE and has received a lot of support from Google.
- IntelliJ IDEA has been gaining in popularity in recent years.
- Android Studio hasn't been released yet, but it is based on the IntelliJ source code and is being jointly developed by IntelliJ and Google. As a result, it has the potential to become a popular IDE for Android development.

Figure J-1   How to install IntelliJ

How to configure IntelliJ for this book

After you install IntelliJ, you set the directory for the Java JDK, the Android SDK, and the workspace as shown in figure J-2. For this book, you should switch the workspace to the directory that stores the source code for this book.

How to configure IntelliJ for this book

1. Start IntelliJ.
2. When you see the Welcome to IntelliJ IDEA screen, select the Configure→Project Defaults→Project Structure item from the menu system. Under the Project Settings category, select the Project item. This should display the general settings for the project.
3. In the "Project SDK" category, click the New button and select the JDK item. Then, navigate to the directory that contains your Java JDK. For a PC, it should be in a directory something like this:

    `C:\Program Files\Java\jdk1.7.0_09`

    On a Mac, it should be in this directory:

    `/System/Library/Java/JavaVirtualMachines/jdk1.6.0`

4. In the "Project SDK" category, click the New button and select the Android SDK item. Then, navigate to the directory that contains your Android SDK. If you installed the ADT bundle, it should be in a directory something like this:

    `C:\adt-bundle-windows-x86_64-20130729\sdk`

    On a Mac, it should be in a directory something like this:

    `/Applications/adt-bundle-mac-x86_64-20130729/sdk`

5. In the Create New Android SDK dialog box, select 1.7 (PC) or 1.6 (Mac) for the Java JDK and select Android 4.2.2 for the build target.
6. In the "Project compiler output" category, click the button to the right of the text box and use the resulting dialog box to select the workspace directory. On a PC, select this directory:

    `C:\murach\android\workspace`

    On a Mac, select this directory:

    `/murach/android/workspace`

Description

- When you use IntelliJ with this book, you typically want to change the workspace to the directory shown above.

Figure J-2   How to configure IntelliJ for this book

How to import an Eclipse project

At this point, you have installed the source code for this book, and you have installed and configured IntelliJ, but you haven't imported any of the projects for this book into IntelliJ. To do that, you can use the procedure shown in figure J-3. This figure shows how to import the Tip Calculator project for chapter 3. However, you can use the same procedure to import any of the projects for this book or to import any of the starting points or solutions for the exercises.

Once you import a project from Eclipse, IntelliJ adds some files to make the Eclipse project also work as an IntelliJ project. As a result, you only need to import an Eclipse project once. After, you can use IntelliJ to close and open the project.

IntelliJ and an emulator after the Tip Calculator app has been run

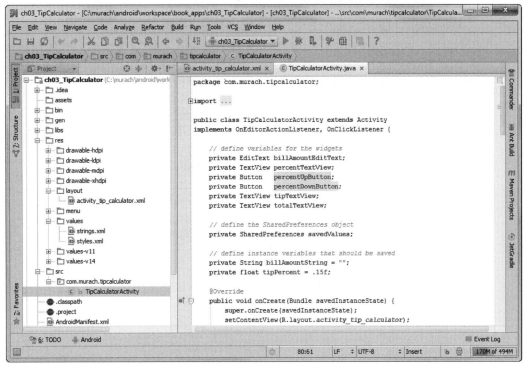

Procedure

1. Start IntelliJ.

2. Select the File→Import Project item from the menu system, or select the Import Project item from the Welcome screen.

3. Browse to the directory that contains the Tip Calculator project for chapter 3. For a PC, this project should be in this directory:

   C:\murach\android\workspace\book_apps\ch03_TipCalculator

4. From the Import Project dialog box, select the "Import project from external model" radio button and select Eclipse.

5. Continue clicking Next and Finish buttons with the default settings to import the project. When you're done, IntelliJ should display the ch03_TipCalculator project in the Projects window.

6. Click the Project tab to display the nodes for the project.

7. Expand the nodes of the project to display its XML and Java files.

8. Double-click on the XML and Java files to open them.

Description

- To get started with IntelliJ, you can import the Tip Calculator app for chapter 3. After that, you can open and close an IntelliJ project without having to import it first.

Figure J-3   How to import an Eclipse project

How to verify that IntelliJ is set up correctly

After importing a project for an app into IntelliJ, you can run the app in an emulator as shown in figure J-4. To start, if you haven't already used the AVD Manager to create an emulator as described in appendix A or B, you start by creating an emulator named Android_4.2_Phone. Then, you can create a run configuration for the project and run the project in that emulator.

In addition, if you have an Android device, you can run the app on it. To start, connect your device to your computer with a USB cable. Then, if you haven't already configured your device for development, you can use appendix A or B to do that. Finally, you can create a run configuration for the project and run the project on your device.

If you are able run the Tip Calculator app on an emulator, IntelliJ is set up correctly for this book. Congratulations! If you are also able to run this app on a device, you have more options for testing and debugging your apps!

Test on an emulator

1. If necessary, create an emulator named Android_4.2_Phone. To do that, click on Tools→Android→AVD Manager from the menu bar to display the AVD Manager. Then, see appendix A or B for details on configuring an emulator.

2. Create a run configuration for the project by clicking on Run→Edit Configurations in the menu bar. Then, select the project. In the Target Device category, select the Emulator radio button, and select the Android_4.2_Phone option from the drop-down menu.

3. In the Toolbar select the Run button. This should run the app on the emulator you selected in the previous step. If necessary, this should start that emulator. Depending on your system, it may take a long time for the emulator to launch, so be patient.

4. If the Tip Calculator app displays correctly, and you can use it to calculate a tip, your system is set up correctly for this book!

Test on a device (optional)

1. Connect the device to your computer with a USB cable. If necessary, configure this device. See appendix A or B for details.

2. Create a run configuration for the project by clicking on Run→Edit Configurations in the menu bar. Then, select the project. In the Target Device category, select the USB Device radio button.

3. In the Toolbar select the Run button. This should run the app on the device.

4. If the Tip Calculator app displays correctly, and you can use it to calculate a tip, your device is configured correctly for testing and debugging!

Figure J-4   How to verify that IntelliJ is set up correctly

Index

.apk file, 14, 15
.arsc file, 14, 15
.class file, 14, 15
.dex file, 14, 15
.java file, 14, 15, 24, 25

A

ACCESS_COARSE_LOCATION permission, 597, 599
ACCESS_FINE_LOCATION permission, 597, 599
ACCESS_NETWORK_STATE permission, 394, 395, 597, 599
Action (broadcast), 412, 413
Action bar, 226, 227
Action constants, 350, 351
Action event, 96, 97
Action key, 96-99
Action overflow icon, 250, 251
Action string (broadcast), 412, 413
ACTION_DOWN constant (MotionEvent), 210, 211
ACTION_MOVE constant (MotionEvent), 210, 211
ACTION_UP constant (MotionEvent), 210, 211
Active state, 104, 105
Activity, 18, 19
    Java code for, 92, 93
    lifecycle, 104, 105
    starting, 258, 259
    working with, 92, 93
Adapter, 342-347
    array, 176-179
    custom, 474-481
    fragment, 480-483
ADB (Android debug bridge), 14, 15
add method (PolylineOptions), 622, 623
addMarker method (GoogleMap), 620, 621
addPolyline method (GoogleMap), 622, 623
addURI method (UriMatcher), 504, 505
ADT (Android Developer Tools), 34, 35
ADT bundle
    install on Mac, 666, 667
    install on PC, 646, 647
ADT plugin
    install on Mac, 666, 667
    install on PC, 646, 647
Advertising (mobile network), 560, 561
AlertDialog.Builder class, 518, 519
Alias (layout), 300, 301
Amazon Appstore, 558, 559
Android API documentation, 108, 109
Android API for Google Maps, 586, 587

Android debug bridge (ADB), 14, 15
Android Developer Tools (ADT), 34, 35
Android Device Chooser dialog box, 52, 53
Android history, 8, 9
Android Manifest, 14, 15, 28, 29
Android Open Source Project (AOSP), 8, 9
Android package, 14, 15
Android package file (APK), 564-567
Android phone, 4, 5
Android runtime, 12, 13
Android SDK
    install on Mac, 666, 667
    install on PC, 646, 647
Android SDK Manager, 650, 651, 670, 671
Android Software Development Kit (SDK), 8, 9
Android stack, 12, 13
Android Studio, 682, 683
Android system architecture, 12, 13
Android tablet, 4, 5
Android versions, 10, 11
Android Virtual Device (AVD), 14, 15,
    create on Mac, 672, 673
    create on PC, 652, 653
AndroidManifest.xml file, 14, 15
animateCamera method (GoogleMap), 618, 619
Animation (Google Maps), 618, 619
Anonymous class,
    as listener, 198, 199
    inner, 198, 199
AOSP (Android Open Source Project), 8, 9
API (Application Programming Interface) 10, 11
API level (Android), 10, 11
APK file (Android package file), 564-567
App, 6, 7
    categories, 6, 7
    deployment, 558-575
    distribution, 558-575
    icon, 534, 535
    run on emulator, 52, 53
    run on physical device, 52, 53
    testing, 54, 55
App widget, 534-551
    add to Home screen, 536, 537
    configure, 548, 549
    create layout, 538-541
    register, 550, 551
    test, 550, 551
appendPath method (Uri.Builder), 508, 509
Application, 6, 7
    framework, 12, 13
    marketplace, 558, 559
Application object, 370-373
Application Programming Interface (API), 10, 11
AppWidgetManager class, 546, 547

onEditorAction method (EditText), 96, 97
onItemClick method (ListView), 346, 347
onPause method (Activity), 106, 107
onPostExecute method (AsyncTask), 324, 325
onPreExecute method (AsyncTask), 324, 325
onProgressUpdate method (AsyncTask), 324, 325
onResume method (Activity), 106, 107
Open Handset Alliance, 8, 9
openFileInput method, 332, 333
openFileOutput method, 330, 331
Open-source code, 4, 5
openStream method, 330, 331
Options menu, 250, 251, 256, 257
orderInCategory attribute (menu item), 254, 255
Orientation (change in an emulator), 106, 107
orientation attribute (linear layout), 158, 159
Original Equipment Manufacturer driver, 654, 655
Outline View (Eclipse), 42, 43
Output stream object, 330, 331

P

Package Explorer (Eclipse), 34, 35
Package, 34, 35
   Android, 14, 15
Padding, 18, 19
   for app widget, 538- 541
Paid app, 560, 561
Palette (Eclipse), 42, 43, 76, 77
Pane, 284, 285
Parent element, 18, 19
Parse an XML file, 332, 333
Passive location listener, 582, 583
Path, 500, 501
Paused state (activity), 104, 105
Pending intent, 388, 389
Perspective (Eclipse), 458, 459
phone attribute (inputType), 170, 171
Phone number (call), 350, 351
Physical device (run app on), 52, 53
Plain old Java object (POJO), 434, 435
POJO (plain old Java object), 434, 435
PolylineOptions class, 622, 623
Popup menu, 252, 253
position method (MarkerOptions), 620, 621
post method, 328, 329
POWER_CONNECTED constant, 412, 413
Preference fragment, 296, 297
Preferences, 260-277
   attributes, 262, 263
   displaying, 264, 265
   enabling/disabling, 274, 275
   getting, 268, 269

Preferences (continued)
   grouping, 272, 273
   using, 270, 271
   with Java, 276, 277
PRIMARY KEY attribute (column), 432, 433
PRIORITY_BALANCED_POWER_ACCURACY
   constant (LocationRequest), 610, 611
PRIORITY_HIGH_ACCURACY constant
   (LocationRequest), 610, 611
PRIORITY_NO_POWER constant (LocationRequest),
   610, 611
Private service, 378, 379
Problems View (Eclipse), 62, 63
progress attribute (SeekBar), 180, 181
Project
   build, 14, 15
   compile, 14, 15
   Eclipse, 34, 35
   run, 14, 15
Properties
   common, 86, 87
   setting, 84, 85
Properties View (Eclipse), 42, 43
protectionLevel attribute (permission), 597, 599
Provider
   content, 500-517
   for app widget, 544, 545
Public service, 378, 379
Publisher account (Google Play), 570, 571
put method (ConvertValues), 450, 451
putExtra method (Intent), 348, 349

Q

Query, 446, 447, 506, 507
query method (SQLiteDatabase), 446, 447
Quick Fix (Eclipse), 62, 63

R

Radio button, 174, 175
Radio group, 174, 175
   event handling, 202, 203
RDBMS, 430, 431
READ_GSERVICES permission, 597, 599
Receiver
   boot completed, 414, 415
   connectivity changed, 416, 417
   custom broadcast, 420, 421
References for widgets, 94, 95
Register
   app widget, 550, 551
   content provider, 512, 513

What software you need for this book

- Java SE 6 (JDK 1.6) or later.
- The ADT (Android Developer Toolkit) bundle. This includes the Android SDK, the Eclipse IDE, and the ADT plugin for Eclipse.
- You can download this software for free and install it as described in appendix A (PC) or appendix B (Mac).

The downloadable files for this book

- Complete source code for the apps presented in this book so you can view the code and run the apps as you read each chapter.
- Starting source code for the exercises presented at the end of each chapter so you can get more practice in less time.
- Source code for the solutions to the exercises that you can use to check your exercise solutions.

How to download the files for this book

- Go to www.murach.com and navigate to the page for *Murach's Android Programming.*
- Click the "download" link. Then, follow the instructions to download the file that contains the source code for the book applications. For Windows, download the self-extracting zip file. For Mac, download the regular zip file.
- Double-click the downloaded file and respond to the dialog boxes that follow.
- For more details, see appendix A (PC) or appendix B (Mac).

www.murach.com

Author:	Joel Murach
Editor:	Ray Halliday
Android Consultant:	Leo Landau
Production Editor:	Ben Murach
Production:	Maria Spera

Books for Java programmers

Murach's Java Programming
Murach's Java Servlets and JSP
Murach's Oracle SQL and PL/SQL

Books for Web developers

Murach's HTML5 and CSS3
Murach's JavaScript and jQuery
Murach's JavaScript and DOM Scripting
Murach's PHP and MySQL
Murach's MySQL

Books for .NET programmers

Murach's C# 2012
Murach's ASP.NET 4.5 Web Programming with C# 2012

Murach's Visual Basic 2012
Murach's ASP.NET 4.5 Web Programming with VB 2012

Murach's SQL Server 2012 for Developers

Books for IBM mainframe programmers

Murach's OS/390 and z/OS JCL
Murach's Mainframe COBOL
Murach's CICS for the COBOL Programmer

For more on Murach books, please visit us at www.murach.com

Printed in the United States of America

10 9 8 7 6 5 4 3 2 1
ISBN: 978-1-890774-71-4

TRAINING & REFERENCE

murach's
Android
programming

Joel Murach

 MIKE MURACH & ASSOCIATES, INC.

4340 N. Knoll Ave. • Fresno, CA 93722
www.murach.com • murachbooks@murach.com